Communications in Computer and Information Science 1209

Commenced Publication in 2007
Founding and Former Series Editors:
Simone Diniz Junqueira Barbosa, Phoebe Chen, Alfredo Cuzzocrea,
Xiaoyong Du, Orhun Kara, Ting Liu, Krishna M. Sivalingam,
Dominik Ślęzak, Takashi Washio, Xiaokang Yang, and Junsong Yuan

More information about this series at http://www.springer.com/series/7899

Sabu M. Thampi · Rajesh M. Hegde ·
Sri Krishnan · Jayanta Mukhopadhyay ·
Vipin Chaudhary · Oge Marques ·
Selwyn Piramuthu · Juan M. Corchado (Eds.)

Advances in Signal Processing and Intelligent Recognition Systems

5th International Symposium, SIRS 2019
Trivandrum, India, December 18–21, 2019
Revised Selected Papers

 Springer

Editors
Sabu M. Thampi
Indian Institute of Information Technology
Trivandrum, India

Rajesh M. Hegde
Indian Institute of Technology Kanpur
Kanpur, India

Sri Krishnan
Ryerson University
Toronto, ON, Canada

Jayanta Mukhopadhyay
Indian Institute of Technology Kharagpur
Kharagpur, India

Vipin Chaudhary
State University of New York at Buffalo
Buffalo, NY, USA

Oge Marques
Florida Atlantic University
Boca Raton, FL, USA

Selwyn Piramuthu
University of Florida
Gainesville, FL, USA

Juan M. Corchado
University of Salamanca
Salamanca, Spain

ISSN 1865-0929   ISSN 1865-0937 (electronic)
Communications in Computer and Information Science
ISBN 978-981-15-4827-7   ISBN 978-981-15-4828-4 (eBook)
https://doi.org/10.1007/978-981-15-4828-4

This Springer imprint is published by the registered company Springer Nature Singapore Pte Ltd.
The registered company address is: 152 Beach Road, #21-01/04 Gateway East, Singapore 189721, Singapore

Preface

These proceedings contain the papers presented at the 5th International Symposium on Signal Processing and Intelligent Recognition Systems (SIRS 2019). SIRS aims to provide the most relevant opportunity to bring together researchers and practitioners from both academia and industry to exchange their knowledge and discuss their research findings. The symposium was held in Trivandrum, Kerala, India, during December 18–21, 2019. SIRS 2019 was hosted by the Indian Institute of Information Technology and Management-Kerala (IIITM-K). The symposium provided a forum for the sharing, exchange, presentation, and discussion of original research results in both methodological issues and different application areas of signal processing, computer vision, and pattern recognition.

In response to the call for papers, 63 papers were submitted for presentation and inclusion in the proceedings of the conference. The papers were evaluated and ranked on the basis of their significance, novelty, and technical quality. A double-blind review process was conducted to ensure that the author names and affiliations were unknown to the Technical Program Committee (TPC). Each paper was reviewed by the members of the TPC and finally, 19 regular papers, 5 invited papers, and 8 short papers were selected for presentation at the symposium.

We thank the program chairs for their wise advice and brilliant suggesstions in organizing the technical program. We would like to extend our deepest appreciation to the Advisory Committee members. Thanks to all members of the TPC, and the external reviewers, for their hard work in evaluating and discussing papers. We wish to thank all the members of the Organizing Committee, whose work and commitment were invaluable. Our most sincere thanks go to all the keynote speakers who shared with us their expertise and knowledge. We wish to thank all the authors who submitted papers and all participants and contributors of fruitful discussions. The EDAS conference system proved very helpful during the submission, review, and editing phases.

We thank IIITM-K for hosting the conference. Our sincere thanks to Dr. Saji Gopinath, Director at IIITM-K, for his continued support and cooperation. Recognition also goes to the Local Organizing Committee members who all worked extremely hard on every detail of the conference programs and social activities. We appreciate the contributions of all the faculty and staff of IIITM-K and the student volunteers who contributed their time to make the conference a great success.

We wish to express our gratitude to the team at Springer for their help and cooperation.

December 2019

Sabu M. Thampi
Rajesh M. Hegde
Sri Krishnan
Jayanta Mukhopadhyay
Vipin Chaudhary
Oge Marques
Selwyn Piramuthu
Juan M. Corchado

Organization

Chief Patron

Madhavan Nambiar IAS (Rtd.) IIITM-K, India

Patron

Saji Gopinath IIITM-K, India

General Chairs

Rajesh M. Hegde IIT Kanpur, India
Sri Krishnan Ryerson University, Canada
Jayanta Mukhopadhyay IIT Kharagpur, India
Vipin Chaudhary State University of New York at Buffalo, USA

General Executive Chair

Sabu M. Thampi IIITM-K, India

Technical Program Chairs

Oge Marques Florida Atlantic University, USA
Selwyn Piramuthu University of Florida, USA
Theodoros Tsiftsis Jinan University, China
Juan M. Corchado University of Salamanca, Spain

Workshop Chairs

Anustup Choudhury Dolby Laboratories, USA
Maheshkumar H. Kolekar IIT Patna, India
Domenico Ciuonzo University of Naples Federico II, Italy

Advisory Committee

Francesco Masulli University of Genoa, Italy
Pramod K. Varshney Syracuse University, USA
Dharma P. Agrawal University of Cincinnati, USA
Soura Dasgupta The University of Iowa, USA
Kuan-Ching Li Providence University, Taiwan
Sushmita Mitra Indian Statistical Institute Kolkata, India

Hideyuki Takagi	Kyushu University, Japan
Deniz Erdogmus	Northeastern University, USA
Sergey Mosin	Kazan Federal University, Russia

Technical Program Committee

Hossein Malekinezhad	Michigan Technological University, USA
V. B. Surya Prasath	Cincinnati Children's Hospital Medical Center, USA
Jun Qin	Southern Illinois University Carbondale, USA
Thakshila Wimalajeewa	Syracuse University, USA
Peng Zhang	Stony Brook University, USA
Zhe Zhang	Geroge Mason University, USA
Sergey Biryuchinskiy	Vigitek, Inc., USA
Anustup Choudhury	Dolby Laboratories, USA
Sharath Pankanti	IBM Research, USA
Vesh Raj Sharma Banjade	Intel Corporation, USA
Yingmeng Xiang	GEIRI North America, USA
Ognjen Arandjelovic	University of St Andrews, UK
Cathryn Peoples	Ulster University, UK
John Soraghan	University of Strathclyde, UK
Quoc-Tuan Vien	Middlesex University, UK
Wenwu Wang	University of Surrey, UK
Hector Zenil	Oxford University, UK, and Karolinska Institute, Sweden
Burhan Gulbahar	Ozyegin University, Turkey
Yasin Kabalci	Nigde Omer Halisdemir University, Turkey
Tarek Bejaoui	University of Paris-Sud 11, France
Elyas Rakhshani	Delft University of Technology, The Netherlands
Thaweesak Yingthawornsuk	King Mongkut's University of Technology Thonburi, Thailand
Grienggrai Rajchakit	Maejo University, Thailand
Bao Rong Chang	National University of Kaohsiung, Taiwan
I-Cheng Chang	National Dong Hwa University, Taiwan
Tzung-Pei Hong	National University of Kaohsiung, Taiwan
Gwo-Jiun Horng	Southern Taiwan University of Science and Technology, Taiwan
Jiachyi Wu	National Taiwan Ocean University, Taiwan
Emilio Jiménez Macías	University of La Rioja, Spain
Juan Pedro Lopez Velasco	Universidad Politécnica de Madrid, Spain
Gema Piñero	Universitat Politècnica de València, Spain
Jithin Ravi	Universidad Carlos III de Madrid, Spain
Addisson Salazar	Universidad Politécnica de Valencia, Spain
Viranjay Mohan Srivastava	University of KwaZulu-Natal, South Africa
Theo Swart	University of Johannesburg, South Africa
Ales Zamuda	University of Maribor, Slovenia
Anna Antonyová	University of Prešov, Slovakia

Chau Yuen	Singapore University of Technology and Design, Singapore
Dragana Baji	University of Novi Sad, Serbia
Ervin Varga	University of Novi Sad, Serbia
Sameera Abar	King Khalid University, Saudi Arabia
Haikal El Abed	Technical Trainers College (TTC), Saudi Arabia
El-Sayed El-Alfy	King Fahd University of Petroleum and Minerals, Saudi Arabia
Kashif Saleem	King Saud University, Saudi Arabia
Justin Dauwels	Nanyang Technological University, Singapore
Salvatore Distefano	University of Kazan, Russia
Andrey Krylov	Lomonosov Moscow State University, Russia
Anh Huy Phan	SKOLKOVO Institute of Science and Technology, Russia
Felix Albu	Valahia University of Targoviste, Romania
Mihaela Albu	Politehnica University of Bucharest, Romania
Valentina Balas	Aurel Vlaicu University of Arad, Romania
Doru Florin Chiper	Gheorghe Asachi Technical University of Iasi, Romania
Bogdan Dumitrescu	Politehnica University of Bucharest, Romania
Radu Vasiu	Politehnica University of Timisoara, Romania
Faycal Bensaali	Qatar University, Qatar
Fernando Perdigão	Instituto de Telecomunicações, Portugal
Adão Silva	Instituto de Telecomunicações, University of Aveiro, Portugal
Anna Bartkowiak	University of Wroclaw, Poland
Andrzej Borys	Gdynia Maritime University, Poland
Tomasz Piotrowski	Nicolaus Copernicus University, Poland
Zbigniew Piotrowski	Military University of Technology, Poland
Ryszard Tadeusiewicz	AGH University of Science and Technology, Poland
Belal Amro	Hebron University, Palestine
Bin Cao	Harbin Institute of Technology Shenzhen Graduate School, China
Jiayu Chen	Wuhan University, China
Mingkai Chen	Nanjing University of Posts and Telecommunications, China
Shiwen He	Central South University, China
Wei-Chiang Hong	Jiangsu Normal University, China
Philip Moore	Lanzhou University, China
Wei Wei	Xi'an University of Technology, China
Jingjin Wu	BNU-HKBU United International College, China
Kui Xu	Army Engineering University of PLA, China
Yuhua Xu	PLA University of Science and Technology, China
Qiang Yang	Zhejiang University, China
Jihong Yu	Beijing Institute of Technology, China, and Simon Fraser University, Canada

Rongqing Zhang	Tongji University, China
Aniruddha Bhattacharjya	Tsinghua University, China
Ramesh Rayudu	Victoria University of Wellington, New Zealand
Raouyane Brahim	University of Hassan II, Morocco
Ahmed El Oualkadi	Abdelmalek Essaadi University, Morocco
César Cárdenas	Tecnológico de Monterrey Campus Guadalajara, Mexico
Rosaura Palma-Orozco	Instituto Politécnico Nacional, Mexico
Badrul Hisham Ahmad	Universiti Teknikal Malaysia, Malaysia
Mohd Ashraf Ahmad	Universiti Malaysia Pahang, Malaysia
Kamarulafizam Ismail	Universiti Teknologi Malaysia, Malaysia
Thinagaran Perumal	University Putra Malaysia, Malaysia
Bahbibi Rahmatullah	Universiti Pendidikan Sultan Idris, Malaysia
Muhammad Rashid	IIUM Gombak Malaysia, Malaysia
Ibrahim Nasir	Sebha University, Libya
Mohammad Abushariah	The University of Jordan, Jordan
Toru Takahashi	Osaka Sangyo University, Japan
Toshio Tsuji	Hiroshima University, Japan
Tomonobu Sato	Hitachi ICT Business Services, Ltd., Japan
Paolo Crippa	Università Politecnica delle Marche, Italy
Gianluigi Ferrari	University of Parma, Italy
Ugo Fiore	University of Naples Federico II, Italy
Stavros Ntalampiras	Università degli studi Milano, Italy
Pisana Placidi	University of Perugia, Italy
Alessandro Testa	Ministry of Economy and Finance, Italy
Kenneth Dawson-Howe	Trinity College Dublin, Ireland
Dheyaa Sabr Al Azzawi	Wasit University, Iraq
Reza J. Alitappeh	Mazandaran University of Science and Technology (MAZUST), Iran
Hamed Mojallali	University of Guilan, Iran
Hugeng Hugeng	Universitas Tarumanagara, Indonesia
Shikha Agrawal	Rajiv Gandhi Proudyogiki Vishwavidyalaya, India
Shamim Akhter	JIIT, India
Shajith Ali	SSN College of Engineering, India
Krishna Battula	Jawaharlal Nehru Technological University Kakinada, India
Chinmay Chakraborty	BIT Mesra, India
Nishant Doshi	PDPU, India
Nithin George	IIT Gandhinagar, India
Surajeet Ghosh	Indian Institute of Engineering Science and Technology, India
Ashish Goel	JIIT, India
Madhu Jain	JIIT, India

Shruti Jain	Jaypee University of Information Technology, India
Raveendranathan K. Chellappan	College of Engineering Thiruvananthapuram, India
Vineet Khandelwal	JIIT, India
Mofazzal Khondekar	B. C. Roy Engineering College, India
Anirban Kundu	Netaji Subhash Engineering College, India
Noor Mahammad Sk	IIITDM Kancheepuram, India
T. C. Manjunath	Dayananda Sagar College of Engineering, India
Joycee Mekie	IIT Gandhinagar, India
Abhishek Midya	NIT Silchar, India
Vinay Mittal	IIT Chittoor, India
Ravibabu Mulaveesala	IIT Ropar, India
Kalavathi Palanisamy	The Gandhigram Rural Institute - Deemed University, India
Dipti Patil	Pune University, India
Munaga Prasad	IDRBT, India
Sathidevi P. Savithri	NIT Calicut, India
Vadivelu R.	Anna University, India
Jaynendra Kumar Rai	Amity University Uttar Pradesh, India
Harikumar Rajaguru	Bannari Amman Institute of Technology, India
Priya Ranjan	Amity University Noida, India
G. Ramachandra Reddy	Vellore Institute of Technology, India
Priti Rege	College of Engineering Pune, India
Andrews Samraj	Mahendra Engineering College, India
Jankiballabh Sharma	Rajasthan Technical University, India
Neeru Sharma	Jaypee University of Information Technology, India
Raghvendra Sharma	Amity University Gwalior, India
Pushpendra Singh	Bennett University, India
Arun Sinha	Vellore Institute of Technology, India
China Sonagiri	Institute of Aeronautical Engineering, India
Seshan Srirangarajan	IIT Delhi, India
Ciza Thomas	College of Engineering Trivandrum, India
Nilesh Uke	University of Pune, India
Madhur Upadhayay	Shiv Nadar University, India
Samudra Vijaya	IIT Guwahati, India
Rajeev Shrivastava	MPSIDC, India
Hrishikesh Sharma	Innovation Labs, Tata Consultancy Services Ltd., India
Jayakumar Singaram	Independent IoT Consultant, India
Kalman Palagyi	University of Szeged, Hungary
Jozsef Vasarhelyi	University of Miskolc, Hungary
Hing Keung Lau	Hong Kong Institute of Vocational Education, Hong Kong
Dimitris Ampeliotis	University of Patras, Greece
Grigorios Beligiannis	University of Patras, Greece

Konstantinos Blekas	University of Ioannina, Greece
Katerina Kabassi	Ionian University, Greece
Michael Vrahatis	University of Patras, Greece
George Tambouratzis	Institute for Language and Speech Processing, Greece
Ilka Miloucheva	Media Applications Research, Germany
Matthias Vodel	Chemnitz University of Technology, Germany
Valerio Frascolla	Intel, Germany
Munir Georges	Intel, Germany
Lotfi Chaari	University of Toulouse, France
Aladine Chetouani	Polytech Orleans, France
Paul Honeine	Université de Rouen, France
Pascal Lorenz	University of Haute Alsace, France
Kester Quist-Aphetsi	University of Brest France, France
Patrick Siarry	University of Paris 12, France
Mohamed Moustafa	Egyptian Russian University, Egypt
Eduardo Pinos	Universidad Politécnica Salesiana, Ecuador
Vladislav Skorpil	Brno University of Technology, Czech Republic
George Dekoulis	Aerospace Engineering Institute, Cyprus
Nizar Bouguila	Concordia University, Canada
Stefka Fidanova	Institute of Information and Communication Technologies, Bulgaria
Minh-Son Dao	Universiti Teknologi Brunei, Brunei
Marcelo Alencar	Federal University of Campina Grande, Brazil
Rodrigo Capobianco Guido	São Paulo State University, Brazil
Joao Paulo da Costa	University of Brasília, Brazil
Marcio Eisencraft	Escola Politécnica da Universidade de São Paulo, Brazil
Alexandre Gonçalves Silva	Federal University of Santa Catarina, Brazil
Waslon Lopes	Universidade Federal da Paraíba, Brazil
Wemerson Parreira	Federal University of Santa Catarina, Brazil
Joel Rodrigues	National Institute of Telecommunications (Inatel), Brazil
Otavio Teixeira	Universidade Federal do Para (UFPA), Brazil
Suryakanthi Tangirala	University of Botswana, Botswana
José Luis Hernandez Ramos	European Commission - Joint Research Centre (JRC), Belgium
Gustavo Fernández Domínguez	Austrian Institute of Technology (AIT), Austria

Organized by

IIITM-Kerala

Contents

Signal and Image Processing

Intelligent Recognition Techniques and Applications

Acoustic Identification of Nocturnal Bird Species

Michelangelo Acconcjaioco and Stavros Ntalampiras[(✉)]

Department of Computer Science, University of Milan,
Via Celoria 18, 20133 Milan, Italy
michelangelo.acconcjaioco@studenti.unimi.it, stavros.ntalampiras@unimi.it

Abstract. Automatic classification of bird species based on their vocalizations is a topic of crucial relevance for the research conducted by biologists, ornithologists, ecologists, and related disciplines. This work in concentrated on nocturnal species; even though the analysis of their population trends is a key indicator, there is a gap in the literature addressing their audio-based identification. After compiling a suitable dataset including six nocturnal bird species, this study employs both supervised (k-Nearest Neighbors, Support Vector Machines) and unsupervised (k-means) methods operating in the feature space formed by Mel-spectrograms. We conclude that automatic classification based on k-Nearest Neighbour and Support Vector Machines provide almost excellent results with a recognition rate in the order of 90%.

Keywords: Bird vocalizations · Bioacoustics · Acoustic ecology · Machine learning · Audio signal processing · Audio pattern recognition

1 Introduction

Acoustic bird monitoring could serve a series of environmental and scientific purposes [13,14], such as:

- reduce the need of volunteers in this kind of biological projects,
- assist the observations made by amateur bird watchers,
- catalog biodiversity automatically,
- identify and count birds in a specific habitat of interest,
- record long-term population trends, etc.

Automatic acoustic classification of bird species is currently attracting the interest of the research community [7,25,26]. Such frameworks are typically composed by signal processing and pattern recognition modules. Signal processing typically involves extraction of features suitable for the problem at hand. Machine learning algorithms use these features to make inferences including prediction, classification and, in general, characterization of novel audio patterns [15,19,20].

Animals use acoustic vocalizations as a very efficient way to communicate since sound does not require visual contact between emitter and receiver, it can

© Springer Nature Singapore Pte Ltd. 2020
S. M. Thampi et al. (Eds.): SIRS 2019, CCIS 1209, pp. 3–12, 2020.
https://doi.org/10.1007/978-981-15-4828-4_1

travel over long distances and can carry the information content, it remains practically unaffected under low visibility conditions, e.g. dense forests, night, rain etc. [3,21]. Animals communicate for reasons that are vital for their existence. Vocalizations are employed for mate attraction, territorial defense and for early warning of other species' members that a dangerous predator is in presence.

Animals produce a variety of sounds to communicate [4,22]. These sounds vary from short simple calls (also called pulses, syllables, or notes) to versatile long songs, which are composed of a complex hierarchy of syllables (very common in singing birds). The communication strategy and the diversity of the frequency structure of the sound pattern depend heavily on the environmental context of the living organism.

Sound recognition is based on the fact that a sound source emits consistent acoustic patterns with a distinctive and characteristic way to distribute its energy over time on its composing frequencies, which constitutes its so-called spectral signature [16]. This spectral signature comprises a unique pattern that can be revealed and subsequently identified automatically by employing statistical pattern classification techniques.

Bird sounds can be broadly classified as songs and calls which can be further divided into hierarchical levels of phrases, syllables and elements or notes. Songs are longer vocalizations which usually include a variety of notes in a sequence while bird calls are short communications which are often the single notes [2]. To the best of our knowledge, the literature does not include solutions targeting nocturnal species specifically as they typically comprise a relatively small part of the dataset [24]. The only work studying such species is a preliminary one [1] addressing flight calls of five species.

With the availability of automated recording units, one can collect data facilitating long term monitoring of remote locations even without human supervision [18]. This is essential in the case of nocturnal species where clean vocalizations can be obtained as less environmental and/or man-made noise is typically present. This work is based on the promise that non-intrusive automatic acoustic monitoring can provide useful information towards explaining population trends and patterns of nocturnal bird species. Such information could complement other modalities, such as video, temperature, humidity, etc.

In this work, we extract the Mel-Frequency Cepstral Coefficients towards characterizing bird calls of the following six species (a) *Bubo bubo*, (b) *Strix aluco*, (c) *Glaucidium passerinum*, (d) *Tyto alba*, (e) *Athene cunicularia*, and (f) *Megascops kennicottii*. Such a compilation of species may serve not only automated monitoring but educational purposes too assisting young students, amateur observers, etc. Based on the related literature, subsequent processing includes the application of both supervised and unsupervised classification methods. More precisely we considered k-Nearest Neighbour, Support Vector Machines and k-means clustering. Encouraging performance was achieved showing the efficacy of the proposed classification scheme.

The rest of this work is organized as follows: Sect. 2 analyzes data collection, pre-processing and feature extraction stages. The next two Sects. (3 and 4) briefly describe the supervised and unsupervised classification algorithms respectively. Finally, Sect. 5 draws our conclusions.

2   Data Collection, Pre-processing and Feature Extraction

This section explains the data collection, audio signal pre-processing, and feature extraction stages of this work.

2.1   Data Collection and Pre-processing

The audio files employed in this paper come from Xeno-Canto database, which is a citizen-science project, i.e. a repository in which volunteers are able to record, upload and annotate recordings of bird songs and calls. Since its birth in 2005, it has collected over 400,000 sound recordings from more than 10,000 species worldwide, and has become one of the largest collections of bird vocalizations in the world. All recordings are published under appropriate Creative Commons licenses[1].

The following six bird species were chosen covering both European and American autochthonous species:

– *Bubo bubo* (Eagle Owl) - Europe
– *Strix aluco* (Tawny Owl) - Europe
– *Glaucidium passerinum* (Owl) - Europe
– *Tyto alba* (Barn Owl) - Europe
– *Athene cunicularia* (Burrowing Owl) - South America
– *Megascops kennicottii* (Western Screech-Owl) - North America

It is interesting to note that nocturne rapacious species were considered in this paper. These were selected because the recordings available on Xeno-Canto repository are practically noise-free, hence facilitating the modeling process. Thus, the considered classification algorithms operate on the structure of bird vocalizations alone both during training and testing phases. Representative spectrograms of the considered species are shown in Fig. 1.

The dataset is composed of 167 files with 25–30 files for each species. Each signal was normalized by removing the corresponding mean value to compensate for potential microphone calibration problems. Each audio file contains 1 to 4 calls while the maximum duration is 8 s. 70% of the dataset was used for model training and validation and the learned models were assessed on the remaining 30%.

---

[1] Xeno Canto, https://www.xeno-canto.org/about/xeno-canto.

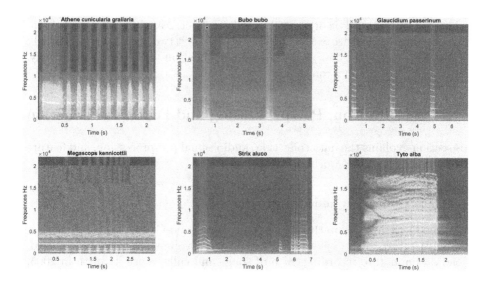

Fig. 1. Representative spectrograms of the considered species.

2.2 Feature Extraction

The audio signal, is voluminous and as such, it is hard to process it directly in any analysis task. We therefore need to transform the initial data representation to a more suitable one, by extracting audio features that represent the properties of the original signals while reducing the volume of data.

Mel Frequency Cepstral Coefficients (MFCCs) are able to describe the spectrum of an audio recording in very compact, yet informative manner. Feature extraction (shown in Fig. 2) can be viewed as a data rate reduction procedure because it reduced the available audio information to a feature set of relatively small dimensionality [5].

An audio signal is constantly changing, so to simplify its processing we assume that on short time scales the audio signal exhibits stationary behavior. We frame the signal into 20–40 ms frames. Shorter frames don't include enough information to get a reliable spectral estimates, while in longer ones the signal changes significantly throughout the frame and important information might be discarded [10,11,17].

This approach is also employed during the feature extraction stage; the audio signal is broken into overlapping frames and a set of features is computed per frame. In this work the frame size is set to 30 ms, and the overlap between subsequent frames is 15 ms.

This type of processing generates a sequence, denoted as F, of feature vectors per audio signal. The dimensionality of the feature vector depends on the nature of the adopted features; in this work, the feature vector is a matrix $n \times 35$ where n is the frames number in which a file is divided and 35 is the number of extracted features. During the specific process, we used the library provided by [5].

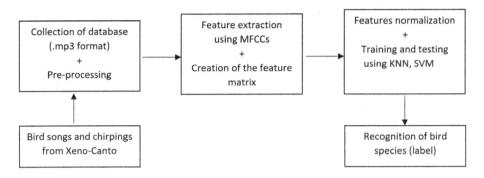

Fig. 2. Pipeline of the audio pattern recognition scheme for identifying nocturnal bird species.

After the feature extraction, we apply silence removal at the frame level during which only frames with energy values above a certain threshold are kept. Standard normalization techniques including mean removal and variance scaling are applied as well. After the transformation, all features included in the training set have zero mean and unit standard deviation.

3   Supervised Methods and Performances

In this work, the features matrix distribution is modeled by two different kinds of supervised machine learning methods used for classification, i.e. a discriminative and an instance-based one. These methods learn from the train feature matrix the main characteristics of a class and generate a specific model that is used for classify the samples included in the test-set.

3.1 k-NN and SVM

In brief, the standard version of k-NN was employed with the Euclidean distance as a metric [27], while the SVM is composed of a radial basis kernel following the line of thought explained in [12]. Both k-NN and SVM operate at the frame level, thus we obtain a prediction for each frame which is useful during online operation, i.e. when the model is applied on incoming streams of audio. Instead, when the target is classifying already archived data, a prediction at the file level is desired. Such a process is carried out by means of a majority voting scheme where all frames of each test file are taken into consideration and the most frequent class is selected to label the unknown file. Following such a scheme is expected to provide improved performance w.r.t the one operating at the frame level since more information is available during testing, while frame predictions which are inconsistent with the predicted flow of classes are discarded.

Table 1. The confusion matrix (in %) achieved by the k-NN approach ($k = 10$). The average recognition rate is 94%. The highest rates per species are emboldened.

Presented	Responded					
	eagle owl	*owl*	*screech owl*	*tawny owl*	*barn owl*	*burrowing owl*
eagle owl	**94.09**	0	5.42	0	0.49	0
owl	0.13	**98.03**	0	1.84	0	0
screech owl	3.23	0.15	**96.62**	0	0	0
tawny owl	0	0.21	0.21	**89.69**	4.74	5.15
barn owl	1.07	1.43	0.12	1.43	**93.44**	2.5
burrowing owl	0.39	0.96	0	0	6.74	**91.91**

Table 2. The confusion matrix (in %) achieved by the SVM approach. The average recognition rate is 88.8%. The highest rates per species are emboldened.

Presented	Responded					
	eagle owl	*owl*	*screech owl*	*tawny owl*	*barn owl*	*burrowing owl*
eagle owl	**78**	0	22	0	0	0
owl	0.07	**96**	1.2	2.8	0	0
screech owl	3.2	0	**97**	0.23	0	0
tawny owl	0.82	1.0	2.1	**85**	3.7	7.2
barn owl	3.7	1.1	0.12	0	**92**	2.6
burrowing owl	0.77	0.19	0	0.19	15.0	**84**

3.2   Analysis of Results for k-NN and SVM

As described in Subsect. 2.1, the dataset was split in two parts (test and train) and the features extraction is performed for both of them. It's important to underline that the train and test sets are the same for all classification methods enabling a reliable comparison.

Tables 1 and 2 include the confusion matrices obtained by k-NN and SVM at the frame level. We can see that both supervised methods achieve high recognition rates. The average rates are 94% and 88.8% for k-NN and SVM respectively. k-NN outperforms SVM in classifying all classes and by 5.2% on average. As we can see the species recognized best is the *common owl* (98.03%) and the worst one is *Tawny owl* (89.69%). The value of k, i.e. 10, was determined during the training phase as the value providing the maximum classification rate on a validation set, which is a non-overlapping part of the training one.

These rates are significantly boosted when entire files are considered. More precisely, the k-NN made only one error where a *Tawny Owl* song is confused a with a *Burrowing Owl* song. SVM made three errors during this classification experiment; indeed, it confuses an *Eagle Owl* song with a *Western Screech Owl* song, a *Tawny Owl* song with a *Burrowing Owl song* and a *Burrowing Owl* song with a *Barn Owl* song.

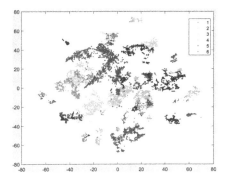
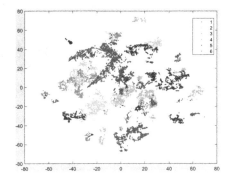

Fig. 3. Clusters in 2 dimension representation. (Color figure online)

Fig. 4. Clusters in 3 dimension representation. (Color figure online)

4   Unsupervised Clustering

The unsupervised method used in this work is k-means clustering. In this setting, a train set is absent and all features are normalized, juxtaposed and grouped in k clusters. Here, the ideal outcome would be that all observations describing the same species belong to the same cluster. Similarly to Sect. 3, the Euclidean distance was used as a similarity metric between the extracted features.

The obtained results are illustrated in Figs. 3 and 4 which provide a interpretative representation in 2 and 3 dimensional planes correspondingly. Suitable dimensionality reduction was carried out using the standard t-distributed stochastic neighbor embedding algorithm [8].

By observing Figs. 3 and 4 we can see that clusters are better distinguished in the 3D plot instead of the 2D. However, the largest parts of almost all clusters are distinguishable which, given the dimensionality reduction carried out for illustration purposes, shows the efficacy of the feature set used in this work. Clusters 4 (cyan) and 5 (blue) are not clearly visible (neither in 2D nor in 3D) and the respective observations are mostly scattered in small groups occupying different parts of the graph. At the same time, these are the clusters with highest number of outliers. In order to assess the quality of the clustering in a quantitative manner as well, the silhouette criterion [23] was computed. Its result is demonstrated in Fig. 5.

Silhouette analysis can be used to study separability between the resulting clusters. This metric lies in a range of $[-1, 1]$, while we are interested in the values corresponding to a 6-cluster setting equal to the number of bird species considered here.

It should be noted that silhouette coefficients (as these values are typically referred to) close to $+1$ indicate that the sample is far away from the neighboring clusters. A value of 0 indicates that the sample is very close to the decision boundary between two neighboring clusters and negative values indicate that those samples might have been assigned to the wrong cluster.

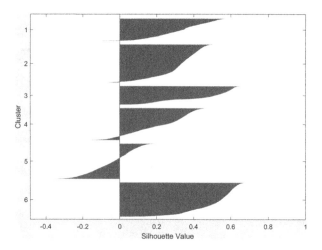

Fig. 5. The silhouette plot considering 6 clusters.

As we can see in Fig. 5 there are lot of outliers (negative value) for cluster 5 and its silhouette coefficient indicates that it isn't well separated from his neighbor clusters. Less misclustered samples are observed in clusters 4, 1, and 2 (descending order). We can argue that the clustering performance is satisfactory as the vast majority of samples is properly clustered.

5   Conclusions

This work evaluated both supervised and unsupervised classification schemes on nocturnal bird species identification based on their vocalizations. We conclude that k-NN and SVM operating in the Mel-filterbank space provide excellent performances. At the same time the results are clearly interpretable by observing the neighbors and separating hyperplanes in k-NN and SVM respectively.

Our future work includes (a) assessing the system's performance in real world conditions following the line of thought explained in [18] and (b) explore more complex classificaion schemes [6,9] while always focusing on interpretability so that interested parties (e.g. biologists, ornithologists, psychologists, computer scientists, etc.) can gain clear insights regarding the problem at hand by analyzing the operation of the modeling approach.

References

1. Bastas, S.A.: Nocturnal bird call recognition system for wind farm applications. Master's thesis, The University of Toledo (2011)
2. Fagerlund, S., Laine, U.K.: New parametric representations of bird sounds for automatic classification. In: 2014 IEEE International Conference on Acoustics, Speech and Signal Processing (ICASSP), pp. 8247–8251, May 2014. https://doi.org/10.1109/ICASSP.2014.6855209

3. Farnsworth, A., et al.: Reconstructing velocities of migrating birds from weather radar - a case study in computational sustainability. AI Mag. **35**(2), 31 (2014). https://doi.org/10.1609/aimag.v35i2.2527

4. Fink, D., et al.: Crowdsourcing meets ecology: hemisphere-wide spatiotemporal species distribution models. AI Mag. **35**(2), 19 (2014). https://doi.org/10.1609/aimag.v35i2.2533

5. Giannakopoulos, T., Pikrakis, A.: Introduction to Audio Analysis: A MATLAB® Approach. Academic Press, Cambridge (2014)

6. Koops, H.V., van Balen, J., Wiering, F.: Automatic segmentation and deep learning of bird sounds. In: Mothe, J., et al. (eds.) CLEF 2015. LNCS, vol. 9283, pp. 261–267. Springer, Cham (2015). https://doi.org/10.1007/978-3-319-24027-5_26

7. Laiolo, P.: The emerging significance of bioacoustics in animal species conservation. Biol. Conserv. **143**(7), 1635–1645 (2010). https://doi.org/10.1016/j.biocon.2010.03.025

8. van der Maaten, L., Hinton, G.: Visualizing data using t-SNE. J. Mach. Learn. Res. **9**, 2579–2605 (2008). http://www.jmlr.org/papers/v9/vandermaaten08a.html

9. Ntalampiras, L.A., et al.: Automatic classification of cat vocalizations emitted in different contexts. Animals **9**(8), 543 (2019). https://doi.org/10.3390/ani9080543

10. Ntalampiras, S.: Hybrid framework for categorising sounds of Mysticete whales. IET Signal Proc. **11**(4), 349–355 (2017). https://doi.org/10.1049/iet-spr.2015.0065

11. Ntalampiras, S.: Moving vehicle classification using wireless acoustic sensor networks. IEEE Trans. Emerg. Top. Comput. Intell. **2**(2), 129–138 (2018). https://doi.org/10.1109/TETCI.2017.2783340

12. Ntalampiras, S.: Audio pattern recognition of baby crying sound events. J. Audio Eng. Soc. **63**(5), 358–369 (2015). https://doi.org/10.17743/jaes.2015.0025

13. Ntalampiras, S.: Bird species identification via transfer learning from music genres. Ecol. Inform. (2018). https://doi.org/10.1016/j.ecoinf.2018.01.006, https://www.sciencedirect.com/science/article/pii/S1574954117302467

14. Ntalampiras, S.: Automatic acoustic classification of insect species based on directed acyclic graphs. J. Acoust. Soc. Am. **145**(6), EL541–EL546 (2019). https://doi.org/10.1121/1.5111975

15. Ntalampiras, S.: Generalized sound recognition in reverberant environments. J. Audio Eng. Soc. **67**(10), 772–781 (2019). https://doi.org/10.17743/jaes.2019.0030

16. Ntalampiras, S.: On acoustic monitoring of farm environments. In: Thampi, S.M., Marques, O., Krishnan, S., Li, K.-C., Ciuonzo, D., Kolekar, M.H. (eds.) SIRS 2018. CCIS, vol. 968, pp. 53–63. Springer, Singapore (2019). https://doi.org/10.1007/978-981-13-5758-9_5

17. Ntalampiras, S., Potamitis, I., Fakotakis, N.: A multidomain approach for automatic home environmental sound classification, Makuhari, Japan, September (2010)

18. Potamitis, I.: Automatic classification of a taxon-rich community recorded in the wild. PLoS ONE **9**(5), e96936 (2014). https://doi.org/10.1371/journal.pone.0096936

19. Raghuram, M.A., Chavan, N.R., Belur, R., Koolagudi, S.G.: Bird classification based on their sound patterns. Int. J. Speech Technol. **19**(4), 791–804 (2016). https://doi.org/10.1007/s10772-016-9372-2

20. Rai, P., Golchha, V., Srivastava, A., Vyas, G., Mishra, S.: An automatic classification of bird species using audio feature extraction and support vector machines. In: 2016 International Conference on Inventive Computation Technologies (ICICT), vol. 1, pp. 1–5, August 2016. https://doi.org/10.1109/INVENTIVE.2016.7823241

21. Riede, K.: Monitoring biodiversity: analysis of Amazonian rainforest sounds. Ambio **22**(8), 546–548 (1993). http://www.jstor.org/stable/4314145

22. Riede, K.: Acoustic monitoring of Orthoptera and its potential for conservation. J. Insect Conserv. **2**(3/4), 217–223 (1998). https://doi.org/10.1023/a:1009695813606

23. Rousseeuw, P.J.: Silhouettes: a graphical aid to the interpretation and validation of cluster analysis. J. Comput. Appl. Math. **20**, 53–65 (1987). https://doi.org/10.1016/0377-0427(87)90125-7

24. Salamon, J., et al.: Towards the automatic classification of avian flight calls for bioacoustic monitoring. PLOS ONE **11**(11), e0166866 (2016). https://doi.org/10.1371/journal.pone.0166866

25. Stowell, D., Benetos, E., Gill, L.F.: On-bird sound recordings: automatic acoustic recognition of activities and contexts. IEEE/ACM Trans. Audio Speech Lang. Process. **25**(6), 1193–1206 (2017). https://doi.org/10.1109/TASLP.2017.2690565

26. Stowell, D., Wood, M.D., Pamuła, H., Stylianou, Y., Glotin, H.: Automatic acoustic detection of birds through deep learning: the first Bird Audio Detection challenge. Methods Ecol. Evol. **10**(3), 368–380 (2018). https://doi.org/10.1111/2041-210x.13103

27. Witten, I.H., Frank, E., Hall, M.A.: Data mining: practical machine learning tools and techniques. In: Morgan Kaufmann Series in Data Management Systems, 3 edn. Morgan Kaufmann, Amsterdam (2011). http://www.sciencedirect.com/science/book/9780123748560

Internet of Assistants: Humans Can Get Assistance Anywhere, Anytime and Any Areas

Minh-Son Dao[1]([✉]) and Mohamed Saleem Haja Nazmudeen[2]

[1] Big Data Analytics Laboratory, National Institute of Information and Communications Technology (NICT), Koganei, Japan
dao.minhson@gmail.com
[2] Universiti Teknologi Brunei, Bandar Seri Begawan, Brunei
mohamed.saleem@utb.edu.bn

Abstract. Internet of Things (IoT) applications such as Smart Cities, Smart Healthcare, etc. are tremendously on the rise which leads us to the research domain of how humans interact with the smart environments in terms of privacy, autonomy, and efficiency. One such research domain is that of intelligent personal assistants that target the participants of the smart application environments. So far, studies in the area of Personal Assistants (PA) mainly contribute towards assisting individuals in interacting with the smart environment with respect to their privacy and autonomy. We would like to take the PAs to the next level so that we could provide customized services to them based on their privacy settings. In this paper, we introduce a new paradigm called the Internet of Assistants (IoA) where assistants from personal and smart environments are interconnected so that they can interact with each other using ontology. We provide the architectural framework called Evolutional Smart Assistant Platform (ESAP) that enables the successful implementation of smart assistant applications. IoA provides numerous possibilities of applications few of which are discussed in this paper. For instance, in a smart eating environment, a PA of a diner can interact with a service assistant (SA) of a restaurant to provide a customized, enjoyable and healthy dining experience.

1 Introduction

Internet of Things (IoT) consists mainly of three layers namely, (1) sensors and actuators, (2) networking and (3) applications (Mahmoud et al. 2015). IoT sensors not only include physical sensors it also consists of virtual sensors where information can be gathered from social networks which are termed as cyber physical world (Poovendran 2010; Rajkumar et al. 2010). Due to tremendous sensing capabilities and the information generated the possibility of effective use of this information is various. This leads to the complexity and concerns for participants in the IoT applications which need to be addressed. Here, we propose a solution to this problem by utilizing the usage analytics, heterogeneous sensing, data retrieval and applied machine learning that opens up a new arena of opportunity which we term it as Internet of Assistants (IoA). IoA also facilitates a human centred approach which focuses on comfort, vitality and high quality of life. To realize the IoA concept we propose two agents namely, Personal Assistant (PA) and Service Assistant (SA). The basic constructs of both the assistants are the same

© Springer Nature Singapore Pte Ltd. 2020
S. M. Thampi et al. (Eds.): SIRS 2019, CCIS 1209, pp. 13–22, 2020.
https://doi.org/10.1007/978-981-15-4828-4_2

nevertheless the difference arises depending on service provided or consumed in the IoA environment.

PAs that makes use of machine learning and artificial intelligence are being developed to address privacy issues and complexity. These PAs are capable of making autonomous decisions on behalf of the participants with respect to smart environments using personal sensors. One such scenario is that the permission for letting the environment use the personal profile of the participant from their smartphone could be managed by PAs based on user preferences (Santos et al. 2016; Chayapathy et al. 2017). However, the contributions so far in this domain are only based on PAs to smart environment. Our proposed (IoA) consists of interconnected PAs and SAs with capabilities to collaborate autonomously within the context of preassigned privacy settings. IoA enables PAs and SAs to communicate with each other through ontologies and they can negotiate based on their preferences and privacy settings to come up with a suitable use case model.

2 Motivation and Purposes

"Humans have always been interested in understanding themselves and their environment. Understanding their relationship with the environment is important to survival as well as thriving in the present situation and planning for the future" (Jain and Jalali 2014).

The above statement kindled a fire of a plan for our current research: *To design a mechanism that can match the needs of the right information required by humans and resources information collected from the environment to understand a human-environment relationship towards having suitable actions*. This mechanism can make a significant change in information retrieval: humans do not need to look for information, but the information comes to humans depending on their needs.

The scope of applying this mechanism is very large and various. Applications built based on this mechanism could vary from personal to societal areas such as personal healthcare, personal assistant, situation management, smart-assistant, smart-environment, and the internet of assistants (IoA). The internet of assistants is the new terminology invented by the candidate and will be discussed later.

The proposed mechanism concerns two types of data: (1) *personal data*: the emergence of wearable sensors and cloud-computing has revolutionized the ability to objectively measuring physical, physiological, and mental activities of humans towards feeding data to help humans particularly understanding themselves, and (2) *environment data*: the non-stop development of social networks and internet of things (IoT) has provided the big data that recorded the environment status and how humans interact and perceive their surrounding environment objectively and subjectively.

Although there are many works focusing on getting insights from these types of data, few of them try to link them together to meet the needs of humans and reasonably explain causal relations among these insights. For example, human behaviour can be understood by using smartphones but there is no cue to answer whether the surrounding environment influences such behaviour or vice versa (Dao et al. 2017).

Another issue is how to choose the right data sources among heterogeneous data sources to link these types of data or to extract reasonable insights by integrating these types of data? In other words, given N sensors and M models needed to be built from N sensors, the problem is how to build M models from N sensors automatically and expert independently? For example, given N physical activities sensors (e.g. accelerometer, gyroscope), M physiological sensors (e.g. ECG, heart pulse, blood pressure), P environment sensors (e.g. rain, humidity, wind), and the context of asthma patient, how to select an optimal subset of sensors to find the correlation/association between asthma attacks, environment changing, and patient's lifestyle? (Dao et al. 2015). Another example is presented in [4] that whether we confirm the causal relations among students' lifestyles, emotions, and in-classroom interaction?

3 Architecture

The strategy of building this proposed system is to start from the generic framework, contextualize this framework to specific categories, get the common factors from these categories' frameworks, refine the generic framework, and continue looping the cycle until getting the concrete and precise framework.

Then, three main categories are built based on the proposed mechanism: (1) Personal assistant, (2) Service assistant, and (3) Internet of Assistants.

3.1 Personal Assistant (PA)

Differ from existing personal assistants, the PA built by the candidate focuses on using data from physical, physiological, and mental sensors to create several models that reflect personal situations according to a spatio-temporal-context constraint. These models will help to store personal data on a semantic level instead of low-level signal data. For example, in (Dao et al. 2015) PHASOR is an instance of PA that can help humans understand their own activities objectively and automatically by feeding a stream of daily activities to their lifelog instead of accelerometer and gyroscope signal data. A PA is abstractly designed as the pair of <personal situations, needs information>_[spatio, temporal, context] where the first component represents a set of models, the second component expresses the needs of extra information from other assistants, and the index denotes the domain where the PA is established.

3.2 Service Assistant (SA)

The SA uses data gathered from the environment including physical, social, and human sensors to generate the environment situation according to a given condition. The terminology of the human sensor is used to express the data generated by humans subjectively but not from their own situation such as multimedia recorded by multimedia devices and published on cyber. The SA is abstractly designed as a pair of <environment situations, required information>_[spatio, temporal, context], where the first component represents a set of models used to understand the environment situation (e.g. weather situations = (cold, hot, wind, snow); natural disaster situations = (flood, landslide, earthquake); hospital situations = (available experts of a certain disease, vacant patient beds)), the second component expresses the required information from other assistants, and the index denotes the domain where the SA is established. For example, in (Jalali et al. 2015), the attempt of creating an instance of the SA is presented. In this work, the environment situations component contains semantic information of weather such as high temperature, strong wind, low humidity, and these states are connected by the extended finite-state automata. The required information component is expressed as a "correlation between asthma attacks and weather situation".

3.3 Internet of Assistants (IoA)

The idea of the Internet of Assistants is to create an internet of assistants including both personal and environment assistants to help humans understand the human-environment relationships towards to survive as well as thrive in the present situation and plan for the future. In order to do this, any pair <personal assistant, service assistant> can be linked together by using the "needs information" component to activate the "required information" component to create a new node of relationships. This node itself could, in turn, be another assistant and contribute to the evolution of IoA. With IoA, independent systems developed by different sources can connect together to exchange, share, and enrich information towards enlarging the IoA. Therefore, humans can get the information they need anywhere and anytime.

4   Use Cases

4.1   A New Way for Shopping – Fast, Secure, Convenience

Mary
Passes by a Shop
Personal Assistant
Recognize a fashion shop
Checks "connection" option set by Mary
If **yes**, start connecting to a service assistant of the shop for sharing Mary's great taste in fashion
Shop Service Assistant
Capture her image
Gather her great taste in fashion
Send to the personal assistant list of products with prices (GAN images generated by using Mary's image and fashion catalogues)
Personal Assistant
Connect to Mary's bank
Display list of products (emphasize of items whose prices she can affordable)
Mary
Decide to buy one item
Personal Assistant
Connect to Mary's bank service assistant and the shop's bank service assistant: transfer money
Connect to Mary's house service bank: gather time to delivery ->send to the shop's service assistant

Shop Service Assistant
Connect to Delivery company's service assistant
Deliver a package to Mary's house

See Fig. 1.

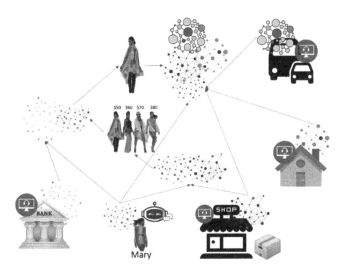

Fig. 1. Shopping with IoA

4.2 A New Way for Smart Eating - Fast, Healthy and Comfortable

John
Passes by a Restaurant
Personal Assistant (PA)
Recognize a restaurant
Checks "connection" option set by John
If **yes**, start connecting to a service assistant of the restaurant for sharing John's eating habit
Restaurant Service Assistant (R-SA)
Gather his eating habit
Send to the PA list of selected meals with names and calories
Personal Assistant
Connect to John's health service assistant (**H-SA**)
Display list of meals he can eat to avoid any damage from food (e.g. allergy, diabetes)
John
Decide to order food and go inside
Personal Assistant
Connect to the R-SA to order a meal

Restaurant Service Assistant
Prepare the meal and a seat
Personal Assistant
Connect to John's gym service assistant (G-SA) (e.g.
workout's methods) and H-SA (e.g. personal health statis-
tics) to make the plan for workout
John
Enjoy his meal and his workout time.

See Fig. 2.

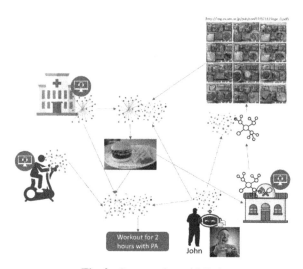

Fig. 2. Smart eating with IoA

4.3 A New Way for a Smart Loan - Fast, Paperless and Convenient

John
Wants to apply for a loan
Personal Assistant (PA)
Activate related personal information portal
Bank's Service Assistant (SA)
Connects to **John's PA**
collect the details of John's business
connects to **government's SA** for collecting further legal
details about John's business
connects to the **revenue department's SA** for collecting
the revenue details of John's business for previous years
connect to **location services' SA** for tracking the move-
ent of John related to his business operations
analyses all the data to provide a verdict for loan ap-
proval which can be immediately processed by the manager
Bank loan to John is approved

See Fig. 3.

Fig. 3. Loan processing using IoA

5 Evolutional Smart Assistant Platform (ESAP)

Figure 4 illustrates the architecture of the evolutional smart assistant platform (ESAP). Green blocks denote methods, technologies, and approaches that are utilized to proceed data while blue blocks describe the output of these green blocks. The data analysis is divided into the following stage,

- **Integration Discovering Analysis:** Gather and aggregate data collected from sensors: In this stage, data collected from sensors or other applications and sources are integrated into a data warehouse. A Big Data system is built to handle such big data that comes from different sources. The integration of discovering analysis should be performed here to aggregate data towards decreasing the complicated diversity of data format.
- **Discover event patterns:** In this stage, event patterns are discovered and extracted by utilizing up-to-date techniques of Machine learning, data mining, data system, and artificial intelligence. Here, the characteristic of the domain where such event patterns are extracted plays an important role to make sure these event patterns are useful for that domain. Discovered vent patterns are stored in the Event Ledger. An event indexing and querying schema are built here to manage events that come from different domains.

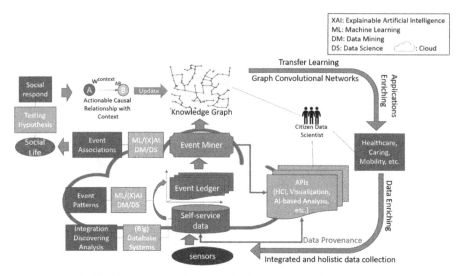

Fig. 4. Evolutional Smart Assistant Platform (ESAP) (Color figure online)

- **Exploit and explore event association:** Machine learning, data mining, data system, and artificial techniques are applied here to build event miner tool. This tool aims to exploit and explore event associations that can represent complicated situations happen in real life. Specific models are built based on these associations towards having an ability to detect and predict complex events.
- **Test hypothesis:** This stage is very important to get rid of complex event models that do not suit real situations. The social testing field is expected to use for determining which complex event models can be useful for human life.
- **Update knowledge graph:** Complex event models that survive after the hypothesis testing stage will be updated to a knowledge graph and be ready for use.
- **Build application:** By reusing, transferring, and sharing knowledge graph and APIs. The collaborative development can be leveraged to fast build a new application. In this stage, APIs mashup should be built to share all developed models so people can utilize it to access the knowledge graph as well as integrate different models to build a new model.

Since ESAP shares knowledge and models via APIs mashup, more and more applications are built quickly towards launching into IoAs more and more data. These data will be the new input data of ESAP. Thus, ESAP can feed, learn, and evolve during its life cycle. The APIs also eliminates the need for experts with high-level programming skills. The applications can easily be built by smartly matching suitable ESAPs to the queried ESAP to answer/recommend to users. A prototype of an assistant is shown in Fig. 5.

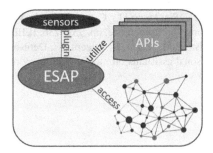

Fig. 5. A universal prototype of an assistant

6 Conclusion

In this paper, we have proposed how usage analytics, heterogeneous sensing, data retrieval and applied machine learning can all be brought into one framework namely ESAP which can pave way for a new paradigm of applications that we term as IoA. We have elaborated on the underlying mechanism how PAs, SAs, and IoAs could be built. Currently we are in the process of developing the ESAP and testing it in various use cases from different business and service sectors such as banking, healthcare, etc. We foresee a lot of challenges in designing the service and personal assistants in such a way that all parties place their trust in the system. Once the system is built there is a need for an empirical study to understand how IoA will be perceived both by end-user and business users in smart city/nation context. In order for this system to be successful proper incentives should be devised for all parties involved in the system.

References

Chayapathy, V., Anitha, G.S., Sharath, B.: IOT based home automation by using personal assistant. In: 2017 International Conference on Smart Technologies for Smart Nation (SmartTechCon), pp. 385–389. IEEE (2017)

Dao, M.S., Nguyen, D., Tien, D., Riegler, M., Gurrin, C.: Smart lifelogging: recognizing human activities using PHASOR (2017)

Dao, M.-S., Zettsu, K., Pongpaichet, S., Jalali, L., Jain, R.: Exploring spatio-temporal-theme correlation between physical and social streaming data for event detection and pattern interpretation from heterogeneous sensors. In: 2015 IEEE International Conference on Big Data (Big Data), pp. 2690–2699. IEEE (2015)

Jalali, L., Dao, M.-S., Jain, R., Zettsu, K.: Complex asthma risk factor recognition from heterogeneous data streams. In: 2015 IEEE International Conference on Multimedia & Expo Workshops (ICMEW), pp. 1–6. IEEE (2015)

Poovendran, R.: Cyber–physical systems: close encounters between two parallel worlds [point of view]. Proc. IEEE **98**, 1363–1366 (2010)

Jain, R., Jalali, L.: Objective self. IEEE Multimedia **21**, 100–110 (2014). https://doi.org/10.1109/MMUL.2014.63

Mahmoud, R., Yousuf, T., Aloul, F., Zualkernan, I.: Internet of things (IoT) security: current status, challenges and prospective measures. In: 2015 10th International Conference for Internet Technology and Secured Transactions (ICITST), pp. 336–341 (2015)

Rajkumar, R., Lee, I., Sha, L., Stankovic, J.: Cyber-physical systems: the next computing revolution. In: Design Automation Conference, pp. 731–736. IEEE (2010)

Santos, J., Rodrigues, J.J., Silva, B.M., Casal, J., Saleem, K., Denisov, V.: An IoT-based mobile gateway for intelligent personal assistants on mobile health environments. J. Netw. Comput. Appl. **71**, 194–204 (2016)

Artificial Intelligence Enabled Online Non-intrusive Load Monitoring Embedded in Smart Plugs

Ruiqi Guo[1,2], Yingmeng Xiang[1(✉)], Zeyu Mao[1,3], Zhehan Yi[1], Xiaoying Zhao[1], and Di Shi[1]

[1] GEIRI North America, San Jose, CA 95134, USA
{yingmeng.xiang,zhehan.yi,xiaoying.zhao,di.shi}@geirina.net
[2] Mechanical Engineering Department, University of California-Berkeley, Berkeley, CA 94720, USA
ruiqiguo@berkeley.edu
[3] Department of Electrical and Computer Engineering, Texas A&M University, College Station, TX 77843, USA
zeyumao2@tamu.edu

Abstract. As an Internet-of-Things (IoT) device for smart homes, smart plugs have been pervasive in households, which enable users to monitor and control their electrical appliances remotely and automatically. It is promising that, the networks of smart plugs in the power system will enable autonomous demand response for optimal grid operation. This benefits power systems from several aspects, e.g., enhancing renewable penetration and reducing the peak load. In order to facilitate energy management and minimize the impact of load shedding on the customers, it is meaningful to know the type of appliances connected to the smart plugs in a real-time and non-intrusive manner. Conventionally, the online non-intrusive load monitoring (NILM) is conducted in a central server, and it requires a large number of measurements transmitted to the cloud, which can impose a huge communication burden. In this paper, using the edge-computing capability of the smart plugs, some lightweight artificial intelligence-based NILM algorithms are developed and implemented inside the smart plugs. These practical algorithms are validated using massive hardware experiments. Case studies indicate that high accuracy can be achieved for NILM with limited measurements and limited storage.

Keywords: NILM · Smart plugs · Edge intelligence · Artificial intelligence

1 Introduction

As one of the most critical infrastructures in the world, electric power systems play a vital role in supporting our human society and economy development. Thus, the security and reliability of power systems should always be guaranteed, while the economy of power systems should be improved. In order to enhance the security, reliability, and economy of power systems, demand response programs are developed around the world. In demand response programs, the end-user customers can change their electricity usage patterns in

© Springer Nature Singapore Pte Ltd. 2020
S. M. Thampi et al. (Eds.): SIRS 2019, CCIS 1209, pp. 23–36, 2020.
https://doi.org/10.1007/978-981-15-4828-4_3

response to the electricity prices or the incentives, and it is an important research topic how to develop and implement the demand response programs in a cost-effective manner [1]. With the rapid development of information and computer technologies, Internet-of-Things (IoT) is emerging as a promising solution to promote demand response [2]. Actually, State Grid Corporation of China (SGCC), which provides most of the electricity supply in China, has just launched an ambitious strategic plan to integrate the IoT technologies with the electric power grids [3].

As a typical IoT device, smart plugs (or smart outlets) [4] have been widely deployed by electricity consumers. A smart plug typically can monitor the power usage of the appliance plugged into it in real-time, transmit the power usage as well as other measurements to the central server, and receive remote commands or settings to control the operation of the connected appliance. Considering these functions, smart plugs can turn the conventionally passive appliances into "smart" appliances, monitor and control the appliances in a more effective manner for demand response. And it is expected that smart plugs might be pervasive in the near future, serving as a critical device that bridges a huge number of distributed appliances with power system operators.

From the perspective of power system operators/administrators, it is of great importance to know the type of appliances connected to the smart plugs. In this way, the power system operators/administrators can more accurately monitor and predict the load demands, develop better strategies to promote the demand response and curtail the less important appliance when load shedding is needed. For example, when a contingency (such as a major generation tripping) happens and load shedding is desired, the power system operators can curtail the less important appliances (such as air-conditioners and refrigerators) and do not affect the normal operation of those more important appliances (like desktop PCs and monitors).

However, despite the merits, smart plugs are not able to automatically detect the type of appliances connected to them. The type of appliances cannot be known by the power system operators/administrators unless the customers manually identify the type and update it on the user interface. This manual detection method will obviously pose a great burden on the customers. Also, the appliances (including laptop PCs, lamps, heaters, vacuums, etc.) connected to a smart plug can change frequently, and it is not realistic to rely on the customers to manually detect the type of appliances.

In recent years, some non-intrusive load monitoring (NILM) methods were proposed [5]. A majority of these NILM methods are for the smart meter, which can disaggregate the power consumption to different types of appliances. A very limited number of NILM methods were developed for smart plugs to detect the type of appliances [6]. However, in those methods, the NILM algorithms are running in a central server, and it requires the smart meters/smart plugs to continuously send the measurements to a central server, which can greatly increase the communication burden, and incur privacy and cybersecurity issues. Thus, it is desired to develop some online NILM methods which can run on-board inside a smart plug.

It should be noted here that such on-board NILM methods are greatly different from those NILM methods that run in a server. First, on-board NILM methods need to be lightweight as smart plugs usually have limited computational capability. Second, they should not require a huge amount of memory or storage, since the microprocessors in the smart plugs usually have quite limited memory or storage. Third, they should be simple and easy to implement. This is because in a computer/server it is easy to implement some complicated methods with machine learning algorithm libraries, but this is quite difficult for a smart plug.

Considering state-of-the-art research, the major novelty of this paper is that some on-board NILM methods are developed for online applications based on the authors' previous research and real-world implementation of the Grid Sense system [7]. Further, those on-board NILM methods are validated using massive hardware experiments.

The rest of this paper is organized as follows. The problem formulation is described in Sect. 2 and the solution algorithms are provided in Sect. 3. Experimental validations are conducted in Sect. 4. Section 5 concludes the paper and suggests future research directions.

2   Problem Formulation

Figure 1 depicts an overview of the hardware architecture for NILM in a smart plug. An electric appliance can be connected to a smart plug, which is capable of precisely measuring the real-time power usage information through its sensor. The microcontroller unit (MCU) inside the smart plug continuously receives the measurements transmitted from its sensor and analyze them with an embedded NILM algorithm. The NILM algorithm can quickly give an identification result to automatically update the type of appliance that is connected to the smart plug. The updated appliance type information will be eventually sent to a cloud server or a smart hub.

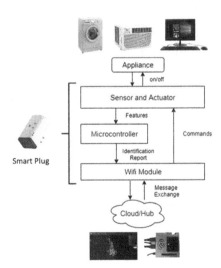

Fig. 1. Hardware architecture

The problem formulation of NILM for smart plugs is shown in Fig. 2. The NILM algorithm takes real-time measurements, including power, current, and power factor, and identify the type of appliance considering a dataset for appliance characteristics. The results of the NILM, i.e., type of the appliance, are sent to the communication module. The database for appliance characteristics should be carefully established offline before applied online.

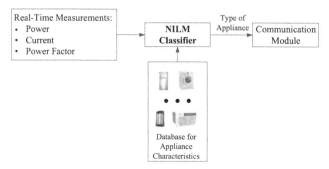

Fig. 2. Illustration of the NILM in a smart plug

3   Solution Approach

This section provides the solution approach to the NILM problem, as well as the metrics to evaluate the performance of the solution approach.

3.1   Setup of Hardware Test System

A general framework for the NILM machine learning algorithm is shown in Fig. 3. It can be divided into two phases: the offline training phase and the online application phase. In the training phase, the database for appliance characteristics is established using given data of the measurements and the corresponding type of appliances. When

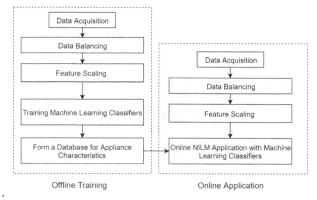

Fig. 3. General framework of machine learning NILM algorithm

the measurement data is acquired, raw data balancing is performed, then feature scaling is conducted. The scaled data is fed to the machine-learning-based NILM classifiers. With sufficient training, a database for appliance characteristics will be achieved, which will benefit the online application.

Data Balancing. The raw input data consists of the real-time measurements of power, current and power factor for each device at each moment. These measurements together with the corresponding states of the appliances connected to the smart plug are stored as samples. The total number of samples for each appliance are different and depends on the frequency of appliance usage: some appliances have longer usage hours such as air conditioner, while the others (such as microwave oven) may have relatively lower usage. The classification categories are not equally represented, and the performance of a classifier, might not be appropriate when the data is imbalanced. As a result, data balancing becomes an essential first step for algorithm development. Usually the class-balanced data can be created by under-sampling or over-sampling. Synthetic minority oversampling technique (SMOTE) is a very popular oversampling method that is usually beneficial with low-dimensional data [8], which is ideal for the NILM of smart plugs since each measurement only has three dimensions (i.e., power, current, power factor).

Feature Scaling. The three dimensions of raw data have different scales and ranges: the current value usually ranges from 0 to 20 A, the power value ranges from 0 to 2000 W, and the power factor value ranges from 0 to 1. While many machine learning algorithms count the Euclidean distance between two data points, if one of the features has a much broader range of values than other features, this feature will domain the distance. Therefore, the range of all features should be normalized so that each feature contributes equally to the final distance. In this research, StandardScaler is applied to standardize the data [9]. The standardized value X of an input x is calculated as:

$$X = \frac{x - u}{s} \tag{1}$$

where u is the mean of the training samples, and s is the standard derivation of training samples.

Machine Learning Classifiers. This step involves passing the training data to different kinds of machine learning classifiers and obtaining the trained classifiers that could be used for load identification in the microcontrollers of smart plugs [10, 11]. The training process can be completed on a computer to handle a large amount of data, but it should be guaranteed that the trained classifiers can be applied in the microcontrollers of smart plugs. Two types of well-known lightweight classification algorithms are evaluated [12].

- **Neural Network.** The neural network (NN) algorithm is first considered. One of the most widely used NN classifiers is the multi-layer perception which is has a robust and simple structure and could handle inputs with noise. The typical structure of a neural network consists of the input layer, the hidden layer, and the output layer. The parameters in a neural network are trained iteratively through a back-propagation process. After training for sufficient epochs, the parameters are optimized and become ready for online application.

- **Decision Tree.** Two classification models, i.e., the random forest (RF) and extra tree (ET), are considered. Different from neural networks, the tree-based models are non-parametric logical (symbolic) learning methods. They are not much affected by the impact of data outliers, and can well deal with linearly inseparable data. Decision trees classify different instances based on the nodes and branches. Each node represents a feature to be sorted with certain strategies, and each branch connected with the nodes represents a criterion that the feature can compare with. Starting from the root node, each instance can be classified into the correct category while passing through all the necessary nodes and branches. Figure 4 demonstrates a typical structure of the decision tree. More details about RT and ET can be found at [13].

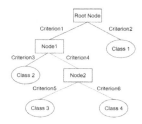

Fig. 4. Overview of decision tree

3.2 Metrics for Evaluating the Performance of NILM

The performances of NILM algorithms are evaluated using a comprehensive set of metrics as follows. Generally, a complete view of a system's performance is usually given by accuracy, precision (or true negative rate), recall (or true positive rate), and F-score [14]. These evaluation criterions are all calculated from the confusion matrix of each classification process. Figure 5 shows an exemplary three-class confusion matrix to demonstrate the contribution of the confusion matrix to the performance evaluation of multi-class classification problems.

	Actual Class		
Predicted Class	1	2	3
1	TP_1	E_{21}	E_{31}
2	E_{12}	TP_2	E_{32}
3	E_{13}	E_{23}	TP_2

Fig. 5. Confusion matrix

TP_1 is the number of samples that are correctly classified from class 1, which is described as True Positive 1. E_{12} is the misclassified samples from class 2 but judged as class 1; similarly, E_{31} is the class 3 samples predicted incorrectly as class 1. The sum of E_{21} and E_{31} is the false negative of class 1 (FN_1), which represents the sum of all the errors in the predicted class 1 row. On the other hand, the samples from class 1 can also be misclassified as class 2 and class 3, the errors are E_{12} and E_{13} while the sum of them is the false positive of class 1 (FP_1). The true positive, false negative and false positive of class 2 and class 3 can be obtained in the same way. Then the accuracy (Acc) of the classifier, overall precision (TNR), recall (TPR) and F-1 score (F1) of each class can be calculated as shown below:

$$\text{Acc} = \frac{\sum_{i=1}^{n} TP_i}{\sum_{i=1}^{n} TP_i + FN_i} = \frac{\sum_{i=1}^{n} TP_i}{\sum_{i=1}^{n} TP_i + FP_i} \tag{2}$$

$$\text{TPR}_i = \frac{TP_i}{TP_i + FN_i} \tag{3}$$

$$\text{TNR}_i = \frac{TP_i}{TP_i + FP_i} \tag{4}$$

$$\text{F1}_i = \frac{2 \times \text{TNR}_i \times \text{TPR}_i}{\text{TNR}_i + \text{TPR}_i} \tag{5}$$

The graphical analysis according to receiver operating characteristics (ROC) curves [15] is also applied in this paper. The advantage of ROC curve is that the shape of the curve will not change with the total number of samples in different classes. For a multi-class problem, there will be one ROC curve for each class. The possibility for each sample to be classified into a specific class is a score. In ROC analysis, the scores are set as the thresholds for classification, which yields different TPR and TNR value combinations. Therefore, TPR represents y-axis and FPR represents x-axis. The area under the ROC curve (AUC) is a quantitate calibration to compare the performance of different classifiers; the higher the AUC value, the closer the classifier will become to an optimal classifier (green line in Fig. 6).

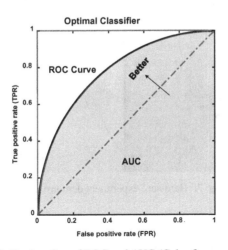

Fig. 6. Explanation of ROC and AUC (Color figure online)

4 Experimental Verification

Hardware experiments are conducted to validate the proposed NILM algorithms, and the performance is analyzed and compared in detail.

4.1 Setup of Hardware Test System

The data collection device for the proposed NILM system is modified from the Sonoff S31 [16], a compact size 1 gang US standard Wi-Fi smart plug with energy monitoring. The energy sensing module for the device is CSE7766 and Wi-Fi module is ESP8266. With the S31 smart plug, we keep track of real-time power, current, and power factor of different home appliances in a laboratory. The energy usage for each appliance is collected every 0.5 s. The data transmission is realized through the MQTT protocol and saved to a computer for training purposes. The machine learning classifiers are trained with scikit.learn [9] and encoded to the ES8266 chip of S31 for appliance identification. The hyperparameter settings for the three kinds of classifiers are obtained through grid search. Finally, the real-time identification results for devices connected to S31 are transmitted and shown on the PC end under the frequency of 1 Hz. The demonstration for the event detection is shown in Fig. 7.

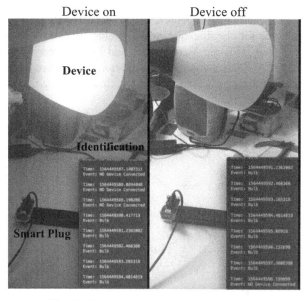

Fig. 7. Hardware experiment demonstration.

4.2 Characterization of Measurements

In practice, we gathered data from 11 different appliances over a three-week period. The class index and device name of the 11 appliances are shown in Table 1. The total dataset

Table 1. List of 11 appliances used in the experiments

Class index	Device name	Class index	Device name
0	Air conditioner	6	TV
1	Bulb	7	Vacuum cleaner
2	Water heater	8	Heat fan
3	Hairdryer	9	Micro-oven
4	Computer	10	Cook-top
5	Monitor		

Table 2. Scaling of measurements

Measurements	u	s
Current (A)	6.20	5.85
Power (W)	698.91	651.61
Power Factor (%)	94.46	109.93

is balanced with SMOTE algorithm and split into training sets to train the machine learning classifiers and testing sets to evaluate the performance. The training sample size is 22638 and the testing sample size is 15092. The StandardScaler is obtained from Eq. (1) with training data, the mean value (u) and standard derivation value (s) are shown in Table 2.

By fitting the training data to the StandardScaler, the standardized training dataset is obtained and boxplots for the standardized training data are drawn to analyze the data distribution.

As shown in Figs. 8, 9, and 10, distributions of features vary in different devices. There are data overlaps among different kinds of devices for each feature, but the overlap effect could be offset by combining three features together, as shown in Fig. 11.

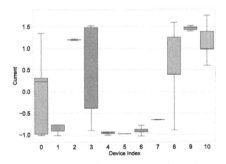

Fig. 8. Current values after standardization

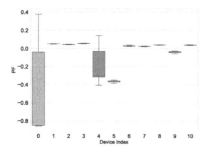

Fig. 9. Power factor values after processing

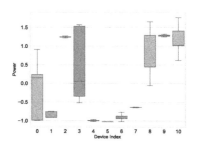

Fig. 10. Power values after processing

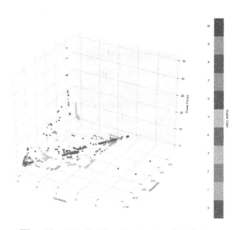

Fig. 11. 3D plot for the standardized data

4.3 NILM Results

The ROC curves for the three models are put together to qualitatively evaluate the effectiveness of the feature combination of current, power, and power factor. As shown in Fig. 12, all of the three models are very close to the optimal classifier. The AUC values of different classes in each classifier are similar and could be approximated as 1, which indicates that all of the three chosen classifiers are performing satisfactorily with the correct feature combination.

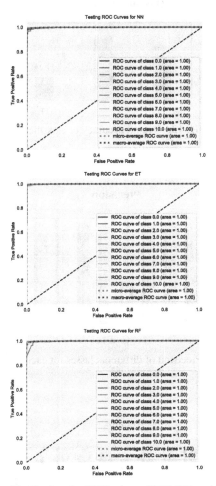

Fig. 12. Testing the ROC curve of NN, ET, and RF

To further compare the performance of these three models, a more quantitative analysis is necessary. Thus, the overall accuracy, recall, precision and F-1 score based on the confusion matrix are compared among the NN, ET and RF models. Overall identification accuracy is shown in Fig. 13, which shows that the random forest model has the highest overall accuracy based on the existing data.

The precision, recall, and F-1 score values of different classes for each classifier are shown in Figs. 14, 15 and 16. From these values, it can be observed that the random forest model and neural network model have better performance than the extra tree model, while random forest has the highest precision, recall and F-1 score values in all classes.

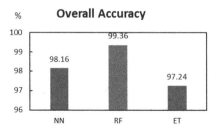

Fig. 13. Overall accuracy of classifiers

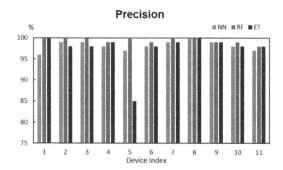

Fig. 14. Precision of different classes for each classifier

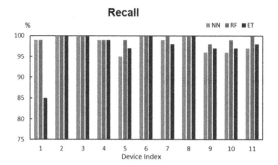

Fig. 15. Recall of different classes for each classifier

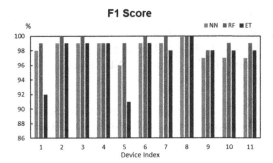

Fig. 16. F-1 score of different classes for each classifier

4.4 Analysis of Storage Requirements for the NILM Algorithms

After implementing the three classifiers in the smart plugs, the occupancy of Read-Only Memory (ROM) and Random-Access Memory (RAM) of the ESP8266 board are further evaluated to make sure it is within the maximum storage limit. The final result is shown in Fig. 17, and it can be found that the NN model has the highest RAM occupancy but the lowest ROM occupancy, the RF model and ET model almost have the same RAM occupancy but RF model has less ROM occupancy. All of the three models are within the limit of storage, which validates the proposed NILM methods are lightweight and effective.

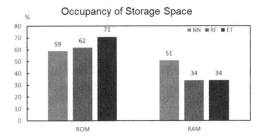

Fig. 17. Occupancy of storage space.

5 Conclusions and Future Work

This paper proposes several artificial intelligence algorithms (including neural network, random forest, and extra tree) for the on-board NILM in smart plugs. First, the hardware architecture is described; second, the NILM algorithms are explained in detail step by step; then, the metrics for evaluating the performance of NILM algorithms are studied. Hardware experiments are carried out, and the results validate that the proposed methods are effective and lightweight. The proposed NILM methods can be easily implemented in general smart plugs with limited computational capability and storage space. The NILM methods can run on-board, and detect the type of appliances plugged to the smart plugs accurately in real-time. Since the NILM algorithms are running in the smart plugs, they do not require a massive amount of measurement transmitted from the smart plugs to a central server, which can greatly reduce the communication burden and provide a practical and cost-effective way for NILM.

In the future, the proposed NILM will be tested considering more uncertainties and noises. And the computational efficiency of the NILM methods will be further improved with guaranteed accuracy.

Acknowledgment. This work is supported by the SGCC Science and Technology Program under project Distributed High-Speed Frequency Control under UHVDC Bipolar Blocking Fault Scenario (SGGR0000DLJS1800934).

References

1. Haider, H.T., See, O.H., Elmenreich, W.: A review of residential demand response of smart grid. Renew. Sustain. Energy Rev. **59**, 166–178 (2016)
2. Bedi, G., Venayagamoorthy, G.K., Singh, R., Brooks, R.R., Wang, K.-C.: Review of Internet of Things (IoT) in electric power and energy systems. IEEE Internet Things J. **5**, 847–870 (2018)
3. South China Morning Post News. https://www.scmp.com/news/china/society/article/3034684/chinas-largest-utility-plans-national-power-grid-integrating
4. Musleh, A.S., Debouza, M., Farook, M.: Design and implementation of smart plug: an Internet of Things (IoT) approach. In: 2017 International Conference on Electrical and Computing Technologies and Applications (2017)
5. Hosseini, S.S., Agbossou, K., Kelouwani, S., Cardenas, A.: Non-intrusive load monitoring through home energy management systems: a comprehensive review. Renew. Sustain. Energy Rev. **79**, 1266–1274 (2017)
6. Barker, S., Musthag, M., Irwin, D., Shenoy, P.: Non-intrusive load identification for smart outlets. In: IEEE International Conference on Smart Grid Communications, pp. 548–553 (2014)
7. Xiang, Y., Lu, X., Yu, Z., Shi, D., Li, H., Wang, Z.: IoT and edge computing based direct load control for fast adaptive frequency regulation. In: IEEE PES General Meeting (2019)
8. Blagus, R., Lusa, L.: SMOTE for high-dimensional class-imbalanced data. BMC Bioinformatics **14**, 106 (2013). https://doi.org/10.1186/1471-2105-14-106
9. Cournapeau, D., et al.: scikits.learn: machine learning in Python. http://scikit-learn.sourceforge.net
10. Duan, J., Xu, H., et al.: Zero-sum game-based cooperative control for onboard pulsed power load accommodation. IEEE Trans. Industr. Inform. **16**, 238–247 (2019)
11. Duan, J., et al.: Deep-reinforcement-learning-based autonomous voltage control for power grid operations. IEEE Trans. Power Syst. **35**, 814–817 (2020)
12. Kotsiantis, S., Kanellopoulos, D., Pintelas, P.: Data preprocessing for supervised learning. Int. J. Comput. Sci. **1**, 111–117 (2006)
13. Singh, A., Thakur, N., Sharma, A.: A review of supervised machine learning algorithms. In: 2016 3rd International Conference on Computing for Sustainable Global Development, pp. 1310–1315 (2016)
14. Goutte, C., Gaussier, E.: A probabilistic interpretation of precision, recall and F-score, with implication for evaluation. In: Losada, D.E., Fernández-Luna, J.M. (eds.) ECIR 2005. LNCS, vol. 3408, pp. 345–359. Springer, Heidelberg (2005). https://doi.org/10.1007/978-3-540-31865-1_25
15. Tharwat, A.: Classification assessment methods. Appl. Comput. Inform. (2018). https://www.sciencedirect.com/science/article/pii/S2210832718301546
16. Itead (2019). https://www.itead.cc/sonoff-s31.html

Using Evaluation Data Analytics
in Environmental Education Projects

Katerina Kabassi(✉), Aristotelis Martinis, and Athanasios Botonis

Department of Environment, Ionian University, 29100 Zakynthos, Greece
{kkabassi,amartinis}@ionio.gr, nasbotonis@gmail.com

Abstract. The importance of evaluating the effectiveness of Environmental Education (EE) programs cannot be underestimated. However, only a few evaluation experiments of these programs are implemented and, therefore, it is not easy to collect data about the evaluation of environmental education projects. As a result, we present the design of a tool that helps users implement environmental education projects by collecting raw evaluation data and processing these data using multi-criteria decision making-theories. More specifically, instructors can add EE projects and/or evaluating existing ones. But most importantly, instructors may choose the projects that they are interested in and the system uses a combination of two Multi-Criteria Decision Making theories in order to evaluate all projects added in the system, rank them and present the best ones to the instructor interacting with the system.

Keywords: Data analytics · Environmental Education · Multi-Criteria Decision Making

1 Introduction

In schools, Environmental Education (EE) is often delivered through an educational program. The objective is to change the learners' cognitive, affective and participatory knowledge, skills and behavior [3]. EE is considered a very important tool for solving many of the environmental problems. This has lead to a wide acceptance of environmental education programs in most parts of the world [2]. Most of the EE initiatives target young people developing their capacity to make important choices and to improve their knowledge and behavior in the natural environment [12]. Environmental education is of great importance for raising awareness among young people and societies in general, so evaluating such programs and actions is a necessary task that may lead to examination, confirmation and eventually, improvement of their outcomes [2].

Although the importance of evaluating the effectiveness of EE programs has been recognized, few systematic evaluations of these programs and their resulting impact on the environment have been conducted [8]. Carleton-Hug and Hug [3] focus on the importance of the evaluation and point out that evaluation should be conducted in the first stages of program implementation. Additionally, evaluation prior to the implementation of the EE project could help the improvement of the projects concerned.

© Springer Nature Singapore Pte Ltd. 2020
S. M. Thampi et al. (Eds.): SIRS 2019, CCIS 1209, pp. 37–47, 2020.
https://doi.org/10.1007/978-981-15-4828-4_4

According to Bitgood [1] the main factors for the scarcity of evaluation studies include: (1) lack of understanding of the evaluation process; (2) failure to give priority to evaluation; (3) concerns about possible negative consequences; (4) the kind of the institution (including management style, authority for decision-making, and leadership personalities); (5) lack of motivation, as well as the lack of consequences for avoiding evaluations. Zint, Dowd, and Covitt [14] add that most environmental educators feel unable to conduct evaluations to inform their practices or to commission quality external evaluation. At the same time, most environmental educators do not have the opportunity to take courses as part of their formal training or subsequent professional development through which they could learn about program evaluation [15].

Given this background, the focus of the reported here research is on designing a tool that collects data on the existing environmental education projects, helps users evaluate these projects to monitor these projects in order to support their evaluation. More specifically, the user selects the projects s/he is interested in and the system collects evaluation data of these projects. The system then uses a combination of two Multi-Criteria Decision Making models to analyze these data and draw conclusions on the quality of the examined projects. The models used are: Analytical Hierarchy Process (AHP) [9] and Fuzzy Technique for the Order of Preference by Similarity to an Ideal Solution (Fuzzy TOPSIS) [4, 7]. Our system is primarily designed to meet the needs of Environmental Educators without program evaluation expertise. The significance of the development of such a system is four-fold: First, this system will help in the reuse of EE programs. Second, the automation of the process simplifies the procedure and makes it accessible to almost anyone. Third, this system collects raw data about the EE evaluation. Fourth, data analytics on EE projects will help users in making a better decision on the EE project selection.

A very interesting effort to automate the evaluation process of EE programs is MEERA [15], which is a free, self-directed learning resource that informs environmental educators about the evaluation of EE programs and facilitates them accessing evaluation resources. The database of MEERA includes summaries and in-depth profiles EE program evaluations. However, the particular system mainly assists the evaluation of one EE program, after its implementation, it does not collect data analytics and combine them to evaluate different EE projects. A quite different approach in helping users evaluate EE projects is that of Fleming and Easton [6] who attempted to improve environmental educators' evaluation capacity through distance education. They describe a distance education course that enables environmental educators and natural resource professionals to conduct evaluations at the local level within the mission and resources of their organizations. However, our approach differs from these systems as it helps formal Environmental Educators with decision management matters. For this purpose, the proposed tool uses Multi-Criteria Decision Making theories.

2   Designing the System

The system collects information about the EE projects as well as the teachers and the students implementing these projects. The system's main functionality is presented in the diagram presented in Fig. 1. Users can:

- insert a new EE project
- evaluate an existing EE project
- select EE projects to compare them and select the best one

More specifically, instructors may evaluate an existing EE project using a questionnaire that has only multiple-answer questions that use linguistic terms. Questions that use linguistic terms are easier for users that are familiar with the evaluation process of EE projects. The questions correspond to specific criteria and evaluation data are stored in a database for future reference. The questions of the questionnaire are the following:

1. How flexible and adaptable to each age group of participants is (the EE project)?
   (a) Very Good (b) Good (c) Fair (d) Poor (e) Very Poor
2. Does the description of (the EE project) cover the topic adequately?
   (a) Very Good (b) Good (c) Fair (d) Poor (e) Very Poor
3. Is (the EE project) based on a pedagogical theory or does it use a pedagogical method?
   (a) Very Good (b) Good (c) Fair (d) Poor (e) Very Poor
4. Are the objectives of (the EE project) are explicitly expressed or stated?
   (a) Very Good (b) Good (c) Fair (d) Poor (e) Very Poor
5. How good is the overall impact of (the EE project), depending on the programming and available support material.
   (a) Very Good (b) Good (c) Fair (d) Poor (e) Very Poor
6. How good is the quantity and quality of a cognitive object offered to students
   (a) Very Good (b) Good (c) Fair (d) Poor (e) Very Poor
7. Are the skills cultivated through activities involving active student participation
   (a) Very Good (b) Good (c) Fair (d) Poor (e) Very Poor
8. How good is (the EE project) in changing the students' attitudes, intentions, and attitudes?
   (a) Very Good (b) Good (c) Fair (d) Poor (e) Very Poor
9. Please rate the enjoyment of the trainees throughout (the EE project)
   (a) Very Good (b) Good (c) Fair (d) Poor (e) Very Poor
10. How successful is (the EE project) in providing many different kinds of activities, interventions, and methods.
    (a) Very Good (b) Good (c) Fair (d) Poor (e) Very Poor

An instructor can also select two or more EE projects that s/he is interested in and wish to conduct a comparative evaluation. After the initiation of the comparison process of these projects by the instructor, the system uses the combination of AHP and Fuzzy TOPSIS as well as the evaluation data analytics stored in the system's database for creating a final ranking of these projects and propose the best one.

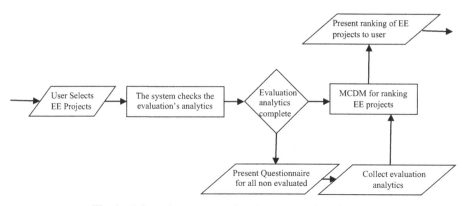

Fig. 1. A flow chart representing the system's functionality.

3   Employing AHP

Analytic Hierarchy Process [9] is one of the most popular MCDM theories. AHP aims to analyze a qualitative problem through a quantitative method. AHP is used for forming the set of criteria and calculating their weights using a group of evaluators. The first step in the implementation of AHP according to [13] is developing a goal hierarchy. For this purpose, the overall goal, the criteria, and the decision alternatives are arranged in a hierarchical structure. The overall goal is the EE programs' evaluation. However, we use only the first steps of the AHP application as AHP is used only for the estimation of the weights. For this reason, the essential steps in the application of AHP are: setting the group of evaluators (Subsect. 3.1), forming the set of criteria (Subsect. 3.2), creating a matrix of the criteria (Subsect. 3.3) in order to make pairwise comparisons of the criteria and then calculate their weights (Subsect. 3.4).

3.1   Setting the Group of Evaluators

A group of experts in EE was used in order to select the set of criteria and comment on the importance of the criteria by making pairwise comparisons of the criteria. The group was made of 10 environmental education experts. The demographics of the group participants were as follows: 50% were female, 60% had a master's degree and 30% a PhD. All of them had experience in the implementation of EE projects. The results of this process are reported next.

3.2   Forming the Set of Criteria

Stern, Powell, and Hill [11] conducted a systematic literature review of peer-reviewed research studies published between 1999 and 2010 that empirically evaluated the outcomes of environmental education (EE) programs in order to study how the characteristics of Environmental Education programs affect the expected results. They also investigated the relationship between the characteristics and the results of the evaluation.

According to this study, some criteria that are important in evaluating EE projects have been extracted. Such criteria are: Knowledge, Awareness, Skills, Intentions, Behavior, Enjoyment, Attitudes, etc. Some of these criteria have been integrated into criteria with wider meaning, e.g. Intentions, Behaviour, and Attitudes have been grouped into the criterion "Attitudes". The group of criteria has been completed by other criteria that were considered important for evaluating EE programs e.g. Completeness, Clarity, Pedagogy, Effectiveness.

As a result, the group of criteria used for evaluating the EE projects is formed to include the following:

- uc1 - Adaptivity: This criterion reveals how flexible the program is and how adaptable it is to each age group of participants.
- uc2 - Completeness: This criterion shows if the available description of the program covers the topic and to what extent.
- uc3 - Pedagogy: This criterion shows whether the EE program is based on a pedagogical theory or if it uses a particular pedagogical method.
- uc4 - Clarity: This criterion represents whether or not the objectives of this program are explicitly expressed or stated.
- uc5 - Effectiveness: This criterion shows the overall impact, depending on the programming and available support material.
- uc6 - Knowledge: The quantity and quality of a cognitive object offered to students.
- uc7 - Skills: This criterion reveals if skills are cultivated through activities involving active student participation.
- uc8 - Attitudes: This criterion reveals the change in student attitudes, intentions, and attitudes through the program.
- uc9 - Enjoyment: This criterion shows the enjoyment of the trainees throughout the EE project.
- uc10 - Multimodality: This criterion represents whether the EE project provides many different kinds of activities, interventions, and methods.

3.3 Setting up a Pairwise Comparison Matrix of Criteria

In this step, a comparison matrix is formed so that the criteria are compared pair wisely. The evaluators must use the nine-point scale developed by Saaty [9] to make comparisons. In the comparison process, a V from the scale that is presented in Table 1 is assigned to the comparison result of two elements P and Q at first, then the value of comparison of Q and P is a reciprocal value of V, i.e. 1/V. The value of the comparison between P and P is 1.

The comparison process is performed by the group of human experts formed in Subsect. 3.1. Each one of the evaluators completed the matrix of the pairwise criteria comparison. Each cell of the final matrix is calculated as a geometric mean of the corresponding cell in the ten (10) matrices collected by the human Environmental Educators. As a result, a final matrix is constructed (Table 2).

Table 1. Matrix for the pairwise comparison of criteria

	P	Q
P	1	V
Q	1/V	1

Table 2. Matrix for the pairwise comparison of criteria using the geometric mean of the values assigned by evaluators

Criteria	uc1	uc2	uc3	uc4	uc5	uc6	uc7	uc8	uc9	uc10
uc1	1.00	1.77	2.71	0.38	0.50	0.62	0.35	0.59	0.48	1.71
uc2	0.57	1.00	2.10	0.66	0.69	0.53	0.43	0.58	0.85	1.45
uc3	0.37	0.48	1.00	0.31	0.36	0.30	0.23	0.33	0.25	0.78
uc4	2.63	1.52	3.18	1.00	0.90	1.07	0.76	1.30	0.70	3.27
uc5	2.00	1.45	2.76	1.11	1.00	1.44	0.86	1.50	1.13	2.35
uc6	1.62	1.90	3.34	0.94	0.69	1.00	0.92	1.65	1.39	2.07
uc7	2.87	2.31	4.27	1.32	1.17	1.09	1.00	2.44	1.54	3.02
uc8	1.69	1.71	3.05	0.77	0.66	0.61	0.41	1.00	1.53	1.40
uc9	2.07	1.18	3.94	1.21	0.89	0.72	0.65	0.65	1.00	2.43
uc10	0.58	0.69	1.29	0.31	0.43	0.48	0.33	0.71	0.41	1.00

3.4 Calculating Weights of Criteria

After making pairwise comparisons, estimations are made that resulted in the final set of weights of the criteria. In this step, the principal eigenvalue and the corresponding normalized right eigenvector of the comparison matrix give the relative importance of the various criteria being compared. The elements of the normalized eigenvector are the weights of criteria or sub-criteria. In order to have the weights easily estimated before the system is fully implemented in a website, we have used the 'Priority Estimation Tool' (PriEst) [10], an open-source decision-making software that implements AHP. The weights of the criteria have been estimated as follows:

$w_{uc1} = 0.072$, $w_{uc2} = 0.071$, $w_{uc3} = 0.036$, $w_{uc4} = 0.129$, $w_{uc5} = 0.133$, $w_{uc6} = 0.127$, $w_{uc7} = 0.171$, $w_{uc8} = 0.099$, $w_{uc9} = 0.111$, $w_{uc10} = 0.051$.

The results would be slightly different if the geometric mean instead of eigenvector would be used as a method in AHP for the calculation of the final set of criteria. The criterion 'Skills' was considered by the experts the most important criterion while evaluating projects of Environmental Education. 'Effectiveness', 'Clarity' and 'Knowledge' were also considered rather important. Less important criteria were the criteria 'Pedagogy' and 'Multimodality'.

4 Combining AHP with Fuzzy TOPSIS for Comparing EE Projects

Fuzzy TOPSIS assumes that the weights of the criteria have been estimated and uses them to evaluate the alternative EE project. The main steps of Fuzzy TOPSIS for the evaluation of the alternative EE projects are the following:

1. **Assigning values to the criteria.** In order to make this process easier for the user, especially for those that do not have experience in multi-criteria analysis, the users could use linguistic terms for characterizing the twelve heuristics presented above. The linguistic terms are presented in Table 3. The values of the criteria are stored in the system's database and have been acquired by the instructors that have answered the questionnaire for each project.
2. **Linguistic terms are transformed into fuzzy numbers.** Each linguistic term is assigned to a fuzzy number, which is a vector $\tilde{a} = (a_1, a_2, a_3)$. The matches are presented in Table 3 [4].

Table 3. Linguistic variables and fuzzy numbers.

Linguistic term	Fuzzy number
Very poor	(0,0,1)
Poor	(0,1,3)
Fair	(3,5,7)
Good	(7,9,10)
Very Good	(9,10,10)

3. **Aggregating values of different instructors.** Each EE project must have only one vector value for each criterion in order to construct the MCDM matrix in the next step. As the values stored for each EE project's criteria are vectors then the geometric mean of these vectors is a vector that has the geometric mean of the corresponding values of these vectors. In order to aggregate all the values of the decision-makers in one single value, the geometric mean is used. The geometric mean of two fuzzy numbers $\tilde{a} = (a_1, a_2, a_3)$ και $\tilde{b} = (b_1, b_2, b_3)$ is calculated as follows:

$$\tilde{c} = (\sqrt{a_1 b_1}, \sqrt{a_2 b_2}, \sqrt{a_3 b_3}) \tag{1}$$

4. **Dynamic Construction of the MCDM matrix.** A fuzzy multi-criteria group decision-making problem can be expressed in a matrix format.

$$\tilde{D} = \begin{array}{c} \\ A_1 \\ A_2 \\ \\ A_m \end{array} \begin{array}{c} C_1 \ C_2 \quad C_{10} \\ \begin{pmatrix} \tilde{x}_{11} & \tilde{x}_{12} & & \tilde{x}_{110} \\ \tilde{x}_{21} & \tilde{x}_{22} & & \tilde{x}_{210} \\ & & \tilde{x}_{ij} & \\ \tilde{x}_{m1} & \tilde{x}_{m2} & & \tilde{x}_{m10} \end{pmatrix} \end{array}, i = 1, 2, \ldots, m; \ j = 1, 2, \ldots, 10, \tilde{x}_{ij} = (a_{ij}, b_{ij}, c_{ij}) \tag{2}$$

where i shows the alternative EE projects that the instructor has selected to evaluate and j shows one of the 10 criteria. Each $\tilde{x}_{ij} = (a_{ij}, b_{ij}, c_{ij})$ is a triangular fuzzy number.

5. **The normalization of fuzzy numbers.** To avoid the complicated normalization formula used in classical TOPSIS, Chen [4] proposes a linear scale transformation in order to transform the various criteria scales into a comparable scale. The particular normalization method aims at preserving the property that the ranges of normalized triangular fuzzy numbers belong to [0,1]. The normalization of a fuzzy number $\tilde{x}_{ij} = (a_{ij}, b_{ij}, c_{ij})$ is given by the formula:

$$\tilde{r}_{ij} = (\frac{a_{ij}}{c_j^*}, \frac{b_{ij}}{c_j^*}, \frac{c_{ij}}{c_j^*}), \text{ where } c_j^* = \max_i c_{ij} \tag{3}$$

6. **Calculating the weighted normalized fuzzy numbers of the MCDM matrix.** Considering the different importance of each criterion, which is imprinted in the weights of the criteria, the weighted normalized fuzzy numbers are calculated: $\tilde{u}_{ij} = \tilde{r}_{ij}(\cdot)\tilde{w}_j$ and these values are used to construct the weighted normalized fuzzy MCDM matrix $\tilde{V} = [\tilde{u}_{ij}]_{M \times N}, i = 1, 2, \ldots, m; j = 1, 2, \ldots, n.$

7. **Determination of the Fuzzy Positive-Ideal Solution (FPIS) and the Fuzzy Negative-Ideal Solution (FNIS).** The FPIS and the FNIS are calculated as follows:

$$\text{FPIS: } A^* = \{\tilde{u}_1^*, \tilde{u}_2^*, \ldots, \tilde{u}_i^*, \ldots, \tilde{u}_n^*\}, \tilde{u}_j^* = (1, 1, 1) \tag{4}$$

$$\text{FNIS: } A^- = \{\tilde{u}_1^-, \tilde{u}_2^-, \ldots, \tilde{u}_i^-, \ldots, \tilde{u}_n^-\}, \tilde{u}_j^- = (0, 0, 0) \tag{5}$$

8. **Calculation of the distance of each alternative from FPIS and FNIS.** The distances $(d_i^*$ and $d_i^-)$ of each weighted alternative $i = 1, 2, \ldots, m$ from FPIS and FNIS is calculated as follows:

$$d_i^* = \sum_{j=1}^{n} d_u(\tilde{u}_{ij}, \tilde{u}_j^*), i = 1, 2, \ldots, m \tag{7}$$

$$d_i^- = \sum_{j=1}^{n} d_u(\tilde{u}_{ij}, \tilde{u}_j^-), i = 1, 2, \ldots, m \tag{8}$$

where $d_u(\tilde{a}, \tilde{b})$ is the distance between two fuzzy numbers $\tilde{a}, \tilde{b}$. The distance of two fuzzy numbers $\tilde{a} = (a_1, a_2, a_3)$ and $\tilde{b} = (b_1, b_2, b_3)$ is calculated as:

$$d(\tilde{a}, \tilde{b}) = \sqrt{\frac{1}{3}[(\alpha_1 - b_1)^2 + (a_2 - b_2)^2 + (a_3 - b_3)^2]} \tag{9}$$

9. **Calculation of the closeness coefficient of each alternative.** The closeness coefficient of each alternative j, is given by the formula $CC_i = \frac{d_i^-}{d_i^* + d_i^-}, 0 \leq CC_i \leq 1.$ According to the values of the closeness coefficient, the ranking order of all the alternatives is determined. The alternative that is closer to FPIS and further from FNIS as CC_i approaches 1. The values of the closeness coefficient of each alternative and the final ranking of the evaluated EE projects are presented to the user.

5 An Example

We have created a scenario to check the system's performance. Let's suppose the system has online the 52 EE projects of Greece that involve paths [ref]. The user is interested in the following five:

- Geomythical – Geo-environmental paths
- «Forest trails»
- Orienteering
- The olive tree's trail
- Environmental routes

But s/he cannot decide which one to implement. The proposed approach can be used to tackle this problem. The user first needs to answer if s/he wants to answer the questionnaire about each project or can just use previous evaluations of the projects. If the system does not have previous evaluation data for an EE project, it informs the user that s/he has to evaluate the EE project by answering the corresponding questionnaire. In the particular example, the system did not have the evaluation of any project as the system had not run yet. Therefore, first the user answered the questionnaire for all 5 alternative EE projects.

Then the linguistic terms of the answers of the questionnaire are transformed into fuzzy number using transformations in Table 3. These values are used for constructing the MCDM matrix. The values of the MCDM table are normalized and then multiplied by the corresponding weight of each criterion to produce the weighted normalised fuzzy numbers of the MCDM matrix.

Taking into account that the FPIS is $A^* = \{(1, 1, 1), (1, 1, 1), \ldots, (1, 1, 1)\}$ and FNIS is $A^* = \{(0, 0, 0), (0, 0, 0), \ldots, (0, 0, 0)\}$, the distances ($d_i^*$ and d_i^-) of each alternative EE project from FPIS and FNIS are calculated and are presented in Table 4. Finally, the closeness coefficient of each alternative EE project is calculated using the formula $CC_i = \frac{d_i^-}{d_i^* + d_i^-}$ (Table 4).

Table 4. The distances d_i^* and d_i^-, closeness coefficient CC_i and the ranking order to the evaluated EE projects

Criteria	d_i^*	d_i^-	CC_i	Rank
Geomythical/geo-environmental paths	9,80	0,24	0,024	5
«Forest trails»	9,07	0,93	0,093	1
Orienteering	9,23	0,78	0,078	2
The olive tree's trail	9,50	0,53	0,070	3
Environmental routes	9,31	0,70	0,052	4

According to the values of the closeness coefficient, the ranking order of all the alternatives is determined. The ranking order of the five EE projects that have been evaluated is presented in Table 4.

6 Conclusions

Fien, Scott, and Tilbury [5] described the "lack of a widespread culture of evaluation in environmental education", calling it "immature" and largely "neglected". And even in the cases, in which limited evaluation experiments have been performed the results are not available to the teacher that could apply the program. Indeed, as Carleton-Hug and Hug [3] point out, it is imperative that evaluators share the results of the EE evaluation and suggest that evaluation should be incorporated in the earlier stages of program development.

The evaluation of EE programs prior to the application is very important because the application of an EE program costs in time, money and effort and, therefore, we would like to implement only a program that is worth being implemented. However, finding the correct EE project to implement is also a very difficult procedure that needs time and resources. The main problem in the selection of the correct EE project is that Environmental Educators do not have the knowledge to perform evaluation experiments for ranking EE projects and select the best one. Evaluation experiments to compare EE programs are difficult to run since competencies for evaluation are lacking among most environmental educators [14].

Given the current state, we propose here a system that stores the EE projects as well as evaluation data analytics of these projects. The evaluation of each EE project is conducted using a questionnaire with linguistic multiple-choice questions and involves the design of each project. This means that one does not have to run the EE project to evaluate it. This information is stored in the system and can be further used by other instructors. More specifically, each instructor may choose the projects s/he is interested in and the system applies MCDM to rank these projects and select the best one. More specifically, each question of the questionnaire corresponds to one criterion and then AHP is used in combination with Fuzzy TOPSIS. AHP is considered ideal for calculating the weights of the criteria because it allows pairwise comparisons. But pairwise comparisons are not suitable when the number of the alternative EE projects that are evaluated is known beforehand. Therefore, Fuzzy TOPSIS is used. Fuzzy TOPSIS uses linguistic terms, which are in compliance with the responses of the questionnaire and, furthermore, the complexity of the method does not increase as the number of the alternative increases. The existence of this successful combination of the Multi-Criteria Decision Making theories provides the theoretical framework that ensures the validity of the results.

It is among our future plans to fully implement the system and evaluate it with real users. Furthermore, we intend to use data analytics about instructors and students to improve this evaluation process.

References

1. Bitgood, S.: Institutional acceptance of evaluation: review and overview. Vis. Stud. **11**(2), 4–5 (1996)

2. Boeve-de Pauw, J.: Moving Environmental Education Forward through Evaluation. Stud. Educ. Eval. **41**(4), 1–3 (2014)
3. Carleton-Hug, A., Hug, J.W.: Challenges and opportunities for evaluating environmental education programs. Eval. Progr. Plan. **33**, 159–164 (2010)
4. Chen, C.T.: Extensions of the TOPSIS for group decision-making under fuzzy environment. Fuzzy Sets Syst. **114**, 1–9 (2000)
5. Fien, J., Scott, W., Tilbury, D.: Education and conservation: Lessons from an evaluation. Environ. Educ. Res. **7**(4), 379–395 (2001)
6. Fleming, L., Easton, J.: Building environmental educators' evaluation capacity through distance education. Eval. Progr. Plan. **33**(2), 172–177 (2010)
7. Hwang, C.L., Yoon, K.: Multiple Attribute Decision Making: Methods and Applications. Springer, New York (1981)
8. Kuhar, C.W., Bettinger, T.L., Lehnhardt, K., Tracy, O., Cox, D.: Evaluating for long-term impact of an environmental education program at the Kalinzu Forest Reserve Uganda. Am. J. Primatol. **72**, 407–413 (2010)
9. Saaty, T.: The Analytic Hierarchy Process. McGraw-Hill, New York (1980)
10. Sirah, S., Mikhailov, L., Keane, J.A.: PriEsT: an interactive decision support tool to estimate priorities from pair-wise comparison judgments. Int. Trans. Oper. Res. **22**(2), 203–382 (2015)
11. Stern, M.J., Powell, R.B., Hill, D.: Environmental education program evaluation in the new millennium: what do we measure and what have we learned? Environ. Educ. Res. **20**(5), 581–611 (2014). https://doi.org/10.1080/13504622.2013.838749
12. Tilbury, D.: Environmental education for sustainability: defining the new focus of environmental education in the 1990s. Environ Educ. Res. **1**(2), 195–212 (1995)
13. Zhu, Y., Buchman, A.: Evaluating and selecting web sources as external information resources of a data warehouse. In: The Third International Conference on Web Information Systems Engineering (WISE 2000), pp. 149–160 (2000)
14. Zint, M.T., Dowd, P.F., Covitt, B.A.: Enhancing environmental educators' evaluation competencies: insights from an examination of the effectiveness of the My Environmental Education Evaluation Resource Assistant (MEERA) website. Environ. Educ. Res. **17**(4), 471–497 (2011). https://doi.org/10.1080/13504622.2011.565117
15. Zint, M.: An introduction to My Environmental Education Evaluation Resource Assistant (MEERA), a web-based resource for self-directed learning about environmental education program evaluation. Eval. Progr. Plan. **33**(2), 178–179 (2010)

Identifying the Influential User Based on User Interaction Model for Twitter Data

C. Suganthini and R. Baskaran[✉]

Department of Computer Science and Engineering, College of Engineering, Anna University, Chennai, Tamil Nadu, India
chinna246@gmail.com, baaski@annauniv.edu

Abstract. In the social network, the user Interaction model is used to measure the interest depending on the interaction behavior between different entities. An entity or node can be people, groups or organizations and the links or edges are shown to represent the relationship between them. Edges are used to identify whether there is communication or social interaction between different users. Based on the user's interest or activities made in a social group, it is used to identify whether the interaction in the network is active or inactive. Social media is one of the fast-growing, dynamic and unpredictable in detecting future influencers which becomes harder in the social group. The features identified from the user interaction model and thus the potential influencers and current influencers are identified. Incidence matrix is used to identify the interaction behavior between different users in a social group. Thus, the incidence edges play an essential role in the vertex-edge incidence matrix interaction to identify the active and inactive user interactions in a social network. The Fraction of Strongly Influential (FSI) users score is used to evaluate the user interaction model by analyzing the influential user's interest and thus the top trending user interests are achieved.

Keywords: Feature extraction · Influential user interaction and Edge Incidence matrix interaction · Fraction of Strongly Influential users

1 Introduction

In a social network sites where one connects with those sharing personal or professional interest in it. Some of the social network sites such as Facebook, Twitter, and LinkedIn which allow users to create or join groups of common interest where they can do some of the activities such as post, comments, photos and videos can be shared and discussion can take place. Social media has become a key media for sharing their opinions, interest and personal information such as messages, photos, videos, news, etc. Users spend several hours on social media daily. Similar to influence analysis, researchers also evaluate information diffusion models on social media to identify key users who increase the diffusion of information. In this work, social influence defines as the effect of users on others that results in sharing information, which is the most common definition of social influence. The users tend to be influential experts on specific topics such as sports, economy, politics, rather than being global experts. This result leads us to explore

© Springer Nature Singapore Pte Ltd. 2020
S. M. Thampi et al. (Eds.): SIRS 2019, CCIS 1209, pp. 48–63, 2020.
https://doi.org/10.1007/978-981-15-4828-4_5

more on user's activities such as sharing information on their interests. In the proposed approach, the features are related to user activities such as ideas, opinions, interest, and thoughts. In this paper, the influential user interaction model of user-related features is introduced. These user's features are classified into core interest and marginal interest of users in the user interaction model. The proposed model is applied in a distributed manner to efficiently analyze the influence of the user's interest.

One of the most important aspects of social media is that it is so dynamic and fast, users and relations may appear and disappear very frequently. It is important to make predictions about who may be influential in the future rather than only identifying current influencers. In this proposed method, the influential users are identified, who would spread the information more, on specific topics. To believe that these users tend to be topic experts and those members are observed as experts on a few numbers of topics.

The remaining part of this paper is organized as follows. In Sect. 2 the existing literature related to the research topic is reviewed. The proposed user influential model based on the vertex-edge incidence matrix for identifying influential users in the social group is presented in Sect. 3. Section 4 describes the experimental data and evaluation results are discussed. Finally, Sect. 5 exposes the conclusion by giving future research challenges.

2 Related Work

2.1 Influential User

David et al. [1] introduce the two basic diffusion models is adopted for spreading an idea or innovation and influence among its members through a social network. Understanding the behavior and comparing its performance for identifying influential individuals. Wei et al. [2] identify the problem to select the initial user who influences the large number of people in the network, i.e. finding the influential individuals in a social network. Michael et al. [3] developed an approach to determine which users have significant effects on the activities of others using the longitudinal records of member's log-in activity. This approach identifies the specific users who most influence the site activity of others. Amit et al. [4] propose the propagation model to maximize the expected influence spread nodes that eventually get activated.

2.2 User Interaction

Francesco [5] proposed a framework to analyze the interaction between social influence and homophily. The features analyzed from social relations and similarity is more important to predict future behavior than either the social influence or similarity feature. Fangshuang et al. [6] Diversified influence maximization which is performed to help the task such as recommending a set of movies with different user's interest and finding a team of experts (answerers) with a comprehensive scope of knowledge. Qindong et al. [7] propose the user interaction process which is one of the most important features in analyzing user behavior, information spreading model, etc. Li et al. [8] propose a framework based on frequent pattern mining to find the influence users as well as the proper time to spread information. To identify influential bloggers to set collective statistics and find users.

Fredrik et al. [9] proposed work analyzes the influential users and their behavior to predict user participation in the social network. Wang et al. [10] propose a dynamic regional interaction model is to evaluate the influential user identification in understanding the information dissemination process according to the interactions between adjacent and non-adjacent users in online social networks. Jiyoung et al. [11] propose topic diffusion in web forms using the epidemic model where the user in online communities spread the diffusion information. Amrita et al. [12] propose the influence of users through the information spreading and diffusion region. Thus finds the number of active users is influenced by the top influential user with higher network reachability holds the larger diffusion in the network region.

Wenzheng et al. [13] propose an influence propagation process the social decision of a user depends more on the network structure and the social structures of different groups of its neighbors rather than the number of its neighbors. Madhura et al. [14] propose a technique to identify the behavior and opinion of influential users in a social network. Yuchen et al. [15] propose the diffusion model and the influence spread under the model. The information diffusion process identifies the k users to maximize the expected number of influenced users in the social network. Mohammed et al. [16] propose an algorithm for the top k-influential users by the selection process. Identification of such influential users helps to identify, understand and discover the underlying interactions of interesting users in the network. Bo et al. [17] propose the most influential node discovery method for discovering influential nodes in the social network to identify the maximum influential node. Zeynep et al. [18] proposed approach include identification authorities on network is used to maximize the spread of information and to recommend other users.

Most of the existing works does not handle the dynamic interaction of users in social network sites. Therefore it is essential to adopt the user interaction model approach which helps in identifying the influential individuals interacted during their time interval in the social network. Incidence matrix interaction is an efficient method which plays an important task in analyzing the influential and non-influential individuals. Thus the influential user interests and influential topics help to identify top trending topics from the user interaction model.

3 Proposed System

The proposed work includes the data extraction, pre-processing, feature extraction and the influential user interaction model is shown in Fig. 1. The first level includes data retrieved from the user timeline tweets. The second level includes the pre-processing and feature extraction. Tweets are pre-processed and the user list is generated to store the user information, both produce the processed tweets. From the processed tweets the hashtag is extracted in the feature extraction level. The third level includes the influential user interaction model, in which the user interest is classified into active and inactive user interaction. Based on the parameters of core interest and marginal interest, the influential and non-influential user interaction is identified respectively. Then the incidence matrix is used to identify the structure of user interaction from the user's timeline. Tweet score is considered to find influential topics from the extracted tweets. Based on the influential topics and the influential users identified the top trending interested topics are analyzed.

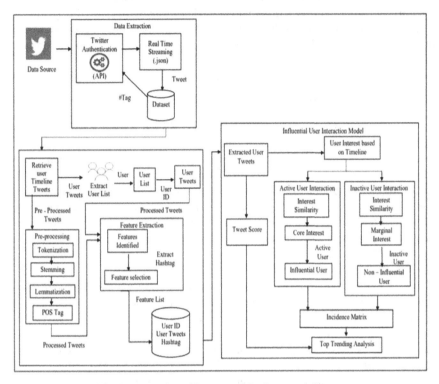

Fig. 1. Overall Architecture of the Proposed System

3.1 Data Extraction

Social site information is not publically available, a password is required to collect data from the user. The Twitter platform provides various API, SDK for developing an application that is used to access the data and data retrieved from the social site's Oauth account of Twitter API. Social Interaction data includes more active users among social media sites. Users retrieve the data which are connected or related to that particular user's post. The information in the retrieved user data includes tweets, followers, Likes, retweets, etc. Twitter users can control photo tagging and sharing of their friend list with the public user can also share the status with specific people [22]. Collecting and analyzing such network sites data allows us to study the structure of the relationship between users in a social network.

3.2 Feature Extraction

The first step of feature extraction is pre-processing. Tweets are tokenized, stemmer and lemmatization applied and analyzed based on part of speech tag. Irrelevant content and

URLs in the tweets are removed. From the extracted tweets the names of users who have shared the message on a specific topic of interest extracted. The extracted information is stored using a user's list which contains the user's names and their hashtag. The processed tweets retrieved from the user's timeline via his/her hashtag by using the Twitter API. Features extracted from the user's activities such as the sharing of information. The hashtag is extracted based on the user who shares their interest from the processed tweets.

3.3 User Interaction Model

Social members with similar features often connected than with more dissimilar ones. If there is any relationship between the members of the social group, there must be some common characteristics between those members which form a social relation. This basic idea leads to introduce the structure of the user interaction model. These characteristics are referred to as an influence. For example, in online social sites such as Twitter, Facebook, LinkedIn, etc., the certain influence is shown as the frequency with which friends tag each other based on common interests (movies, songs, books, etc.), in their posts. These influences suggest us how similar are the members of the social network [23]. Users are more willing to make friend with the one who has the same interest than the person who holds different attitudes [24].

The user interaction model is classified into two types such as active and Inactive user interaction based on the user's interest which further grouped into core interest and marginal interest respectively. In this paper, the two main problems by identifying how user interests change over time and whether user interests have a preference. Thus to define user interest based on core and marginal interest and the incidence edge matrix is used to capture user interests change over time and user interest structure is also identified. The user Interaction model generates the interesting topics of users from the tweets retrieved from the user timeline. This module first identifies entities that are can be directly extracted from the user's tweets and then scores them to reflect the influence of each topic of interested users. The time-based dynamics of user interest and how the topics of individual user's interest change over time. The proposed work deals with the user's interest from user tweets as a set of influenced concepts where concept may refer to an arbitrary entity and the influence indicates how important the concept is for the user's interests.

User Interest

The topics of interest of a user u is a set of influenced concepts where a concept c is represented via the entity.

$$UI(u, t) = \{(c_u, I(c_u, t, tweet_u)) \mid c_u \in C_{UE}\} \tag{1}$$

Where $I(c_u, time, tweet_u)$ is an influence which is estimated for the concept c by the given user u based on user tweets denoted as $tweet_u$ posted by u and based on the given time t. C_{UE} is consists of a set of entities.

Tweet Score

The topic score of each tweet T, posted by given a user u, can be measured by frequency and confidence for an entity e, where a tweet t is represented via a set of entities E.

$$TS(u, T) = \sum_{i=1}^{n} f_{ei} \, c_{ei} \text{ where } f_{ei} \in F_E \text{ and } c_{ei} \in C_E \tag{2}$$

Interest Similarity

The tweets retrieved from the user timeline is considered with a vector. Assume a user u interests is the vector $S(u, t_1)$ at time t_1 and $S(u, t_2)$ at time t_2, where $t_1 < t_2$, which are represented via topics sharing of interests using the same vector representation that is used for a given user interests and the similarity score is used to identify the users interest change degree according to their cosine similarity.

$$SimI(S(u, t_1), S(u, t_2)) = \frac{(S(u, t_1) \cdot S(u, t_2))}{\|S(u, t_1)\| \quad \|S(u, t_2)\|} \tag{3}$$

Active User Interaction

Identifying the core interest among like, share and comment with a similar interest in a group. User's interest among the shared information in social media may not change depends on the trending information in the network. Active user interaction in the social group is formed due to frequent interaction among the user who shares the common interest leads to strong tie strength between them. This type of interaction is also referred to as a strong interaction. The user interest score is calculated from the given Eq. (1), if the score is high then they are identified as an active user, then the interest similarity score is calculated from the given Eq. (3), if the value of SimI is high, it indicates the interests of the same user have less change over time i.e. their interest is stable for a long period thus more amount of users has interacted.

Core Interest (CI)

The number of times that the user interacts with each other and thus they tend to have a closer relationship between them. The individual core interests will not change in a short period will stay for the long-term. The User's core interest is calculated which depends on the interaction among them and they are stable. The active users will hold the same core interest based on the user's interaction which belongs to a similar interest. The core interest score is calculated from the given Algorithm 1.

Inactive User Interaction

Inactive user interaction in the social network is formed by fewer activities or dissimilar interests among them which leads to a weak tie strength between them. This type of interaction is also referred to as a weak interaction. Users may not share the related information or most likely topic in social media. Inactive user interaction in the social network is formed by fewer activities or dissimilar interests among them which leads to a weak tie strength between them. This type of interaction is also referred to as a weak interaction. Users may not share the related information or most likely topic in social media. User's interest among the shared information in social media may change depends on the shared information in the network. Interaction between the users is an irrelevant topic. Weak ties between the users indicated the interaction between them is weak/less and inactive. The user interest score is calculated from the given Eq. (1), if the score is low then they are identified as an inactive user, then the interest similarity score is calculated from the given Eq. (3), if the value of SimI is low, it indicates the interests of the same user has a large change over time.

Algorithm 1: User Interaction Algorithm
Input: User Interest
Output: Active and inactive user identification
S: Active and Inactive User Interaction
M: CI of each Category in Aa
R: MI of each Category in IAa
Aa: Active interested user who shares the common interest
IAa: Inactive interested user who shares the dissimilar interest
1. $S := \emptyset$
2. $M := \emptyset, R := \emptyset$
3.   for i = 1, 2, 3, … , k users do
4.   Evaluate UI according to Equation 1
5.   Evaluate SimI according to Equation 3
6.   If UI and SimI value is high, then update to Aa
7.   If UI and SimI value is less, then update to IAa
8. Pa_i = Total number of users in the interaction
9.   Threshold value = 0.5
10.   for j = 1,2,3 , …, m in Aa
11.   Core Interest CI = $|Aa_j| / |Pa_i|$
12.   CI > Threshold value
13.   Active user is obtained by CI
14.   append M
15.   for j = 1,2,3, … r in IAa
16.   Marginal Interest MI = $|IAa_j| / |Pa_i|$
17.   MI < Threshold value
18.   Inactive user is obtained by MI
19.   append R
20.   S = { M, R }
21. return S

Marginal Interest (MI)
The temporary or marginal interest is the interaction among the user's interest will change over a short period and thus the marginal interest is unstable. Inactive users will have dissimilar or different interests which change in a short period. Thus the interaction among them is distinguished with dissimilar interest. The marginal interest score is calculated from the given Algorithm 1.

3.4 Incidence Matrix Interaction

A graph with no parallel edges, m vertices represented as v_1, v_2, v_3, v_4,..., v_m, and t edges denoted as e_1, e_2, e_3 e_4,..., e_t. An Incidence matrix of an undirected graph shows the interaction between two nodes. The representation of matrix which includes nodes or vertex for each 'm' row and edges is the relation between them for each 't' column. The incidence matrix of 'm x t' order is denoted by $[m_{ij}]$ is shown in Eq. (4).

$$M(G) \quad m_{ij} = \begin{cases} 1 & \text{if } m_{ij} \text{ if } e_i \text{ edge is incident to } v_j \text{ vertex} \\ 0 & \text{otherwise} \end{cases} \tag{4}$$

Incidence matrix is also called a Vertex-edge Incidence matrix. If a vertex v incident upon edge e, then the pair of (v,e) will be equivalent to one otherwise it is zero. Thus the incidence graph GU_1 and GU_2 for user u_1 and user u_2. The vertices denote the users and the edges represent the relations among the users. The incidence graph U_1 and U_2 as GU_1 (VU_1, EU_1) and GU_2 (VU_2, EU_2) where VU_1 and VU_2 are the vertices corresponding to the users u_1 and u_2 and EU_1 and EU_2 are the edges corresponding to social interaction among the users u_1 and u_2.

The user interaction model shows the interaction between Users and their tweets and the incidence matrix represents the structure of the user interaction. Users are represented as vertex and the interaction between them is represented as edges. Multiple users post or share the information for certain hashtags. The interested user who tweets for the particular hashtag and their relationship between them exist through the interaction. The incidence matrix interaction is used to represent the structure of active and inactive user interaction. The topmost frequency distribution of hashtag is measured to identify the active user interaction by using a core interest method. The least frequency distribution of hashtag is measured to identify the inactive user interaction by the marginal interest method.

The incidence matrix interaction is a more efficient way to detect the user interest change over time and identify the user interest structure. Incidence matrix for an active user shows the number of users who frequently interacted with other users in the social group, whereas the inactive user shows fewer users who interact with other users over a long duration of the period. Incidence matrix is an efficient method to identify the influential user interest on the user's timeline from the given Eq. (4). From the given Eq. (2), the tweet score is calculated and identifies the influential topics. Thus based on the influential user interest and influential topics identified the Top trending interested topics is achieved by the fraction of strongly influential users score.

4 Performance Evaluation

In this section, the evaluation of the system and experimental results are discussed for the real-world datasets. Mainly, users in social networks will interact by sharing opinions or exchanging messages with each other. Everyone is interested in understanding user behavior and comparing their performance for identifying influential individuals. Social influence and user's similarity are both shows the users own interest and are more predictive for finding future user behavior. Tweets are extracted based on the user's time-line interaction for 10 different categories. The interaction among the users is dependent on how they share their interests among them. The number of tweets extracted in each category and the hashtag frequency distribution for 10 different categories is evaluated based on their extracted hashtag is shown in Table 1.

Table 1. Tweets extracted and hastag frequency distribution

Category	No. of Tweets	Frequency
Game	5347	1001
Education	5879	1007
Entertainment	9200	2022
Environment	13578	5079
Food	11303	3130
Music	12428	4454
News	10829	4975
Sports	14588	9868
Politics	7409	1464
Travel & tourism	8030	2462

4.1   User Interaction Evaluation

From the results of the frequency distribution, the most influential users identified using the user interaction method. For each category, the top 10 influential users are identified and compared to the number of influenced users at a different timeline. The top frequency count is calculated to identify influential users. According to each user participating in the tweeting and retweeting process, the user interaction model divides the user influence into two different types as the active and inactive user influence according to the user interaction among them. The core interest and marginal interest score are evaluated to find the highest and least frequency count respectively for each category from the extracted tweets. The user interest (UI) and Interest Similarity (SimI) score is calculated from the Eqs. 1 and 3 for active and inactive user as shown in the Table 2.

The most influential user leads to active user interaction which is predicted from the top frequency calculated from core interest for the frequently interacted users. The non-influential user leads to inactive user interaction is predicted from the least frequency calculated from marginal interest for not frequently interacted users. The active and inactive user interaction is shown in Table 3.

Consider any three categories from the extracted 10 different categories of tweets to explain the difference between active and inactive user participation. Consider some of the different scenarios such as environment, music and news categories respectively.

Table 2. UI and SimI score for Active and Inactive User

S. No.	Category	Active User		Inactive User	
		UI Score	SimI	UI Score	SimI
1.	Game	0.667864	0.781384	0.1069	0.0692
2.	Education	0.561072	0.575867	0.1609	0.1246
3.	Entertainment	0.540059	0.627811	0.0772	0.0925
4.	Environment	0.888758	0.849076	0.0382	0.0549
5.	Food	0.623323	0.671565	0.0834	0.1035
6.	Music	0.598788	0.524465	0.0903	0.1166
7.	News	0.636181	0.516462	0.0482	0.0512
8.	Sports	0.760438	0.61923	0.0636	0.0764
9.	Politics	0.588362	0.732885	0.1034	0.0905
10.	Travel & tourism	0.706742	0.675854	0.0646	0.0942

Active User Interaction

To assume that if user u1 tagged in most of the tweets where user u2 is tagged as well, there is a high chance of user u1 will also have more interest in a new topic where u2 is already active because the interest of u2 influences u1 to be active and they both have similar interests. For active user interaction for different categories as shown in Fig. 2.

Case 1: In the environment category the number of tweets extracted as 13578 among them the frequency distribution is calculated as 5079. The topmost topics preferred by the core interest of active user interaction with their top frequency score is 4675.
Case 2: In the Music category the no. of tweets extracted as 12428 among them the frequency distribution is calculated as 4975. The topmost topics preferred by the core interest of active user interaction with their top frequency score is 2667.
Case 3: In the News category the no. of tweets extracted as 10829 among them the frequency distribution is calculated as 4454. The topmost topics preferred by the core interest of active user interaction with their top frequency score is 3165.

Inactive User Interaction

If user u1 does not tag the tweets where user u2 is tagged, then there is less chance of user u1 is interested in a new topic where u2 is active because the interest of u2 does not influence u1 to be active and they both have a dissimilar interest. For inactive user interaction for different categories as shown in Fig. 3.

Case 1: In the environment category, the least topics preferred by the marginal interest of inactive users interaction with their least frequency score is 322.
Case 2: In the music category, the least topics preferred by the marginal interest of inactive users interaction with their least frequency score is 402.

Table 3. Active and Inactive User Interaction

S. No.	Category	Top	Least
1.	Game	451	107
2.	Education	690	162
3.	Entertainment	1092	156
4.	Environment	4675	322
5.	Food	1951	261
6.	Music	2667	402
7.	News	3165	240
8.	Sports	7504	628
9.	Politics	768	118
10.	Travel & tourism	1740	159

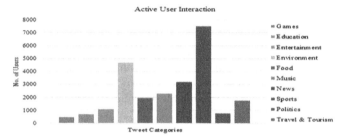

Fig. 2. Active User Interaction based on different categories

Case 3: In the news category, the least topics preferred by the marginal interest of inactive users interaction with their least frequency score is 240.

The timeline interaction graph is considered for a better understanding of the user interaction shown in the different scenarios. Timeline interaction graph which is used to measure the evolution changes of the user interaction in the social network is achieved by the incidence matrix interaction. The active and inactive user interaction is measured based on the interaction with each other is shown over the period.

Strong interaction deals with the common interest shared among the users who frequently interact for a long period. From the results, the interaction among the users is frequent and the user's distribution over the period is very dense. Weak interaction shows the deviation of user interest over a period. The user does not interact frequently and they are inactive users. From the results, the interaction among the users is not frequent and the distribution looks sparse for a certain period is very sparse. The user interaction for the different scenarios shown in Figs. 4 and 5.

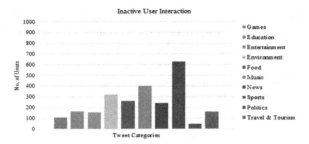

Fig. 3. Inactive User Interaction based on different categories

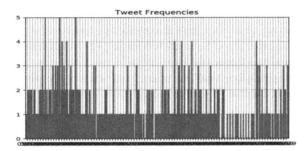

Fig. 4. Active User Interaction for Environment Category

4.2 Evaluation Metrics

Adamic/Adar measure is the inverted sum of degrees of common neighbors among the interaction of two users. A value of 0 indicates that two users are not close to each other, while higher values indicate users are closer.

$$A(x, y) = \sum_{u \in N(x) \cap N(y)} \left(\frac{1}{\log |N(u)|} \right) \tag{5}$$

where $N(u)$ is the number of users adjacent to u.

This measure is used to identify the neighbor's user in the social group. The incidence matrix interaction is used to represent the structure of the user interaction and the values

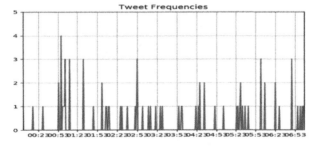

Fig. 5. Inactive User Interaction for Environment Category

are evaluated from the given Eq. (5). This measure is used to know the strong relationships among the user's interaction. If they share a similar interest, then the users are related to each other. If they share the dissimilar interest, then the users are not closer to each other. If two users have a significant number of common interest than the number of their total interest. The measure is higher and the interaction between them is stronger refers to influential user interaction [21].

4.3 Fraction of Strongly Influential (FSI)

The Fraction of Strongly Influential is the value of the number of users interacted among other users over a period. A user is strongly influenced by an active user who frequently shares the common interest during the long time interval. A user is weakly influenced with inactive user among which they have less interaction based on the dissimilar interest with long duration. The fraction of Strongly Influential (FSI) users score will lie between 0 and 1, with 1 being the frequently strongly influenced whereas 0 leads to weakly influenced users. The precision, recall, accuracy, and F-measures evaluated based on the Fraction of Strongly influential (FSI) users score as shown in Table 4.

Table 4. User Interaction based on FSI

	Similar interest	Dissimilar interest
Active User	Most Influential user with similar interest (MIUSI)	Influential user with a dissimilar interest (IUDSI)
Inactive User	Non-influential user with similar interest (NIUSI)	Non-Influential user with a dissimilar interest (NIUDSI)

MIUSI = MIUSI is active with frequent interaction. This is a correct positive prediction.
IUDSI = IUDSI is active with frequent interaction. This is an incorrect negative prediction.
NIUSI = NIUSI is inactive with non-frequent interaction. This is an incorrect positive prediction.
NIUDSI = NIUDSI is inactive with non-frequent interaction. This is a correct negative prediction.

Precision
Precision is calculated as the number of Influential similar user predicated divided by the total number of similar user predictions. Thus it is also called a positive predictive value. Precision is calculated from Eq. 6.

$$Precision = \frac{\sum MIUSI}{(\sum MIUSI + \sum IUDSI)} \tag{6}$$

Recall

The recall is known as true positive value or sensitivity. It is defined as the number of correct results divided by the number of relevant results. It is also called a true positive rate. The Recall is calculated from Eq. 7.

$$Recall = \frac{\sum MIUSI}{(\sum MIUSI + \sum NIUSI)} \tag{7}$$

Accuracy

Accuracy measures a ratio of correctly predicted observation referred to as influential users to the total observations referred to as the total number of users. Accuracy is calculated from Eq. 8.

$$Accuary = \frac{(\sum MIUSI + \sum NIUDSI)}{(\sum MIUSI + \sum IUDSI + \sum NIUSI + \sum NIUDSI)} \tag{8}$$

F measure

F1 score is used to consolidate precision and recall into one measure, the F1 measure is calculated from Eq. 9.

$$F = 2X \frac{Precision * Recall}{Precision + Recall} \tag{9}$$

The fraction of strongly influential users score is evaluated from the given Eq. (6), Eq. (7), Eq. (8) & Eq. (9) and shown in Fig. 6.

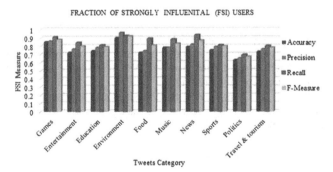

Fig. 6. User Interaction based on FSI Score

5 Conclusion

In many Social network sites such as Facebook, Twitter or LinkedIn the users find new friends or imitate a real-life relationship with friends and establish new friendship relations that cannot exist in real life due to distance or other factors. This study represents the social phenomenon of how users influence the establishment of a new friend's relationship in social network sites. In this proposed work, the user interaction model is used to identify an active user or inactive user interaction in a social group. To characterize a particular relationship between the users by using incidence matrix interaction i.e. an edge plays the important role in identifying the interaction, by the strength of relationship based on common interest among the users in a social group. Thus it is used to identify the influential user based on the vertex–edge incidence matrix interaction. Thus the Fraction of strongly influential Users measure is an efficient way to evaluate the Influential user interaction. As a part of future work, it would be interesting to see whether these model helps to detect the communities in the social network.

Acknowledgement. This publication is an outcome of the R&D work undertaken in the project under the Visvesvaraya Ph.D. Scheme of Ministry of electronics of Info. Technology, Government of India, being implemented by Digital India Corporation (formerly media Lab Asia).

References

1. Kempe, D., Kleinberg, J., Tardos, E.: Maximizing the spread of influence through a social network. In: Proceedings of the Ninth ACM SIGKDD International Conference on Knowledge Discovery and Data Mining – KDD 2003 (2003)
2. Chen, W., Wang, Y., Yang, S.: Efficient influence maximization in social networks. In: Proceedings of the 15th ACM SIGKDD International Conference on Knowledge Discovery and Data Mining - KDD 2009 (2009)
3. Trusov, M., Bodapati, A.V., Bucklin, R.E.: Determining influential users in internet social networks. J. Market. Res. **47**(4), 643–658 (2010). ISSN: 0022-2437
4. Goyal, A., Bonchi, F., Lakshmanan, L.V.S.: A data-based approach to social influence maximization. Proc. VLDB Endowment **5**(1), 73–84 (2011)
5. Bonchi, F.: Influence propagation in social networks: a data mining perspective. In: 2011 IEEE/WIC/ACM International Conferences on Web Intelligence and Intelligent Agent Technology, vol. 12, no. 1, pp. 8–16 (2011)
6. Tang, F., Liu, Q., Zhu, H., Chen, E., Zhu, F.: Diversified social influence maximization. In: 2014 IEEE/ACM International Conference on Advances in Social Networks Analysis and Mining (ASONAM 2014), pp. 455–459 (2014)
7. Sun, Q., Wang, N., Zhou, Y., Wang, H., Sui, L.: Modeling for user interaction by influence transfer effect in online social networks. In: 39th Annual IEEE Conference on Local Computer Networks, pp. 486–489 (2014)
8. Kao, L.-J., Huang, Y.-P.: Mining influential users in social network. In: 2015 IEEE International Conference on Systems, Man, and Cybernetics, pp. 1209–1214 (2015)
9. Erlandsson, F., Brodka, P., Borg, A., Johnson, H.: Finding influential users in social media using association rule learning. Entropy **18**(5), 164 (2016)
10. Nan, W., Sun, Q., Zhou, Y., Shen, S.: A study on influential user identification in online social networks. Chin. J. Electron. **25**(3), 467–473 (2016)

11. Woo, J., Chen, H.: Epidemic model for information diffusion in web forums: experiments in marketing exchange and political dialog. Springer Plus **5**(1), 1–19 (2016)
12. Namtirtha, A., Gupta, S., Dutta, A., Dutta, B., Coenen, F.: Algorithm for finding influential user: base on user's information diffusion region. In: 2016 IEEE Region 10 Conference (TENCON) - International Conference, pp. 2734–2738 (2016)
13. Xu, W., Liang, W., Lin, X., Yu, J.X.: Finding top-k influential users in social networks under the structural diversity model. Inf. Sci. **355–356**, 110–126 (2016)
14. Kaple, M., Kulkarni, K., Potika, K.: Viral marketing for smart cities: influencers in social network communities. In: 2017 IEEE Third International Conference on Big Data Computing Service and Applications (Big Data Service), pp. 110 –126 (2017)
15. Li, Y., Fan, J., Wang, Y., Tan, K.-L.: Influence maximization on social graphs: a survey. IEEE Trans. Knowl. Data Eng. **30**(10), 1852–1872 (2018)
16. Alshahrani, M., Zhu, F., Zheng, L., Mekouar, S., Huang, S.: Selection of top-K influential users based on radius-neighborhood degree, multi-hops distance and selection threshold. J. Big Data **5**(1), 1–20 (2018). Springer
17. Zhang, B., Zhang, L., Mu, C., Zhao, Q., Song, Q., Hong, X.: A most influential node group discovery method for influence maximization in social networks: a trust-based perspective. Data Knowl. Eng. **121**, 71–87 (2019)
18. Zengin, Z., Şule, A., Oguducu, G.: Influence factorization for identifying authorities in Twitter. Knowl. Based Syst. **163**, 944–954 (2019)
19. Gowsikhaa, D., Abirami, S., Baskaran, R.: Automated human behavior analysis from surveillance videos: a survey. Artif. Intell. Rev. **42**(4), 747–765 (2012)
20. Suganthini, C., Sridhar, R.: A survey on community detection in social network analysis. Int. J. Appl. Eng. Res. **10**(75), 1–6 (2015). ISSN 0973-4562
21. Sridhar, R., Kumar, A., Roshini, S.B., Sundaresan, R.K., Chinnasamy, S.: Feature based community detection by extracting Facebook profile details. ICTACT J. Soft Comput. **8**(4), 1706–1713 (2018)
22. Montreal, D.S.: All graphs in which each pair of distinct vertices has exactly two common neighbors. Mathematica Bohemica **1**, 101–105 (2005)
23. Verma, A.K., Pal, M.: Evolving social networks via friend recommendations. In: 11th International Conference on Signal-Image Technology & Internet-Based Systems (SITIS), IEEE Computer Society, pp. 379–383 (2015)
24. Malliaros, F.D., Vazirgiannis, M.: Clustering and community detection in directed networks: a survey. Phys. Rep. **533**(4), 95–142 (2013)

Rank Level Fusion of Multimodal Biometrics Based on Cross-Entropy Monte Carlo Method

Shadab Ahmad[1,2(✉)], Rajarshi Pal[2], and Avatharam Ganivada[1]

[1] University of Hyderabad, Hyderabad, India
shadab.ahmad013@gmail.com, avatharg@uohyd.ac.in
[2] Institute for Development and Research in Banking Technology (IDRBT),
Hyderabad, India
iamrajarshi@gmail.com

Abstract. In unimodal biometric systems, there are several limitations like, non-universality, noisy data and other security risks. To overcome these, multimodal biometric systems are increasingly adopted. Multimodal biometric systems fuse information from multiple biometric traits. Rank level fusion is one of the approaches of information fusion for multimodal biometrics. In this paper, rank level fusion is considered as an optimization problem. Its aim is to minimize the distances between an aggregated rank list and each input rank list from individual biometric trait. A solution of this optimization problem has been proposed using cross-entropy (CE) Monte Carlo method. The proposed CE method uses two distance measures - namely, Spearman footrule and Kendall's tau distances. Superiority of the proposed CE method based on above two distance measures over several existing rank level and score level fusion schemes is achieved on two different datasets.

Keywords: Multimodal biometrics · Rank level fusion · Optimization

1 Introduction

Biometric recognition systems are increasingly being used to recognize an individual in her office, government departments, banks and even for getting necessary commodities. In biometric based identification system, the biometric traits of an individual (i.e., a probe) are matched with those of every enrolled user (i.e., subject) in the system. On the contrary, in a verification system, the biometric traits of an individual are compared with the stored biometric traits of her. Traditionally, biometric system has been unimodal, where a single biometric trait is used to establish one's identity. But the unimodal biometric system faces several challenges as follows: non-universality, noisy data, inter-class similarities, intra-class variation, interoperability issues and susceptibility to circumvention. In order to overcome these challenges, multimodal biometric system has been

ⓒ Springer Nature Singapore Pte Ltd. 2020
S. M. Thampi et al. (Eds.): SIRS 2019, CCIS 1209, pp. 64–74, 2020.
https://doi.org/10.1007/978-981-15-4828-4_6

adopted. Apart from fulfilling the gaps of a unimodal biometric system, multi-modal biometric system is more secure and more accurate which has been proved over the period of time.

Multimodal biometrics combine information from multiple biometric traits (fingerprint and face [1], different finger surfaces [2], face and palm-print [3], etc.) or multiple representations of same biometric trait (various feature descriptors of palm-print biometrics [1]). This fusion of information can occur at sensor level, feature level, score level, rank level and decision level. Scope of the current paper is limited to rank level fusion. Rank level fusion [1–6] method is useful when there is less information (similarity or dissimilarity scores between the probe and the enrolled subjects) available to perform the fusion. Scores from different modalities can also be fused using a score level fusion [7–10] method. But these scores need to be normalized [10] in order to perform the score level fusion. On the contrary, these scores can be sorted to produce a rank list. Multi-level fusion schemes [11] can be found in literature.

A rank level fusion method combines several rank lists as obtained from multiple biometric modalities. A rank list, in this context, is a list of ranks of each subject as compared with a probe (based on a similarity or a dissimilarity score). Several rank level fusion methods have been proposed in the literature. Borda count, weighted borda count, highest rank and logistic regression are basic rank level fusion strategies [12] available in literature. For example, a rank level fusion method based on logistic regression and Borda count for combining kinetic gait and face biometrics has been proposed in [4]. In [3], the fusion of face and palm-print has been performed at rank level using highest rank, Borda count and logistic regression.

Apart from the above methods, several non-linear rank level fusion methods also exist. In [2], few such non-linear weighted rank methods have been used for rank level fusion for finger surface biometrics. These are, hyperbolic tangent, hyperbolic arc sinus, hyperbolic arc tangent, division exponential, and logarithm. In [1], fusion of three rank lists (as obtained from three different features descriptors of palm-print) has been carried out using two nonlinear methods, namely exponential and weighted exponential methods. In [5], rank lists as generated through multiple rank level fusion methods have been consolidated either serially or parallelly. Serial combination is obtained by a combining functions as $f_2(f_1(x))$, where $f_1(x)$ and $f_2(x)$ are two different rank level fusion methods. Similarly, parallel combination has been performed by combining all the rank lists as generated using various rank level fusion methods using a hyperbolic tangent rank level fusion method. Fusion of multimodal biometrics involving face, iris and ear is done at rank level [6]. A Markov chain is used for this task. In this method, a Markov chain has been established on the enrolled subjects. Transitions in this Markov chain represent an order relation among these enrolled subjects. A rank list is obtained using stationary distribution of this Markov chain.

In a completely different approach, the present paper perceives the rank level fusion of multimodal biometrics as an optimization problem. In this context, the goal is to minimize the distances between an aggregated rank list and the input rank lists. Two widely used distance measures in the domain of rank aggregation problems - (i) Spearman footrule [13] and (ii) Kendall's tau [13] distances - are considered in the proposed method. In this work, cross-entropy Monte Carlo algorithm [14] based approach has been proposed to solve the above optimization problem in the context of multi-modal biometrics. The proposed approach has been experimentally studied on two different datasets: (i) NIST BSSR1 [15] and (ii) OU-ISIR BSS4 [16,17]. Experimental results justify the suitability of the proposed approach of rank level in the context of multimodal biometrics.

The rest of this paper is organized as follows: A detailed formulation of rank level fusion of multimodal biometrics as an optimization problem is presented in Sect. 2. Section 3 describes the proposed rank level fusion method using cross-entropy Monte Carlo method. Section 4 reports the results of applying the proposed rank level fusion method as well as several state-of-the-art fusion methods on two multimodal biometric datasets. Finally, Sect. 5 draws the concluding remarks.

2   Formulation of Rank Level Fusion as an Optimization Problem

Let P_1, P_2, ... , P_N be various biometric traits to identify a person. Let the matching score Q_i^j be associated with each such biometric trait P_i for a jth person (subject) for an input probe. A rank list L_i of those subjects can be generated from an ordering of these matching scores. Considering a high value of Q_i^j as good (for a similarity score), the following is true about the rank list L_i: $Q_i^j > Q_i^k$ implies $L_i^j < L_i^k$. Here, L_i^j indicates the rank of the jth subject in the list L_i.

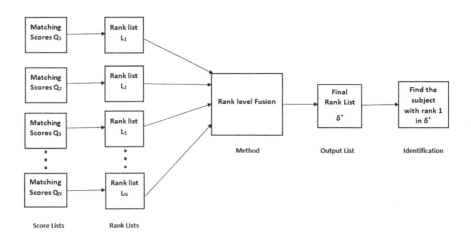

Fig. 1. Fusion of multimodal biometrics at rank level

Therefore, N rank lists are created as L_1, L_2, ..., L_N for biometric traits P_1, P_2, ... , P_N, respectively. A combination of these N rank lists generates an aggregated rank list δ^* as shown in Fig. 1.

$$\delta^* = aggregate(L_1, L_2, ...L_N) \tag{1}$$

The objective, here, is to generate the aggregated rank list δ^* having minimum distances from the input rank lists L_1, L_2, ..., L_N. Hence, the objective function to generate an aggregated rank list can be defined as:

$$minimize \ \Phi(\delta) = \sum_{i=1}^{N} w_i \times d(\delta, L_i) \tag{2}$$

δ denotes an aggregated rank list.

L_i is the ith input rank list (as obtained from biometric modality P_i).

N denotes number of modalities.

w_i denotes the associated weight for rank list L_i. For the reported experiments in this paper, each input rank list has been assigned same weight.

d denotes a distance metric between two lists.

The goal is to obtain an aggregated rank list δ^*, which minimizes the objective function $\Phi(\delta)$, among the set of all candidate rank lists.

Spearman footrule [13] and Kendall's tau [13] distances are applied here to calculate the distance between two rank lists.

In order to estimate the distance between two rank lists, Spearman footrule distance [13] considers summation of the absolute differences between ranks of each subject in two lists as:

$$d(\delta, L_i) = S(\delta, L_i) = \sum_{t \in L_i \cup \delta} |r^\delta(t) - r^{L_i}(t)| \tag{3}$$

Here, $r^\delta(t)$ represents the rank of subject t in the list δ. $r^{L_i}(t)$ represents the rank of subject t in the input rank list L_i.

As per Kendall's tau distance [13], the distance between the aggregated rank list δ and the input rank list L_i is estimated by counting the number of disagreements in the rank ordering by considering every pair of subjects between these two lists.

$$d(\delta, L_i) = K(\delta, L_i) = \sum_{t,u \in L_i \cup \delta} K_{tu}^p \tag{4}$$

$$where, K_{tu}^p = \begin{cases} 0 & \text{if } r^\delta(t) < r^\delta(u), r^{L_i}(t) < r^{L_i}(u), \\ & \text{or } r^\delta(t) > r^\delta(u), r^{L_i}(t) > r^{L_i}(u), \\ 1 & \text{if } r^\delta(t) > r^\delta(u), r^{L_i}(t) < r^{L_i}(u) \\ & \text{or } r^\delta(t) < r^\delta(u), r^{L_i}(t) > r^{L_i}(u), \\ p & \text{if } r^\delta(t) = r^\delta(u) \text{ or } r^{L_i}(t) = r^{L_i}(u) \end{cases} \tag{5}$$

Here, p is penalty and its value is set to 0.5.

If subject t is not present in one of the two lists (either δ or L_i), the rank of the subject ($r^\delta(t)$ or $r^{L_i}(t)$) in the list is considered as one more then the size of the list.

3 Cross-Entropy Monte Carlo Algorithm

In this paper, the cross-entropy Monte Carlo algorithm has been presented to solve the above optimization problem (Eq. 2). The cross-entropy (CE) Monte Carlo algorithm [14] is an iterative method to solve combinatorial problems. The cross-entropy Monte Carlo method is presented in this section.

A rank list δ can be represented as a matrix X of size $n \times k$ having values $x_{jr} \epsilon \{0,1\}$. n is number of subjects in the list and k represents total number of rank positions. In this matrix X, each row represents a subject (person) and column represents each rank. So, this matrix will have its entries as 0's and 1's, while satisfying the following constraint on each row and each column. The summation of x_{jr} values in each row and in each column has to be 1. The places in the matrix X, where 1 is present for each subject, will define the rank of the subject. The matrix X defines a rank list of size k uniquely using the placement of 1's in each row. For example, a set of subjects A, B, C and D having ranks 3, 1, 2 and 4, respectively, can be represented using a 4×4 matrix X as:

$$
X = \begin{bmatrix} 0\,0\,1\,0 \\ 1\,0\,0\,0 \\ 0\,1\,0\,0 \\ 0\,0\,0\,1 \end{bmatrix}
$$

The solution space χ can be defined for the proposed optimization problem as a collection of all such feasible matrices. Here, the aim is to find the rank list (i.e., the matrix X) with minimum objective function value over the solution space. This will generate the aggregated rank list.

X follows the probability mass function (pmf) $P_{v^t}(X)$ which is reflected in a parameter matrix $v_{n \times k} = p_{jr}$.

$$
P_{v^t}(X) \propto \prod_{j=1}^{n} \prod_{r=1}^{k} (p_{jr})^{x_{jr}}
$$

$$
\times I(\Sigma_{r=1}^{k} x_{jr} \leq 1, 1 \leq j \leq n; \Sigma_{j=1}^{n} x_{jr} = 1, 1 \leq r \leq k) \quad (6)
$$

Here, the element at the jr^{th} position of the X matrix is denoted by x_{jr}. Each element in the parameter matrix v is denoted as p_{jr} which indicates the probability of jth subject having r^{th} rank.

The steps of the CE algorithm [14] is given below. These steps are repeated until the algorithm converges.

1. **Initialization:** Let t denote the current iteration number, which is initialized to 0. A parameter matrix v^0 of size $n \times k$ is initialized to have same value for every element of the matrix. In this matrix, row corresponds each subject (person) and column corresponds individual ranks. Initially, the probability of jth subject having r^{th} rank p_{jr}^0 (i.e., jr^{th} element in v^0 matrix) is assigned a value $1/n$. Here, total number of subjects (persons) in the list is n and k represents the total number of rank positions. So, initially every subject has

an equal opportunity for being in the candidate list. The objective function $\Phi(\delta)$ in Eq. 2 is evaluated for such a candidate list.

2. **Sampling:** N_S samples (candidate lists) are generated from the probability mass function (pmf) $P_{v^t}(X)$ (Eq. 6) at each iteration t. For each of the N_S candidate lists δ_i's, its objective function value $\Phi(\delta)$ is estimated using Eq. 2. An ascending order sorting of these candidate lists helps to obtain the ρ-quantile $y^t = \Phi_{\lceil \rho N \rceil}$. The value of ρ is assumed as 0.1 for the reported experiments.

3. **Updation:** The probability values p_{jr}^{t+1} are updated as:

$$p_{jr}^{t+1} = (1 - w)p_{jr}^t + w\frac{\sum_{i=1}^{N_S} I(\Phi(\delta_i) \leq y^t)x_{jr}^{(i)}}{\sum_{i=1}^{N_S} I(\Phi(\delta_i) \leq y^t)}, \tag{7}$$

where, the jr^{th} position of the ith sample (denoted by matrix X_i) is $x_{jr}^{(i)}$. For the experiments, N_S is set to 1000, k is set to 5, w is a weight parameter. The value of w is set to 0.25.

4. **Convergence:** This algorithm stops if changes in the optimal list are less than a threshold for a specified number of iterations. The number of iteration is set to 5 for the reported experiments.

4   Experimental Results

Performance of the proposed cross-entropy Monte Carlo method (independently using Spearman footrule and Kendall's tau distances) is experimentally compared against several existing fusion methods (both at rank level and score level). These existing state-of-the-art methods have been mentioned in Sect. 4.1. Moreover, this experimental comparison has been carried out for two different tasks of multimodal biometrics. Hence, two different datasets have been used in these experiments. These datasets along with the corresponding performance measures of all these comparing methods have been reported in Sects. 4.2 and 4.3.

4.1   State-of-the-Art Methods for Performance Comparison

Performance of the proposed rank level technique has been compared against the existing state-of-the-art linear and non-linear rank level fusion methods. Borda count, weighted Borda count and highest rank methods belong to linear rank level fusion methods [12]. Similarly, non-linear rank level fusion methods are exponential [1], weighted exponential [1], division exponential [2] and logarithm [2] methods. The proposed method has also been compared against few state-of-the-art score level fusion methods, sum-rule [9], product-rule [18], max-rule [18] and min-rule [18].

Some of these existing methods (weighted Borda count, exponential, weighted exponential, division exponential and logarithm) require weights to be assigned for various biometric modalities. The performances of these methods depend on these selected weights. An elitist genetic algorithm has been used to search for the set of weights for various modalities. Recognition accuracy (as defined as the percentage of probe for which correct matching subject has been found) is considered as a fitness criteria for this genetic algorithm.

Moreover, the set of selected weights depends on the dataset on which the weights are being trained. Hence, k-fold cross validation (using $k = 5$) is used to eliminate this dependency. The dataset is splitted into k parts. $k - 1$ parts have been considered as training set to learn the set of weights using the elitist genetic algorithm and remaining one partition is used as test set to report the recognition accuracy. This is repeated k-times to get different set of weights and corresponding accuracies on the test sets. Finally, average accuracy from k such test sets has been presented in the result.

4.2   Fusion of Multimodal Biometrics Involving Face and Fingerprint

In this paper, the first dataset namely (BSSR1) is from the data repository of NIST [15]. This dataset has been widely used to study fusion of multimodal biometrics [1,7,8,19]. In this dataset, four biometric modalities have been considered. Two of these modalities are for face biometrics (using two different matchers, termed as G and C in the dataset). Fingerprints of the right index finger and left index finger are the other two modalities. These above four biometric modalities were acquired for each of the 517 persons (subjects) during enrollment phase. The dataset contains similarity scores of each of these subjects as a probe with all 517 subjects as per two different face matchers (termed as G and C) and fingerprint matchers for right and left index fingers. These similar-

Table 1. Cumulative recognition accuracies in % for NIST BSSR1 dataset using various comparative methods

Method			Rank 1	Rank 2	Rank 3
Unimodal	Face Matcher G		83.37	86.28	88.40
	Face Matcher C		88.78	90.52	91.50
	Left fingerprint Matcher		85.70	87.04	87.81
	Right fingerprint Matcher		92.07	93.23	93.62
Existing rank level fusion	Borda count		92.07	93.04	94.00
	Weighted borda count		92.50	94.20	95.36
	Highest rank		79.70	99.81	98.26
	Nonlinear methods	Exponential	89.16	90.13	91.30
		Weighted exponential	87.81	89.16	90.71
		Division exponential	99.23	**99.42**	**99.42**
		Logarithm	98.45	99.03	99.23
Existing score level fusion	Sum		79.50	80.85	81.24
	Max rule		79.88	94.00	98.45
	Min rule		94.78	95.40	95.60
	Product		97.87	98.30	98.70
Proposed CE methods	**CES**		**99.26**	**99.42**	**99.42**
	CEK		98.26	98.84	99.23

ity scores from various biometric modalities are fused using existing score level fusion methods (as mentioned in Sect. 4.1).

Moreover, rank lists are generated based on the given similarity scores. This provides four rank lists for each probe. These rank lists are combined using the proposed rank level fusion method (Sect. 3) and other existing methods as discussed above (Sect. 4.1). Table 1 presents the recognition accuracies of various comparing methods (in %) for the probe subjects within top 1, top 2 and top 3 ranks (cumulative). It also presents the recognition accuracies for each of the unimodal biometrics in this dataset.

From Table 1, the results clearly show that the proposed cross-entropy Monte Carlo algorithm based on Spearman footrule (CES) and Kendall's tau (CEK) distances performs better than most of the comparing methods. The reason for this superiority of the proposed method is that the method considers minimization of the distances between aggregated and input rank lists. Only the performance of the division exponential method is equal to the proposed CE method with Spearman footrule distance. It is also noticed that the proposed method is performing better than each unimodal matcher justifying the need for multi-biometric system.

4.3   Fusion of Multimodal Biometrics for Various Gait Features Representations

Additionally, the second dataset (BSS4) is form the Institute of Scientific and Industrial Research (ISIR), Osaka University (OU) [16,17]. This dataset has also been used for fusion of multimodal biometrics in [20,21]. In this dataset, input image sequence from gait has been processed using five different feature extraction methods: (i) Gait energy image (GEI), (ii) Frequency-domain feature (FDF), (iii) Gait entropy image (GEnI), (iv) Chrono-gait image (CGI), (v) Gait flow image (GFI). The dataset is composed of dissimilarity scores of each of these 3249 subjects as probe with all 3249 subjects for above mentioned features. These dissimilarity scores from above gait features are fused using existing score level fusion methods. Details of these gait feature extraction methods can be found [17,20].

Moreover, rank lists are generated based on the given dissimilarity scores. This provides five rank lists for each probe. These rank lists are combined using the proposed rank level fusion method (Sect. 3) and other existing rank level fusion methods as discussed in Sect. 4.1. Table 2 presents the recognition accuracies of various methods (in %) for the probe subjects within top 1, top 2 and top 3 ranks (cumulative). The table also presents the recognition accuracies for each of the unimodal biometrics in the dataset.

From Table 2, the results clearly show that the proposed cross-entropy Monte-Carlo algorithm based on Spearman footrule (CES) and Kendall's tau (CEK) distances has superior performance over other methods except score level fusion with sum and rank level fusion with division exponential method. The justification made for the previous dataset BSSR1 is also applicable here. It is also

Table 2. Cumulative recognition accuracies in % for OU-ISIR BSS4 dataset using various comparative methods

Method			Rank 1	Rank 2	Rank 3
Unimodal	CEnI		80.95	85.50	87.50
	CGI		83.35	87.44	89.04
	FDF		85.90	89.87	91.23
	GEI		85.72	89.54	91.20
	GFI		74.92	79.93	82.12
Existing rank level fusion	Borda count		83.63	87.47	88.77
	Weighted borda count		84.58	88.34	89.47
	Highest rank		77.41	88.15	91.54
	Nonlinear methods	Exponential	83.56	87.47	88.83
		Weighted exponential	81.60	85.29	87.32
		Division exponential	86.40	89.94	91.51
		Logarithm	85.47	89.20	90.90
Existing score level fusion	Sum		**86.61**	89.72	91.23
	Max rule		85.38	88.27	89.60
	Min rule		77.41	88.15	91.60
	Product		77.41	88.21	91.60
Proposed CE methods	**CES**		86.21	**90.21**	91.51
	CEK		86.30	89.87	**91.63**

noticed that the proposed method is performing better than each unimodal matcher justifying the need for multi-biometric system.

5 Conclusion

Rank level fusion has been studied, in this paper, for multimodal biometrics. The manifold contributions of this paper are highlighted here: The rank level fusion in multimodal biometrics is formulated as an optimization problem. In order to solve this optimization problem, cross-entropy Monte Carlo method (using two distant measures, namely Kandall tau and Spearman footrule distances) is proposed. The proposed method is tested for two different multi-biometric datasets: BSSR1 and BSS4. The proposed method using both of the distance metrics provides better performance in identifying the subjects than most of the existing methods of fusion at rank level (e.g., Borda count, weighted Borda count, highest rank, exponential, weighted exponential, division exponential, logarithm) and score-level (product-rule, sum-rule, max-rule and min-rule) for multimodal biometric systems. Experiments also justify the usefulness of multimodal biometric system over unimodal biometric system. Similarly, the proposed model can be applied for any other multimodal biometrics. Moreover, initial success for the reported experiments is encouraging enough to try out other meta-heuristic search and optimization strategies (like genetic algorithm, particle swarm optimization etc.) in the context of rank level fusion of multimodal biometrics.

References

1. Kumar, A., Shekhar, S.: Personal identification using multibiometrics rank-level fusion. IEEE Trans. Syst. Man Cybern. **41**(5), 743–752 (2011)
2. Jemaa, S.B., Hammami, M., Ben-Abdallah, H.: Finger surfaces recognition using rank level fusion. Comput. J. **60**(7), 969–985 (2017)
3. Devi, D.V.R., Rao, K.N.:A multimodal biometric system using partition based dwt and rank level fusion. In: IEEE International Conference on Computational Intelligence and Computing Research (ICCIC) (2016)
4. Rahman, M.W., Zohra, F.T., Gavrilova, M.L.: Rank level fusion for Kinect gait and face biometric identification. In: IEEE Symposium Series on Computational Intelligence (SSCI) (2017)
5. Sharma, R., Das, S., Joshi, P.: Rank level fusion in multibiometric systems. In: Fifth National Conference on Computer Vision, Pattern Recognition, Image Processing and Graphics (NCVPRIPG) (2015)
6. Monw, M.M., Gavrilova, M.: Markov chain model for multimodal biometric rank fusion. SIViP **7**, 137–149 (2013)
7. Wasnik, P., Raghavendra, R., Raja, K., Busch, C.: Subjective logic based score level fusion: combining faces and fingerprints. In: 2018 21st International Conference on Information Fusion (FUSION), pp. 515–520(2018)
8. Sharma, R., Das, S., Joshi, P.: Score-level fusion using generalized extreme value distribution and DSMT, for multi-biometric systems. IET Biometrics **7**(5), 474–481 (2018)
9. Soltanpour, S., Wu, Q.J.: Multimodal 2D–3D face recognition using local descriptors: pyramidal shape map and structural context. IET Biometrics **6**(1), 27–35 (2016)
10. Kabir, W., Ahmad, M.O., Swamy, M.: Normalization and weighting techniques based on genuine-impostor score fusion in multi-biometric systems. IEEE Trans. Inf. Forensics Secur. **13**(8), 1989–2000 (2018)
11. Kabir, W., Ahmad, M.O., Swamy, M.: A multi-biometric system based on feature and score level fusions. IEEE Access **7**, 59437–59450 (2019)
12. Ross, A.A., Nandakumar, K., Jain, A.K.: Handbook of Multibiometrics. Springer, Boston (2006). https://doi.org/10.1007/0-387-33123-9
13. Pihur, V., Datta, S., Datta, S.: Rankaggreg, an r package for weighted rank aggregation. BMC Bioinform. **10**(1), 62 (2009)
14. de Boer, P.T., Kroese, D.P., Mannor, S., Rubinstein, R.Y.: A tutorial on the cross-entropy method. Ann. Oper. Res. **134**(1), 19–67 (2005). https://doi.org/10.1007/s10479-005-5724-z
15. NIST: Dataset NIST: Biometric scores set (bssr1). https://www.nist.gov/itl/iad/image-group/nist-biometric-scores-set-bssr1
16. OU: The OU-ISIR biometric score database (BSS4). http://www.am.sanken.osaka-u.ac.jp/BiometricDB/BioScore.html
17. Iwama, H., Okumura, M., Makihara, Y., Yagi, Y.: The ou-isir gait database comprising the large population dataset and performance evaluation of gait recognition. IEEE Trans. Inf. Forensics Secur. **7**(5), 1511–1521 (2012)
18. Arun, A., Karthik, N., Anil, K.J.: Handbook of Multibiometrics. Springer, Boston (2006). https://doi.org/10.1007/0-387-33123-9
19. Chia, C., Sherkat, N., Nolle, L.: Towards a best linear combination for multimodal biometric fusion. In: 2010 20th International Conference on Pattern Recognition, pp. 1176–1179 (2010)

20. Makihara, Y., Muramatsu, D., Iwama, H., Yagi, Y.: On combining gait features. In: 2013 10th IEEE International Conference and Workshops on Automatic Face and Gesture Recognition (FG), pp. 1–8 (2013)
21. Mansouri, N., Issa, M.A., Jemaa, Y.B.: Gait features fusion for efficient automatic age classification. IET Comput. Vision **12**(1), 69–75 (2018)

Multilingual Phone Recognition: Comparison of Traditional versus Common Multilingual Phone-Set Approaches and Applications in Code-Switching

K. E. Manjunath[1,3($\boxtimes$)], K. M. Srinivasa Raghavan[1], K. Sreenivasa Rao[2], Dinesh Babu Jayagopi[1], and V. Ramasubramanian[1]

[1] International Institute of Information Technology - Bangalore (IIIT-B), Bangalore, India
{manjunath.ke,srinivasaraghavan.km}@iiitb.org,
{jdinesh,v.ramasubramanian}@iiitb.ac.in
[2] Indian Institute of Technology Kharagpur, Kharagpur, India
ksrao@iitkgp.ac.in
[3] U. R. Rao Satellite Centre, ISRO, Bangalore, India

Abstract. We propose a multilingual phone recognition system using common multilingual phone-set (Multi-PRS) derived from IPA based labelling convention, which offers seamless decoding of the code-switched speech. We show that this approach is superior to a more conventional front-end language-identification (LID)-switched monolingual phone recognition (LID-Mono) trained individually on each of the languages present in multilingual dataset. The state-of-the-art i-vectors are used to perform LID. We address the problem of efficient speech recognition for bilingual code-switching. We analyse the differences between LID-Mono and proposed Multi-PRS, by showing that the LID-Mono approach suffers due to a trade-off between two conflicting factors - the need for short windows for detecting code-switching at a high time resolution and the need for long windows needed for reliable language identification - which limits the overall performance of the LID-Mono system that suffers with high PERs at small windows (poor LID performance) and mismatched decoding conditions at long windows (due to poor code-switching detection time resolution). We show that the Multi-PRS, by virtue of not having to do a front-end LID switching and by using a multilingual phone-set, is not constrained by these conflicting factors and hence performs effectively on code-switched speech, offering low PERs than the LID-Mono system.

Keywords: Code-switched ASR · Indian language ASR · Multilingual ASR · LID-switched monolingual ASR

© Springer Nature Singapore Pte Ltd. 2020
S. M. Thampi et al. (Eds.): SIRS 2019, CCIS 1209, pp. 75–86, 2020.
https://doi.org/10.1007/978-981-15-4828-4_7

1   Introduction

The traditional approach for multilingual phone recognition uses front-end language-identification (LID)-switched monolingual phone recognition (LID-Mono) trained individually on each of the languages present in the multilingual dataset. The traditional approach has several disadvantages: (i). Complex two-stage architecture, (ii). Failure of LID block leads to the failure of entire system, (iii). Developing monolingual phone recogniser is not feasible for all languages. We propose to use a common multilingual phone-set approach to build Multilingual Phone Recognition System (Multi-PRS).

We address the problem of efficient techniques for speech recognition of code-switched speech. In code-switching, two or more languages are mixed and spoken as if they are one language [15,29]. Code-switching (or code-mixing or language-mixing) involves switching between multiple languages either inter-sententially and intra-sententially [17]. Bilingual code-switching is more common compared to the mixing of more than two languages [26]. The reasons for code-switching include (i) availability of a better word or phrase in another language to express a particular idea, (ii) certain words or phrases are more readily available in the other language, (iii) to show expertise in multiple languages. Code-switching is a common practice across the world in multilingual societies, where a speaker has spoken proficiency in more than one language. In this study, we have considered intra-sentential code-switching between two Indian languages, namely, Kannada (KN) and Urdu (UR), with Kannada sentence being the primary language within which switching occurs to Urdu words and phrases.

The proposed Multi-PRS is faced with the specific difficulty of having to arrive at the appropriate phone set based on which such a phonetic decoding can be done on input speech from any of the languages of interest. Such a common phone set has to have a coverage of all the phones occurring across the multiple languages while also ensuring that the individual language's phones are accurately mapped to the phones in the common phone set. We propose the use of International Phonetic Alphabet (IPA) chart to derive common multilingual phone-set. IPA has strict one-to-one correspondence between symbols and sounds which makes it to be able to accommodate all the world's diverse languages. Few notable works based on common multilingual phone-set approach are reported in [32,33,37,38]. Although there are significant efforts to develop multilingual speech recognizers using Indian languages [2,8,34], not many studies have explored the use of IPA based common multilingual phone-set to develop multilingual phone recognizers using Indian languages. The most recent works on multilingual speech recognition using DNNs are reported in [14,21,22,25,42]. Few notable works on code-switched speech recognition using multilingual speech recognisers are reported in [1,13,16,30,36].

The focus of our work here is to compare two approaches of multilingual speech recognition: (i) one involving using a front-end language-identification stage to detect the language spoken in short intervals of speech and then use the recognized language's phone recogniser to decode the speech; we refer to this as a LID-switched monolingual approach (LID-Mono), and (ii) using a Multilingual Phone Recognition System based on common multilingual phone-set (Multi-PRS). We further extend the comparison between LID-Mono and Multi-PRS to code-switching scenario. Because the Multi-PRS can seamlessly decode the code-switched speech without regard to the code-switched instances, since the Multi-PRS is designed using a common phone-set between several languages (from which the code-switched speech could switch between any pair of languages) with the corresponding common phone acoustic-models being trained from shared-data from the multiple languages. The rest of the paper is organized as follows: Sect. 2 describes our experimental setup. Section 3 describes and compares the two approaches of multilingual phone recognition. Section 4 extends the comparison to code-switching scenario. Section 5 provides the summary of the paper.

2   Experimental Setup

2.1   Multilingual Speech Corpora

We describe here the details of the speech corpora of the 6 Indian languages used in this work: Kannada (KN), Telugu (TE), Bengali (BN), Odia (OD), Urdu (UR), and Assamese (AS). The speech corpora was collected as a part of consortium project titled *Prosodically guided phonetic engine for searching speech databases in Indian languages* supported by DIT, Govt. of India [12]. Speech corpora contains 16 bit, 16 KHz speech wave files along-with their IPA transcription [39]. The wave files contain read speech sentences of size between 3 to 10 s. Detailed description of the speech corpora is provided in [4,19,23,28]. We have used a split of 80:20 for train and test data, respectively. 10% of training data is held out from the training and used as development set. Table 1 shows the statistics of the speech corpora.

Table 1. Statistics of multilingual speech corpora

Language	# Speakers		Duration (in hours)			
	M	F	Train	Dev	Test	Total
Kannada (KN)	7	9	2.80	0.33	0.76	3.89
Telugu (TE)	9	10	4.05	0.47	1.07	5.59
Bengali (BN)	20	30	3.42	0.40	0.99	4.81
Odia (OD)	14	16	3.58	0.36	0.97	4.91
Urdu (UR)	53	6	4.12	0.46	1.04	5.64
Assamese (AS)	8	8	2.39	0.23	0.53	2.39

2.2   Testing Set Speech Corpora for Code-Switching Scenario

We have selected 320 code-switched sentences having Kannada as the primary language and code-switching to Urdu words and phrases. The sentences are carefully chosen to cover all the phonetic units of Kannada and Urdu. *Four* male and *four* female speakers who are bilinguals of Kannada and Urdu are made to read 40 sentences each. The speakers are proficient in both spoken Urdu and Kannada, and tend to produce KN-UR utterances. These sentences were transcribed using IPA symbols and then mapped to the common multilingual phone-set to generate the ground-truth transcription for calculation of PER in the decoding.

Figure 1 shows the distribution of durations of KN and UR languages in code-switched KN/UR test sets. The duration of each utterance range from 3.5 s to 11 s in the data set, within which the Urdu words and phrases occur at durations, from which it can be noted that Kannada segments in an utterance are the longer ones, interspersed with Urdu segments of relatively shorter durations, importantly ranging from 0–0.5 s to 3–3.5 s which typically correspond to short words (<500 ms) and multi-word phrases (of the order of 1–3.5 s). This kind of Urdu segments in the code-switched data plays an important factor in determining how the LID-Mono works, particularly in the choice of the speech interval size on which the front-end LID has to operate.

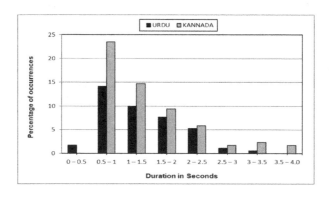

Fig. 1. Distribution of durations of KN and UR languages in code-switched KN/UR test sets.

2.3   Training DNNs

Context dependent DNNs with tanh non-linearity at hidden layers and softmax activation at the output layer are used. DNNs are trained using greedy layer-by-layer supervised training. Initial learning rate was chosen to be 0.015 and was decreased exponentially for the first 15 epochs. A constant learning rate of 0.002 was used for the last 5 epochs. Once all the hidden layers are added to the network, shrinking is performed after every 3 iterations, so as to separately scale the parameters of each layer. Mixing up was carried out halfway between the completion of addition of all the hidden layers and the end of training. Stability of the training is maintained through preconditioned affine components. Once the final iteration of training completes, the models from last 10 iterations are combined into a single model. Each input to DNNs uses a temporal context of 9 frames (4 frames on either side). The number of hidden layers of DNNs used in the development of Phone Recognition Systems (PRS) are tuned by adjusting the width of the hidden layers. It is found that the DNNs with 5 hidden layers are suitable for building PRSs. Bi-phone (phoneme bi-grams) language model is used for decoding. The language model weighting factor and acoustic scaling factor used for decoding the lattice are optimally determined using the development set to minimize the PER. DNNs training used in this study is similar to the one presented in [41]. All the experiments are conducted using the open-source speech recognition toolkit - Kaldi [11].

2.4 Extraction of i-vectors

The i-vectors are one of the most widely used features for language recognition. They are fixed dimension feature vectors that are derived from the variable length sequence of front-end features [24]. A DNN is trained for automatic speech recognition using the labelled speech data from Switchboard (SWB1) and Fisher corpora (about 2000 h). Training uses hidden layers with ReLU activation with layer-wise batch normalization. Mel-frequency cepstral coefficients (MFCCs) are extracted from each input utterance and fed to DNNs. The bottleneck features (80 dimension) are extracted from the bottleneck layer of trained DNN [5,6]. The extracted bottleneck features are the front end features. A Gaussian Mixture Universal Background Model (GMM-UBM) is obtained by pooling the front end features from all the utterances in the train dataset. The means of the GMMs are adapted to each utterance using the Baum-Welch statistics of the front-end features. The i-vectors (400 dimension) are computed based on each adapted GMM mean supervector. Since the *SWB1* and *Fisher* corpora used for training the DNNs have the sampling rate of 8 KHz, we have down-sampled the multilingual speech corpora and the testing datasets (see Sects. 2.1 and 2.2) from 16 KHz to 8 KHz for extracting the i-vectors. Detailed description of extraction of i-vectors is given in [7].

3 Approaches for Multilingual Phone Recognition

The following subsections describe the development and comparison of multilingual phone recognizers using two approaches: (i). LID-switched monolingual phone recognition (LID-Mono) approach, and (ii). Multilingual phone recognition using common multilingual phone-set (Multi-PRS) approach.

3.1 LID-switched Monolingual Phone Recognition (LID-Mono) Approach

LID-Mono is a traditional approach for multilingual phone recognition and is shown in Fig. 2. It consists of two stages. In the first stage, the language of the input speech is determined using a language identification block. In the second stage, the input speech utterance is routed to the monolingual phone recognizer of the language identified in stage-1 and the phones present in the input speech are determined. Monolingual phone recognizer is a conventional PRS developed using the data of single language.

We briefly outline the LID system here. There are two approaches for LID, namely, implicit LID and explicit LID. The explicit LID requires phonetic transcription and language models for each language [3,27], whereas the implicit LID does not need either phonetic transcription or language models [10,35]. Since, we do not have language models for the languages considered in this study, we have carried out implicit LID to perform LID. Support Vector Machines (SVM) [20,40] are used to train the LID classifiers. Multi-class SVM is constructed using one-against-one approach (Max-win voting). The radial basis function is used as a kernel. The *LIBSVM* library is used for building SVM models [9]. We have explored both MFCCs and i-vectors as features for building LID systems. The 13-dimensional MFCCs [18] along-with their first and second order derivatives are computed using a frame-length of 25 ms with a frame-shift of 10 ms. The i-vectors are extracted using the procedure described in Sect. 2.4. Table 2 shows the LID accuracy (%) for various language sets using SVMs. Since, the performance of LID using i-vectors outperforms MFCCs, we have considered only the i-vector based LID systems in all our experiments. LID accuracy decreases as the number of languages increase.

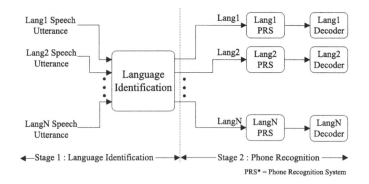

Fig. 2. Multilingual phone recognition using LID-Mono approach.

Table 2. LID accuracy for various language sets using MFCCs and i-vectors.

Languages	LID accuracy (%)	
	MFCCs	i-vectors
KN-BN-OD-UR	91.16	97.98
KN-TE-BN-OD-UR	74.76	96.22
KN-TE-BN-OD-UR-AS	71.19	96.00

3.2 Multilingual Phone Recognition Using Common Multilingual Phone-Set Approach (Multi-PRS)

Figure 3 shows the schematic representation of the Multi-PRS. Unlike Fig. 2, which has two stages, Fig. 3 has a single stage - irrespective of any language, Multi-PRS accepts speech input from any language and decodes it into a sequence of phonetic units [31] using a common phone-set.

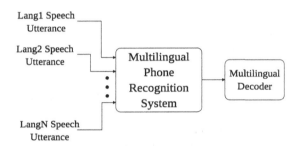

Fig. 3. Multilingual phone recognition using common multilingual phone set approach.

We have developed Multi-PRSs using six Indian languages - KN, TE, BN, OD, UR and AS. The common multilingual phone-set is derived by grouping the acoustically similar IPAs across the languages together and selecting the phonetic units which have sufficient number of occurrences to train a separate model for each of them. The IPAs which do not have sufficient number of occurrences will be mapped to the closest linguistically similar phonetic units present in the common multilingual phone-set. The common multilingual phone-set thus derived contained 44, 46, 46 phones for 4, 5, and 6 languages, respectively. We have also developed monolingual Phone Recognition Systems (Mono-PRSs) for KN, TE, BN, OD, UR, and AS languages using 36, 35, 34, 36, 35, and 32 phones, respectively. Mono-PRSs are used in second stage of LID-Mono systems as shown in Fig. 2. The Mono-PRSs and Multi-PRSs are trained using CD DNNs.

3.3   Comparison of LID-Mono and Multi-PRS

Phone Error Rate (PER) is determined by comparing the decoded phone labels with the reference transcriptions by performing an optimal string matching using dynamic programming.

Table 3. Phone error rates of multilingual phone recognition systems.

Languages	Approach	
	Multi-PRS	LID-Mono (i-vectors)
KN, BN, OD, UR	31.5	35.5
KN, TE, BN, OD, UR	32.5	35.6
KN, TE, BN, OD, UR, AS	31.9	37.9

Table 3 shows the PERs of LID-Mono and Multi-PRS approaches. It is found that the Multi-PRS systems based on common phone set approach outperform the traditional LID-Mono systems. As the number of languages increase the benefits of Multi-PRSs will be more. Higher the number of languages more the benefit from Multi-PRSs compared to LID-Mono. The use of Multi-PRS has an additional advantage of decoding more number of phones compared to the LID-Mono. This would help the language models to recognise the words more accurately.

We show in Fig. 4, the performance of the LID-Mono and Multi-PRSs on test data drawn from 4, 5 and 6 languages in terms of % LID accuracy and % PER. It can be noted that the LID-Mono has an inherently poor performance marked by decreasing %LID accuracy as the number of language classes increase from 4 to 6, which in turn impacts the % PER to increase in going from 4 to 6 languages. When the LID system makes an error, the LID-switched monolingual phone recognition chooses the wrong language phone acoustic models to decode the input speech and naturally incurs an higher %PER. The %PERs of Multi-PRS

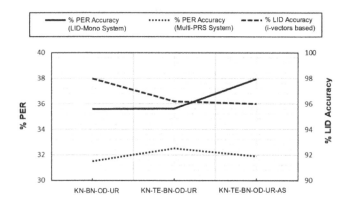

Fig. 4. LID accuracy (%) and PERs (%) of LID-Mono and Multi-PRS approaches.

system, in contrast, have a robust constant performance across the multiple languages clearly arising from its not depending on a front-end LID decision making.

4   LID-Mono and Mutli-PRS Approaches in Code-Switching Application

For comparison of results, in addition to the multilingual phone recognisers based on 4, 5, and 6 languages, we have also developed bilingual phone recognisers using KN and UR languages using both LID-Mono and Multi-PRS approaches. Further, we have analysed how the duration of windows (in seconds) used for performing LID effects the performance of various multilingual phone recognisers. Figure 5 shows a composite display of the performance of LID-Mono and Multi-PRS for different number of test languages, different sizes of the intervals over which the LID makes a decision (called the LID-interval in the x-axis from 500 ms to 5 s and the full utterance), the resulting %LID accuracy and the overall %PER (in the two y-axes).

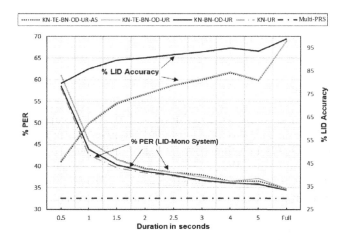

Fig. 5. Comparison of LID-Mono and Multi-PRS systems in code-switching scenario.

Considering the LID-Mono system curves, for small LID-intervals, the LID accuracy is low (45% to 85% for different number of languages considered 6 to 4), with the LID-Mono having to use a wrong language acoustic-model for decoding the corresponding interval. This naturally leads to a very high % PER (of the order of 60%). As the LID-interval size increases the %LID improves, reaching 80–90% for durations of 3 s and 85–95% for longer durations of 5 s (or the full utterance). The corresponding %PER also shows a marked decrease, since the LID-Mono makes a phone-decoding of the given intervals with the correct language acoustic models with increasing accuracy, reaching 40% at 2 secs and down to 35% at 5 secs and more. What is important to note here is that the LID-Mono's performance is dictated by the LID-interval size - smaller sizes are good for detecting code-switching instances at high time-resolution, but have inherently poor LID accuracies and corresponding poor PER; larger sizes are good for yielding high LID accuracy, with corresponding lower PER, but the code-switching instances are missed due to poor time resolution of the LID decision intervals, i.e. for instance, at a 2 s LID-interval, a good proportion of Urdu segments would have occurred 'within' the 2 s interval (as evident from the code-switching duration distribution in Fig. 1). These segments would then be decoded by the LID decision, which, has no question of being 'correct' since the interval in question is 'mixed' in its ground truth, and the LID has to yield a single language decision, potentially resulting in a mismatch in the language(s) in the 2-sec interval and the single monolingual phone-recognizer that would have been brought into to decode the speech in the 2 s interval. This problem becomes more acute as the LID-interval size increases.

On the contrary, the KN-UR Multi-PRS using common multilingual phone-set approach offers a consistent performance of 32.7% without having any LID-interval in its pipe-line, and hence is robust to arbitrary code-switching durational distributions (as in Fig. 1). This makes the Multi-PRS the natural choice to recognize code-switched speech, with practically no particular merit to choose the LID-Mono system, which suffers from the trade-off discussed above, higher design complexity of having to design a LID system (to recognize multiple language classes), and having to design multiple monolingual phone recognition systems.

5   Conclusions

We have developed and compared LID-Mono and Multi-PRS approaches of multilingual phone recognition. We have extended the same study to code-switched speech recognition scenario using code-switched utterance of two Indian languages (Kannada and Urdu). We have studied the performance characteristics of LID-Mono and Multi-PRS approaches with respect to several underlying parameters, such as the interval over which the LID makes a decision and the number of languages on which the LID is designed for the LID-Mono systems, the means of arriving at a common phone-set for the Multi-PRS and shown that while the LID-Mono system suffers from inherent trade-off's between interval sizes, the Multi-PRS offers a robust performance for arbitrary code-switching data.

References

1. Vu, N.T., et al.: A first speech recognition system for Mandarin-English code-switch conversational speech. In: ICASSP, pp. 4889–4892 (2012)
2. Mohan, A., Rose, R., Ghalehjegh, S.H., Umesh, S.: Acoustic modelling for speech recognition in Indian languages inan agricultural commodities task domain. Speech Commun. **56**, 167–180 (2014)
3. Sai Jayram, A.K.V., Ramasubramanian, V., Sreenivas, T.V.: Language identification using parallel sub-word recognition. In: IEEE International Conference on Acoustics, Speech, and Signal Processing (ICASSAP), vol. 1 (2003)

4. Sarma, B.D., Sarma, M., Sarma, M., Prasanna, S.R.M.: Development of Assamese phonetic engine: some issues. In: IEEE INDICON, pp. 1–6 (2013)
5. Jiang, B., Song, Y., Wei, S., Liu, J.H., McLoughlin, I., Dai, L.: Deep bottleneck features for spoken language identification. PLoS ONE **9**(7), e100795 (2014)
6. Jiang, B., Song, Y., Wei, S., Wang, M., McLoughlin, I., Dai, L.: Performance evaluation of deep bottleneck features for spoken language identification. In: International Symposium on Chinese Spoken Language Processing, pp. 143–147 (2014)
7. Padi, B., Ramoji, S., Yeruva, V., Kumar, S., Ganapathy, S.: The LEAP language recognition system for LRE 2017 challenge - improvements and error analysis. In: The Speaker and Language Recognition Workshop, Odyssey (2018)
8. Kumar, C.S., Mohandas, V.P., Haizhou, L.: Multilingual speech recognition: a unified approach. In: INTERSPEECH, pp. 3357–3360 (2005)
9. Chang, C.C., Lin, C.J.: LIBSVM: a library for support vector machines. ACM Trans. Intell. Syst. Technol. **2**, 1–27 (2011). software available at http://www.csie.ntu.edu.tw/ cjlin/libsvm
10. Nandi, D., Pati, D., Rao, K.S.: Implicit processing of LP residual for language identification. Comput. Speech Lang. **41**, 68–87 (2017)
11. Povey, D., et al.: The Kaldi speech recognition toolkit. In: IEEE Workshop on ASRU (2011). http://kaldi-asr.org/
12. Development of Prosodically Guided Phonetic Engine for Searching Speech Databases in Indian Languages. http://speech.iiit.ac.in/svldownloads/pro_po_en_report/
13. Yilmaz, E., Heuvel, H.V.D., Leeuwen, D.V.: Code-switching detection using multilingual DNNs. In: IEEE Workshop on Spoken Language Technology for Under-Resourced Languages, pp. 159–166 (2016)
14. Heigold, G., et al.: Multilingual acoustic models using distributed deep neural networks. In: IEEE International Conference on Acoustics, Speech and Signal Processing (ICASSP) (2013)
15. Kroll, J.F., De Groot, A.M.B. (eds.): Handbook of Bilingualism: Psycholinguistic Approaches. Oxford University Press, New York (2005)
16. Bhuvanagiri, K., Kopparapu, S.K.: Mixed language speech recognition without explicit identification of language. Am. J. Sig. Process. **2**(5), 92–97 (2012)
17. Jorschick, L., Quick, A.E., Glasser, D., Lieven, E., Tomasello, M.: German-English-speaking children-s mixed NPs with 'correct' agreement. Bilingualism: Language and Cognition **14**(2), 173–183 (2011)
18. Rabiner, L., Juang, B., Yegnanarayana, B.: Fundamentals of Speech Recognition. Pearson Education, New Delhi (2008)
19. Madhavi, M.C., Sharma, S., Patil, H.A.: Development of language resources for speech application in Gujarati and Marathi. In: IEEE International Conference on Asian Language Processing (IALP), vol. 1, pp. 115–118 (2014)
20. Li, M., Suo, H., Wu, X., Lu, P., Yan, Y.: Spoken language identification using score vector modeling and support vector machine. In: Interspeech, pp. 350–353 (2007)
21. Muller, M., Waibel, A.: Using language adaptive deep neural networks for improved multilingual speech recognition. In: International Workshop on Spoken Language Translation (IWSLT) (2015)
22. Muller, M., Stuker, S., Waibel, A.: Towards improving low-resource speech recognition using articulatory and language features. In: International Workshop on Spoken Language Translation (IWSLT), pp. 1–7 (2016)
23. Shridhara, M.V., Banahatti, B.K., Narthan, L., Karjigi, V., Kumaraswamy, R.: Development of Kannada speech corpus for prosodically guided phonetic search engine. In: Sixteenth International Oriental COCOSDA (2013)

24. Dehak, N., Carrasquillo, P.A.T., Reynolds, D., Dehak, R.: Language recognition via i-vectors and dimensionality reduction. In: Twelfth Annual Conference of the International Speech Communication Association (2011)
25. Vu, N.T., et al.: Multilingual deep neural network based acoustic modeling for rapid language adaptation. In: ICASSP (2014)
26. Heredia, R.R., Altarriba, J.: Bilingual language mixing: why do bilinguals code-switch? Curr. Dir. Psychol. Sci. 10, 164–168 (2001)
27. Santosh Kumar, S.A., Ramasubramanian, V.: Automatic language identification using ergodic-HMM. In: ICASSP, pp. 609–612 (2005)
28. Kumar, S.B.S., Rao, K.S., Pati, D.: Phonetic and prosodically rich transcribed speech corpus in Indian languages: Bengali and Odia. In: O-COCOSDA, pp. 1–5 (2013)
29. Ford, S.: Language mixing among bilingual children. http://www2.hawaii.edu/~sford/research/mixing.htm
30. Kim, S., Seltzer, M.L.: Towards language-universal end-to-end speech recognition. In: IEEE International Conference on Acoustics, Speech and Signal Processing (ICASSP), pp. 4914–4918 (2018)
31. Siniscalchi, S.M., Lyu, D., Svendsen, T., Lee, C.: Experiments on cross-language attribute detection and phone recognition with minimal target-specific training data. IEEE Trans. Acoust. Speech Signal Process. 20(3), 875–887 (2012)
32. Stuker, S., Schultz, T., Metze, F., Waibel, A.: Multilingual articulatory features. In: ICASSP, vol. 1, pp. 144–147 (2003)
33. Stuker, S., Metze, F., Schultz, T., Waibel, A.: Integrating multilingual articulatory features into speech recognition. In: INTERSPEECH, pp. 1033–1036 (2003)
34. Gangashetty, S.V., Sekhar, C.C., Yegnanarayana, B.: Spotting multilingual consonant-vowel units of speech using neural network models. In: Faundez-Zanuy, M., Janer, L., Esposito, A., Satue-Villar, A., Roure, J., Espinosa-Duro, V. (eds.) NOLISP 2005. LNCS (LNAI), vol. 3817, pp. 303–317. Springer, Heidelberg (2006). https://doi.org/10.1007/11613107_27
35. Nagarajan, T., Murthy, H.A.: A pair-wise multiple codebook approach to implicit language identification. In: Workshop on Spoken Language Processing, pp. pp. 101–108 (2003)
36. Schultz, T.: Multilingual automatic speech recognition for code-switching speech. In: The 9th International Symposium on Chinese Spoken Language Processing (2014)
37. Schultz, T., Waibel, A.: Language independent and language adaptive acoustic modeling for speech recognition. Speech Commun. 35, 31–51 (2001)
38. Schultz, T., Kirchhoff, K.: Multilingual Speech Processing. Academic Press, Amsterdam (2006)
39. The International Phonetic Association: Handbook of the International Phonetic Association. Cambridge University Press (2007). https://www.international phoneticassociation.org/
40. Campbell, W.M., Singer, E., Torres-Carrasquillo, P.A., Reynolds, D.A.: Language recognition with support vector machines. In: Proceedings of the Odyssey: The Speaker and Language Recognition Workshop, pp. 41–44 (2004)
41. Zhang, X., Trmal, J., Povey, D., Khudanpur, S.: Improving deep neural network acoustic models using generalized maxout networks. In: ICASSP, pp. 215–219 (2014)
42. Miao, Y., Metze, F.: Improving low-resource CD-DNN-HMM using dropout and multilingual DNN training. In: INTERSPEECH, pp. 2237–2241 (2013)

Estimation of Bone Mineral Density Using Machine Learning Approach

Bharti Joshi[1], Shivangi Agarwal[2(✉)], Leena Ragha[1], and Navdeep Yadav[3]

[1] Department of Computer Engineering, Ramrao Adik Institute of Technology,
Nerul, Navi Mumbai 400706, India
[2] Department of Electronics Engineering, Ramrao Adik Institute of Technology,
Nerul, Navi Mumbai 400706, India
`agarwal.shivangi@gmail.com`
[3] Department of Instrumentation and Control Engineering, NSIT,
University of Delhi, Sec-3, Dwarka, New Delhi 110078, India
`https://www.rait.ac.in, http://www.nsit.ac.in/`

Abstract. Osteoporosis means porous bone. If a healthy bone is looked under a microscope, it will look like a honeycomb. In case of osteoporosis, the holes and spaces in the honey comb are much larger. This means bone has lost its density. As bones become less dense they tend to break easily, the bones of the spine are at great risk. This paper is the part of the project which designs and develops a bone mineral density (BMD) detecting ultrasound machine. The unit is fully compact and portable and enables the BMD to be evaluated in real time by determining a parameter known as net time delay (NTD). Computer simulations are used to compare dependencies of NTD and BMD. It shows that the dependencies are very high. Hence concluding that NTD is highly proportional to bone mass and machine learning approach can be used to estimate the BMD of the subject. Looking to the severity of the disease and availability of expertise with ultrasonic system, it is proposed to develop low cost Ultrasonic based Bone Densitometer system to find the NTD thus the density of the bone and then find the T-Score in medical terms using Linear regression model.

1 Introduction

Bone density, or bone mineral density (BMD), is the quantity of bone mineral in bone tissue. The concept is of mass of mineral per volume of bone. Improper BMD can cause Osteoporosis i.e. "porous bone". Artificial intelligence has been on the Radar of technology leaders for decades and these autonomous machines can help us making decisions or carry out complex tasks. Machines can do complex and stressful work faster than Human multiple task at a time. So in order to get appropriate data and accurate result, AI has been used. Current existing system for bone densitometer scanning uses X-rays, hence it is costly. Also, the size and weight of such a system is massive. Using X-rays might have some side effects on a person's body. So, to avoid all these problems a new idea of using

© Springer Nature Singapore Pte Ltd. 2020
S. M. Thampi et al. (Eds.): SIRS 2019, CCIS 1209, pp. 87–93, 2020.
https://doi.org/10.1007/978-981-15-4828-4_8

Ultrasonic sound for measuring the bone density is used. Bone density can be easily calculated using the difference between sound intensity at two different points. This system can be made handy, as size of the equipment needed isn't large. Also, it's setup is easy as compared to X-ray based system. So, in order to provide neat and elegant solution, we decided to use Ultrasonic sound based scanning system.

1.1 Purpose and Problem Definition

Osteoporosis is a major health related issue, endangering human lives, in turn causing collateral damage to various entities. Bone density importantly influences our lives. Apriori accurate detection of low bone density is a key that could prevent any further damage due to Osteoporosis. Powerful computational tools are needed for predicting the accurate result of low bone density. There is a need of compact and reliable system for predicting the proper bone density related issues. A branch of AI i.e. Machine Learning has many advantages like experiential prediction to produce optimal results in reasonable amount of time, suitable for predicting output by taking past results into account, so proposed system is based on Machine Learning to predict and categorize the bone density.

1.2 Scope

Bone density index prediction is a crucial issue when fighting with Osteoporosis. A direct prediction of person to have Osteoporosis or not cannot be determined by seeing results, as there are some cases in which even if bone density is low, chances of Osteoporosis are less. So, to neatly predict close to accurate results, machine Learning algorithms like Linear Regression are used. This reduces the uncertainty in output prediction and reduces false predictions. An intelligent system based on Machine Learning algorithms has been implemented for prediction of Osteoporosis.

2 Literature Survey

In [1], Normative information on BMD and body structure measured with dual energy x ray absorptiometry (DXA) were acquired from early age to adolescence. The most easily available and most frequently uses method for measuring BMC and BMD in kids is DXA. The major benefits of this method are its brief time, very small dose of radiation, high precision, and the capacity to evaluate BMC and BMD at both axial and appendicular skeletal locations. This article proposed that before the end of the second span of 10 years, there is a slight rise of the skeletal mass in the lumbar spine and complete body. This research offers reference value for kids and young adults for bone density and body composition measured with DXA.

In [2], The article writer has introduced a device that is completely mobile, light and allows the BMD to be assessed in live time by determining a variable known as NTD. The NTD is described as the difference between the travel period of an ultrasound signal via the heel and the travelling time through an equally thick (to the heel) hypothetical item but comprising only soft tissue. Author's information shows that the fresh instrument and its related ultrasound parameter, the NTD, are highly susceptible to bone mass as measured by DXA scanners. On the other hand, the information demonstrate that the velocity of the ultrasound is much less correlated with BMD, this is due to differences in the quantity of overlying soft tissue density and calcaneus density between individuals. In short, a fresh instrument has been defined which, in specific, has the ability to widen the range of ultrasound application and bone screening in general.

In [3], The writer outlined the bone mineral density test in detail. X rays are utilized to evaluate the mineral quantity in bone. This test is essential for individuals having high potential of osteoporosis, particularly females and older adults. Also called as DXA. It is an significant test for osteoporosis.

In [4], Many efforts have been taken towards enhancing the quality. However, the presence of reliable BMD prediction techniques is still scarce, and they hardly include a real-time analysis simulation module. The main reason for this lack of confidence in prediction is uncertainty in prediction. Past attempts have made a direct prediction of person to have Osteoporosis or not by seeing indexes, which cannot be determined like that, as there are some cases in which even if bone density is low, chances of Osteoporosis are less. Also, as X-ray based systems are tedious to setup and are massive due to X-ray emitter systems, it is troublesome to take them far remote and rural areas. Similarly many researchers have worked in this area [5–8].

2.1 Benefits of the Proposed System

The solution to the problem defined above requires a system based around a model that can predict and categorize the bone density with improved accuracy despite the uncertainty in the input parameters involved. ML does exactly that. Using only data related to bone densities under analysis and also past data, algorithm can produce results in a reasonable amount of time, and is also suitable for parallel processing. The main advantages of the Proposed System are:

1. Improved accuracy of Results: Algorithm has the ability to produce "closer to optimal" results. This leads to a better intuition of the possible category of bone density.
2. Quicker Predictions: The system can make predictions in a "reasonable amount of time" while also "supporting parallel processing", which results in quicker predictions and timely actions to respond to the situation with a proper strategy.

3   Proposed Methodology

3.1   Module Description

As Net Time Delay (NTD) of ultrasound and Bone Density (BD) are correlated, a Linear Regression model can be used for predicting BD according to NTD. Linear Regression is a method to predict the dependent variable (Y) based on the values of independent variables (X). It can be used for the cases where we want to predict some continuous quantity (Fig. 1).

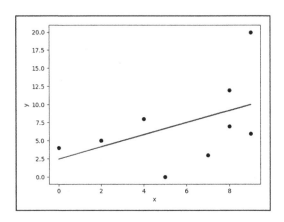

Fig. 1. Representation of linear regression (Color figure online)

Here the X axis is called independent axis, which will be plotted with NTD data. Y axis or dependent axis will be plotted with BD data. Regression line (as shown in blue) will be plotted optimally for all points. As it is assumed that the two variables are linearly related. Hence, we try to find a linear function that predicts the response value (y) as accurately as possible as a function of the feature or independent variable (x). The equation of the regression line is represented as: **y = b0 + b1 * x.**

Linear Regression to convert the Net Time Delay to Bone Density:

```
1  import numpy as np
2  def estimate_coef(x, y):
3      n = np.size(x)
4      m_x, m_y = np.mean(x), np.mean(y)
5      SS_xy = np.sum(y*x - n*m_y*m_x)
6      SS_xx = np.sum(x*x - n*m_x*m_x)
7      b1 = SS_xy / SS_xx
8      b0 = m_y - b1*m_x
9      return(b0, b1)
10
11 def Predict(NTD):
```

```
12   # observations sample random
13   x = np.array([0, 0.5, 1, 1.5, 2, 2.5, 3, 3.5, 4])   #
     taken test data [2]
14   y = np.array([0, 0.4, 0.6, 0.8, 1, 1.4, 1.6, 1.8, 2])
     #taken test data [2]
15   b0,b1 = estimate_coef(x, y)
16   print("The estimated value of b0 = " + str(b0))
17   print("The estimated value of b1 = " + str(b1))
18   BD = b0 + b1 * NTD
19   print("The estimated value of Predicted Bone Density =
     " + str(BD))
20 Predict(0.04)
```

Listing 1.1. Linear Regression

Output:

```
1   The estimated value of b0 = -0.004395604395604602
2   The estimated value of b1 = 0.5355311355311357
3   The estimated value of Predicted Bone Density =
    0.017025641025640827
4
```

Thus now, we have got the prediction about bone density for given NTD. The data used is of a male's lumbar spine bone density, age between 18–23 [1] (Fig. 2).

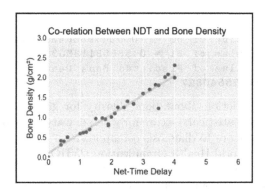

Fig. 2. Correlation between NTD and Bone Density

We have simulated a population mean here, after analyzing the real data we will get the real population mean.

```
1 t_score = (BD - popu_mean)/(std_dev/math.sqrt(sample_size
  -1))
```

Where BD is predicted value of BMD and **popu_mean** is the mean of population or data set taken for analysis.

```
1 t_score = -3.851155020921751
```

3.2 Hardware and Software Requirements

1. Software Requirements
   - Operating System: Any OS with Python support
   - Platform: Python 3.5 or above pre-installed with data analysis modules
   - Python libraries used: numpy, scipy, matplotlib
2. Hardware Requirements
   The hardware required for data collection is out of the scope of the project
   Computational hardware required
   - Minimum 4 GB RAM on system
   - Processor with high clock rate
   - USB input peripheral should be available

We will collect the data via a digitizer using PyUsb, do the computing on the NTD and calculate the bone density to get the T-Score.

4 Project Results

The system is trained with data obtained from a male's lumbar spine bone density, age between 18–23. [1]

For value 0.04 as test value, we get

Predict(0.04)

Output:

```
The estimated value of b0 = -0.004395604395604602
The estimated value of b1 = 0.5355311355311357
The estimated value of Predicted Bone Density =
    0.017025641025640827
```

We've got the prediction about bone density for given NTD. We will use this BD to obtain the statistical T-score for given density. A healthcare provider looks at the lowest T-score to diagnosis osteoporosis.

According to the World Health Organization (WHO):

Normal bone density range has T-score of −1.0 or above.

Low bone density or osteopenia has T-score between −1.0 and −2.5.

Diagnosis of osteoporosis has T-score of −2.5 or below.

Based on T-score, we can predict the category of bone density found for given bone sample, hence can predict chance of Osteoporosis.

```
t_score = (BD - popu_mean)/(std_dev/math.sqrt(sample_size
    -1))
```

Where BD is predicted value of BMD and popu_mean is the mean of population or data set taken for analysis.

```
t_score = -3.851155020921751
Your t-score is below -2.5, you have Osteoporosis
```

5 Benefits of Proposed System

The system in its current form can accurately categorize whether the person is having Osteoporosis or not, depending upon the NTD value obtained from Ultrasonic Transceiver system.

1. Accuracy: As the relation is found to be linearly regressive, it's easy to get required BMD index and hence t-score from it, to predict.
2. Speed: Once trained with given dataset, system can map any NTD value to BMD value and can predict Osteoporosis very easily.
3. Cost Effectiveness: As this system uses Ultrasonic transceiver instead of classical X-Ray based heavy systems, not only the maneuvering cost, but the cost of operation also gets reduced.

6 Conclusion

The aim of the proposed system has been achieved. A well trained system can now predict Osteoporosis accurately in low cost operations. The ease of calculation is possible due to Linear Regression algorithm used for prediction. Trained well enough for accurate dataset, it provides neat and accurate basis for calculation of BMD value. Apriori detection of Osteoporosis can now be done easily to improvise doctor's diagnostics.

References

1. van der Sluis, I.M., de Ridder, M.A., Boot, A.M., Krenning, E.P., de Muinck Keizer-Schrama, S.M.: Reference data for bone density and body composition measured with dual energy x ray absorptiometry in white children and young adults. Arch. Dis. Child. **87**(4), 341–347 (2002). https://doi.org/10.1136/adc.87.4.341
2. Kaufman, J.J., Luo, G., Siffert, R.S.: A portable real-time ultrasonic bone densitometer. Ultrasound Med. Biol. **33**(9), 1445–1452 (2007). https://doi.org/10.1016/j.ultrasmedbio.2007.04.007
3. Krans, B.: Bone Mineral Density Test. https://www.healthline.com/health/bone-mineral-density-test. Accessed 25 May 2018
4. Bochud, N., et al.: Predicting bone strength with ultrasonic guided waves. Sci. Rep. **7**, 43628 (2017). https://doi.org/10.1038/srep43628
5. Jethe, J.V., Ananthakrishnan, T.S., Lakhe, A.S., Patkar, D.P., Parlikar, R.S., Jindal, G.D.: Development of ultrasonic pulser-receiver for bone density assessment. Int. J. Sci. Res. **8**(9) (2019)
6. Chen, Y., Xu, Y., Ma, Z., Sun, Y.: Detection of bone density with ultrasound. Procedia Eng. **7**, 371–376 (2010)
7. Stein, E.M., et al.: Clinical assessment of the 1/3 radius using a new desktop ultrasonic bone densitometer. Ultrasound Med. Biol. **39**(3), 388–395 (2013)
8. Suzuki, T., et al.: Factors affecting bone mineral density among snowy region residents in Japan: analysis using multiple linear regression and Bayesian network model. Interact. J. Med. Res. **7**(1), e10 (2018)

CNN Based Periocular Recognition Using Multispectral Images

Vineetha Mary Ipe$^{(\boxtimes)}$ ⓘ and Tony Thomas ⓘ

Indian Institute of Information Technology and Management, Kerala, India
{vineetha.mary,tony.thomas}@iiitmk.ac.in

Abstract. Over the recent years, the periocular region has emerged as a potential unconstrained biometric trait for person authentication. For a biometric identification scenario to operate reliably round the clock, it should be capable of subject recognition in multiple spectra. However, there is limited research associated with the non-ideal multispectral imaging of the periocular trait. This is critical for real life applications such as surveillance and watch list identification. The existing techniques for multispectral periocular recognition rely on fusion at the feature level. However, these handcrafted features are not primarily data driven and there even exists possibilities for more novel features that could better describe the same. One possible solution to address such issues is to resort to the data driven deep learning strategies. Accordingly, we propose to apply the attributes extracted from pretrained CNN for subject authentication. To the best of our knowledge, this is the first study of multispectral periocular recognition employing deep learning. For our work, the IIITD Multispectral Periocular (IMP) database is used. The best classification accuracy reported for this dataset is 91.8%. This value is not precise enough for biometric identification tasks. The off-the-shelf CNN features employed in our work gives an improved accuracy of 97.14% for the multispectral periocular images.

Keywords: Multispectral periocular recognition · Deep learning · Biometrics · Convolutional neural network (CNN)

1 Introduction

Periocular recognition refers to the automated process of recognizing individuals using periocular images. Periocular is the facial region around the eye including eyelids, eyelashes, and eyebrows [2]. The typical elements of the periocular region are as shown in Fig. 1. Periocular (periphery of ocular) region is unique for each individual because of the distinctive markings of the iris, sclerotic blood vessels and skin texture [3].

The significance of periocular biometrics lies in the fact that it represents a trade-off between using face and iris [13] which can be elucidated as follows:

© Springer Nature Singapore Pte Ltd. 2020
S. M. Thampi et al. (Eds.): SIRS 2019, CCIS 1209, pp. 94–105, 2020.
https://doi.org/10.1007/978-981-15-4828-4_9

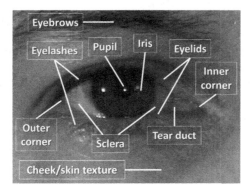

Fig. 1. Elements of the periocular region [2]

- While trying to capture the entire face from a distance, iris information will be of poor quality or low resolution. On the contrary, when the iris is captured from close distance, the whole face may not be visible thereby forcing the recognition system to rely only on the iris. In both these cases, periocular images may be captured with sufficient clarity over a wide range of distances.
- When portions of the face such as the nose and mouth are occluded, the periocular region may be used to determine the identity.
- In cases where the iris cannot be reliably obtained, such as blinking or off-angle poses, the skin in periocular region may be used to recognize the identity.

This implies that the periocular biometric requires less subject cooperation compared to other ocular biometrics which in turn makes the periocular region a potential trait for unconstrained biometrics. However, there is limited research focusing on periocular recognition in multispectral non-ideal imaging scenario. This is critical for real life applications such as tracking, watch list identification, surveillance and security, where it is required to track or authenticate individuals based on the biometric traits in varied illumination conditions. Thus, there is a need for such biometric applications to operate in multiple spectra round the clock. Also, these multispectral images provide rich visual information for authentication and is tougher to be spoofed. This has paved the way for multispectral periocular recognition.

Existing implementations for multispectral periocular recognition rely on fusion of periocular features in different spectra (feature-level fusion) [1]. But, these handcrafted features are not primarily data driven and are based on human assumptions to be the best descriptors for the multispectral periocular trait. There even exists possibilities for more novel or unrevealed features that could better describe the same. Moreover, different handcrafted features exhibit differently with respect to different spectra and this can give rise to compatibility issues while fusing such features for different spectrums. To address such issues, the data driven deep learning strategies can be used.

For our work, we empoly the multispectral images from the IIITD Multispectral Periocular (IMP) database [17], which provides the relevant multispectral ocular images belonging to three spectrums namely near infrared, visible and night vision. The best classification accuracy reported with this database in literature is 91.8% and such works rely on a number of feature descriptors. However, this value is not precise enough to settle with the claim that the features or methodology involved in such tasks are the best approaches for biometric recognition using multispectral periocular images. One possible technique to improve the recognition performance is to resort to the recent trends in deep learning.

The advantage of deep learning strategies lies in the fact that they are primarily data driven i.e. they are able to learn features automatically based on the input data rather than by human hypotheses as in the case of conventional features. Even though deep learning techniques call for huge amount of training data and time for designing and training new CNN architectures for specific applications, deploying off-the-shelf CNN features from pretrained existing architectures help resolve these demands. Accordingly, we propose to apply the extracted attributes from pretrained convolutional neural network (CNN), Alexnet for subject authentication.

Alexnet is a CNN that is trained on 1.2 million high resolution images from the ImageNet database [6]. Hence, the network has learned rich feature depictions for a wide variety of images. This representational power of pretrained deep networks can be effectively utilized by means of feature extraction. In our work, we extract the off-the-shelf CNN features from Alexnet so as to train an SVM based image classifier. We present the efficiency of our method by means of classification accuracy, which is the ratio of the number of exact classifications to the total number of prediction outputs. The off-the-shelf CNN features used in our work provide an improved 97.14% accuracy. Our research work is novel in the prospect that this is the first study for multispectral periocular recognition employing deep learning.

2   Related Works

The pioneering work in the field of periocular biometrics was reported in the literature by Park et al. in 2009 [14]. They made a feasibility study of whether periocular can be used as a biometric trait for recognition and confirmed the same [13]. Bharadwaj et al. also supported the concept of using periocular biometrics when iris recognition fails [4]. Park et al. used conventional methods for attribute extraction, such as scale invariant feature transform (SIFT) [7], local binary patterns (LBP) [12] and histograms of oriented gradients (HOG) [21] in order to assess performance in the presence of various performance degradation factors such as occlusions, scale, pose, and gaze.

Deep learning approaches especially CNNs have gained boundless popularity for computer pattern analysis and computer vision tasks. However, surveys on periocular biometrics [2] suggest that few studies have taken advantage of deep learning methods for improving periocular recognition rates. Zhao et al. [22] presented a semantics assisted convolutional neural network (SCNN) method which uses semantic information apart from learning identities. This technique involves learning from scratch which is computationally expensive and does not have a significant boost in the recognition performance. Proenca et al. [15] proposed a deep learning framework for periocular recognition without iris and sclera (Deep-PRWIS). Though this is an attempt to highlight the importance of periocular region alone, research suggests that use of entire periocular region without masking provides better recognition performance [13], which is the crux of the biometrics. Luz et al. [8] proposed a periocular region recognition (PRR) using VGG (visual geometry group) transfer learning. Kevin et al. [5] performed periocular recognition using off-the-shelf CNN features. Compared against the conventional periocular features such as LBP, HOG, and SIFT, they show an EER (equal error rate) reduction up to 0.4. Considering the pace at which research in deep learning is progressing,we can anticipate much more precise and fast algorithms.

Research in periocular biometrics has paved its way to emerge as a potential trait for unconstrained biometrics. For such a biometric system to operate 24×7, it should be capable of person identification/authentication in multiple spectra. This has led to the development of multispectral biometrics. The multispectral recognition has been performed on the biometric traits such as face, fingerprint, iris, sclera and palmprint [20] and these images have shown to provide integral information owing to the fact that these traits exhibit different physical characteristics in different spectra. Combining the images belonging to different spectrums improves the strength of the biometric systems against spoofing or counterfeiting thus increasing the robustness of the system.

Very few works are reported in literature for multispectral periocular recognition. Tapia et al. [18] presented a gender classification method from multispectral periocular images using features extracted from shape, intensity and texture of periocular images. Algashaam et al. [1] presented a multispectral periocular classification with multimodal compact multi-linear pooling in which complementary data from multispecral images are fused. This work provides the best classification accuracy with the IIITD Multispectral Periocular (IMP) database, the only publically available database for multispectral periocular images. The highest accuracy reported is 91.8% which is not lucrative enough to accept the above approach as the perfect methodology for multispectral periocular recognition. One possible remedy for performance improvement is to incorporate deep learning techniques. Based on these facts, we extract the off-the-shelf CNN features from Alexnet so as to train an SVM based image classifier.

3 Proposed SR Based Periocular Recognition

The framework for periocular recognition is as shown in Fig. 2. This recognition framework is one of its kind whereby off-the-shelf CNN features are extracted from multispectral periocular images of the IMP database.

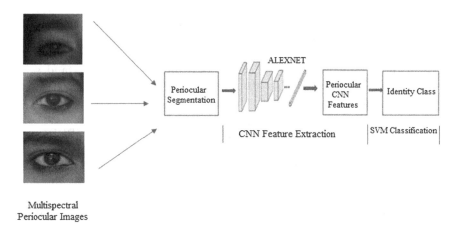

Multispectral
Periocular Images

Fig. 2. Experimental framework

3.1 Periocular Segmentation

The periocular region is segmented from these images by modifying the Viola-Jones algorithm [19]. Such a segmentation routine is precise ensuring the incorporation of vital periocular information such as eye, eyebrow, and surrounding skin texture. After segmentation, the periocular images belonging to distinct individuals are grouped together.

3.2 CNN Feature Extraction

The segmented periocular images are fed into the CNN feature extraction module. The CNN used for feature extraction in our work is Alexnet. It consists of 5 convolutional (conv.) layers and 3 fully connected (f_c) layers. In addition, there are ReLU and max-pooling layers. Multiple convolutional kernels (filters) extract features in an image. The first convolutional layer of Alexnet contains 96 kernels of size $11 \times 11 \times 3$. The architecture of Alexnet is shown in Fig. 3.

Convolutional layer is the integral element of CNN. The convolutional layers comprise a series of kernels or learnable filters which derives the local attributes from the periocular images [9]. Each kernel generates a kernel map or feature map which is produced by convolution operation ($*$). The first convolutional layer extracts low-level features and the succeding convolutional layer extracts higher-level features.

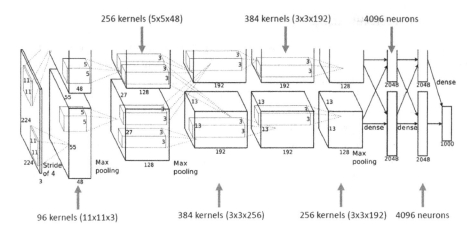

Fig. 3. Alexnet architecture [6]

The non-linear layers introduce non-linearity into the CNN thus enabling the learning of complex models. In Alexnet the non-linear activation function used is ReLU. It applies the following function

$$f(x) = max(0, x) \tag{1}$$

The pooling or subsampling layer reduces the dimensionality of the feature maps. In Alexnet, max pooling is used. The max pooling returns the maximum output inside a rectangular neighbourhood thus making the features robust to minute variations for formerly learned features.

The feature maps of the preceeding layer are convolved with learnable kernels at a convolution layer and passed to the activation function to form the output feature map. Each output combines convolutions with many input maps. This can be represented by Eq. (2).

$$x_j{}^L = f\left(\sum_{i \epsilon M_j} x_i{}^{L-1} * k_{ij}^L + b_j^L\right) \tag{2}$$

where

$$x_j{}^L - j^{th} \ output \ of \ L^{th} \ layer$$
$$k_{ij}^L - kernel \ for \ the \ L^{th} \ layer$$
$$b_j^L - additive \ biases \ for \ the \ L^{th} \ layer$$
$$M_j - set \ of \ input \ maps$$

The CNN feature extraction of periocular images consists of multiple similar steps and each step is made of cascading three aforementioned layers.

3.3 SVM Classification

The features extracted from Alexnet are then fed into the classification module. The periocular recognition task being a multiclass identification problem, we choose to perform classification using simple multi-class SVM [16], which is quite popular in image classification tasks because of its robustness, accuracy, and effectiveness while using small training set. Since SVM is mostly used for the binary classification problem, ECOC (error-correcting output codes) based SVM is used for implementing the multiclass problem, whereby the direct multiclass task is split into several binary classification tasks and the results from these indirect classifiers are then combined to give correct predictions of identity class.

4   Experimental Results

4.1 Database

For our experiments IIITD multispectral periocular (IMP) database [17] is used. It is the only publically available database for multispectral periocular images. It contains 1240 multispectral images of 62 persons in three spectrums namely visible, night vision and near infrared. The visible wavelength images are captured with a digital camera, the near infrared dataset with a iris scanner and the night vision dataset with a Sony camcorder in night vision mode.

4.2 Performance Metric and Baseline Method

To report the performance of our work, we rely on classification accuracy. It is a popular metric for assessing the classification models. It is defined as the ratio of the number of correct predictions to the total number of predictions.

The baseline methodology we used for comparison is the multimodal compact multi-linear pooling by Algashaam et al. [1]. This baseline work achieved periocular recognition accuracy of 91.8%, the best reported in the literature with IMP dataset.

4.3 Experimental Setup

The periocular images obtained from the IMP database are grouped into three categories corresponding to the visible, night vision and near infrared spectra. Each category comprises 62 distinct classes with each class containing the periocular region of a particular individual. Thus these 62 classes correspond to 62 distinct individuals. Additionally, we also group the multispectral images into another 62 classes such that each class contains the images of all the three spectra of an individual. For each of these four categories, we choose 70% of the data corresponding to each class on a random basis for training and the remaining 30% for testing.

In this work, we use Alexnet to extract features from the periocular images. This deep CNN is the ILSVRC 2012 winner with a top-5 error rate of 15.3%, compared to 26.2% achieved by the runner up [6]. This network is trained on more than one million images of the ImageNet database and can classify images into 1000 object categories. This deep network is thus capable of capturing and encoding complex attributes of the periocular images.

For feature extraction, we deploy the pretrained Alexnet model in MATLAB R2018b. With the output from each layer used as an attribute vector to characterise the multispectral periocular images, the performance of each layer is assessed using classification accuracy.

The pretrained CNN model Alexnet is not altered at all using the training data. Only features are extracted from CNN. These activations corresponding to the training images, in turn, are used to train the multiclass SVMs. The multiclass SVM used in our work is equivalent to combining multiple binary SVM classifiers and the number of class depends on the count of subjects in the periocular database.

4.4 Performance Analysis

Different levels of visual content are encoded by each layer of CNN. With the output from each layer used as an attribute vector to characterise the multispectral periocular trait, the performance of each layer is assessed using classification accuracy. The accuracies obtained for different layers of the CNN model are shown in Fig. 4.

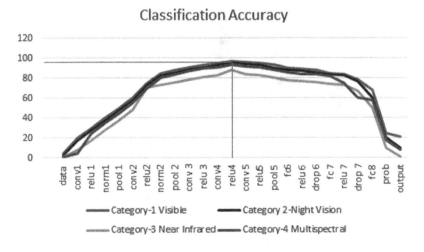

Fig. 4. Performance comparison

For each category, we get a maximum classification accuracy with 'relu4' (fourth rectified linear unit layer) in Alexnet model in Matlab R2018b. The highest classification accuracies obtained are 97.14% (category 1-visible), 95.86% (category 2-night vision), 88% (category 3-near infrared) and 93.55% (category

Classification Accuracy at relu4

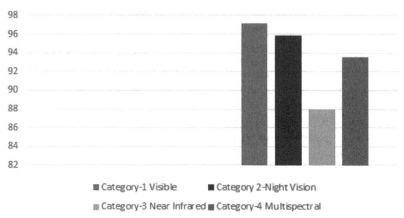

■ Category-1 Visible   ■ Category 2-Night Vision
▨ Category-3 Near Infrared ■ Category-4 Multispectral

Fig. 5. Performance at relu4

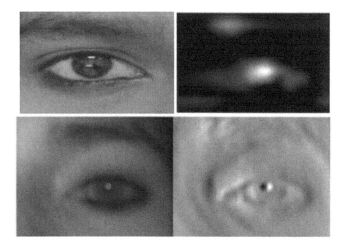

Fig. 6. Activations in Alexnet model

4-multispectral). This is illustrated in Fig. 5. A comparatively low accuracy for near infrared images owes to the fact that these images in the IMP database are blurred and rich visual details are hence not available for recognition.

The difference in accuracy in the distinct layers of Alexnet owes to the distinct levels of features or activations learnt by them. The channels in earlier layers learn simple features like color, edges etc., while those in deeper layers learn much more complex features like that of eyes. The low accuracy in the initial layers is due to the fact that they retain the majority of the input as such or only learn simple features which are not quite efficient for distinguishing individuals. The final layers may not activate at all since there is nothing more to learn beyond a particular layer. Consequently, the recognition accuracy falls towards the last layers.

For visualisation, we normalise the activations to gray-scale such that minimum activation is 0 and the maximum is 1. Such a mapping leads to white pixels representing strong positive activations and gray representing weak activations. Figure 6 shows activations for periocular image in the visible spectrum and in the near infrared spectrum from the layer that gives the maximum accuracy i.e. 'relu4'. From these activations it is evident that the one corresponding to visible spectrum has strong white pixels around the center of the periocular region which is absent in the other case and this strong activation leads to high accuracy.

Comparison of the proposed method with other approaches for multispectral periocular recognition on IMP dataset is shown in Table 1. From this, it is evident that the proposed approach has much better recognition accuracy when compared to the existing methods.

Table 1. Comparison with existing approaches

Methods	Recognition accuracy (%)
Concatenation [20]	86.3
Element-wise multiplication [20]	87.1
Canonical correlation analysis [23]	89.5
Discriminant correlation analysis [24]	90.1
Multimodal compact multi-linear pooling [1]	91.8
Proposed approach	97.14

5 Conclusion

In this work, we have addressed the approach of multispectral periocular recognition from a deep learning perspective. Our experiment shows that there is significant improvement in recognition performance by the application of off-the-shelf CNN features. There are various open questions and challenges regarding the deployment of deep learning to the problem of periocular recognition, even though CNN such as Alexnet is effective in encoding discriminative features for periocular recognition. The computational intricacy of CNNs used for recognition task can be addressed by using model reduction techniques such as pruning and compression [5]. The low accuracy for near infrared images can be addressed using super-resolution [11], a technique for improving the resolution of images. Also, this deep learning approach can also be extended to cross-spectral periocular recognition. Recent developments in Deep Reinforcement Learning and Evolution Theory allow the networks to adapt themselves [10] and achieve better results for the task of periocular recognition. There are several other frameworks in the field of deep learning such as Recurrent Neural Network (RNN), Stacked Auto-Encoder (SAE) and so on. To improve the representation capability of periocular region, these architectures can be used independently or combined with conventional CNNs.

Acknowledgement. This work is done as a part of the project CEPIA (Centre of Excellence in Pattern and Image Analysis) 2019–20, which is funded by the Kerala state planning board.

References

1. Algashaam, F.M., Nguyen, K., Alkanhal, M., Chandran, V., Boles, W., Banks, J.: Multispectral periocular classification with multimodal compact multi-linear pooling. IEEE Access **5**, 14572–14578 (2017)
2. Alonso-Fernandez, F., Bigun, J.: A survey on periocular biometrics research. Pattern Recogn. Lett. **82**, 92–105 (2016)
3. Bakshi, S., Sa, P.K., Majhi, B.: A novel phase-intensive local pattern for periocular recognition under visible spectrum. Biocybern. Biomed. Eng. **35**(1), 30–44 (2015)
4. Bharadwaj, S., Bhatt, H.S., Vatsa, M., Singh, R.: Periocular biometrics: when iris recognition fails. In: 2010 Fourth IEEE International Conference on Biometrics: Theory, Applications and Systems (BTAS) (2010)
5. Hernandez-Diaz, K., Alonso-Fernandez, F., Bigun, J.: Periocular recognition using CNN features off-the-shelf. In: 2018 International Conference of the Biometrics Special Interest Group (BIOSIG) (2018)
6. Krizhevsky, A., Sutskever, I., Hinton, G.E.: ImageNet classification with deep convolutional neural networks. Commun. ACM **60**(6), 84–90 (2017)
7. Lowe, D.G.: Distinctive image features from scale-invariant keypoints. Int. J. Comput. Vis. **60**(2), 91–110 (2004). https://doi.org/10.1023/B:VISI.0000029664.99615.94
8. Luz, E., Moreira, G., Junior, L.A.Z., Menotti, D.: Deep periocular representation aiming video surveillance. Pattern Recogn. Lett. **114**, 2–12 (2018)
9. Namatēvs, I.: Deep convolutional neural networks: structure, feature extraction and training. Inf. Technol. Manag. Sci. **20**(1), 40–47 (2017)
10. Nguyen, K., Fookes, C., Ross, A., Sridharan, S.: Iris recognition with off-the-shelf CNN features: a deep learning perspective. IEEE Access **6**, 18848–18855 (2018)
11. Nguyen, K., Fookes, C., Sridharan, S., Tistarelli, M., Nixon, M.: Super-resolution for biometrics: a comprehensive survey. Pattern Recognit. **78**, 23–42 (2018)
12. Ojala, T., Pietikainen, M., Maenpaa, T.: Multiresolution gray-scale and rotation invariant texture classification with local binary patterns. IEEE Trans. Pattern Anal. Mach. Intell. **24**(7), 971–987 (2002)
13. Park, U., Jillela, R., Ross, A., Jain, A.: Periocular biometrics in the visible spectrum. IEEE Trans. Inf. Forensics Secur. **6**(1), 96–106 (2011)
14. Park, U., Ross, A., Jain, A.K.: Periocular biometrics in the visible spectrum: a feasibility study. In: 2009 IEEE 3rd International Conference on Biometrics: Theory, Applications, and Systems (2009)
15. Proenca, H., Neves, J.C.: Deep-PRWIS: periocular recognition without the iris and sclera using deep learning frameworks. IEEE Trans. Inf. Forensics Secur. **13**(4), 888–896 (2018)

16. Scholkopf, B., Smola, A.J.: Learning with Kernels: Support Vector Machines, Regularization, Optimization, and Beyond. MIT Press, Cambridge (2001)
17. Sharma, A., Verma, S., Vatsa, M., Singh, R.: On cross spectral periocular recognition. In: 2014 IEEE International Conference on Image Processing (ICIP), pp. 5007–5011. IEEE (2014)
18. Tapia, J., Viedma, I.: Gender classification from multispectral periocular images. In: 2017 IEEE International Joint Conference on Biometrics (IJCB) (2017)
19. Viola, P., Jones, M.: Rapid object detection using a boosted cascade of simple features. In: Proceedings of the 2001 IEEE Computer Society Conference on Computer Vision and Pattern Recognition, CVPR (2001)
20. Zhang, D., Guo, Z., Gong, Y.: Multispectral biometrics systems. In: Zhang, D., Guo, Z., Gong, Y. (eds.) Multispectral Biometrics, pp. 23–35. Springer, Cham (2015). https://doi.org/10.1007/978-3-319-22485-5_2
21. Zhang, S., Wang, X.: Human detection and object tracking based on histograms of oriented gradients. In: 2013 Ninth International Conference on Natural Computation (ICNC) (2013)
22. Zhao, Z., Kumar, A.: Accurate periocular recognition under less constrained environment using semantics-assisted convolutional neural network. IEEE Trans. Inf. Forensics Secur. **12**(5), 1017–1030 (2017)
23. Haghighat, M., Abdel-Mottaleb, M., Alhalabi, W.: Fully automatic face normalization and single sample face recognition in unconstrained environments. Expert Syst. Appl. **47**, 23–34 (2016)
24. Haghighat, M., Abdel-Mottaleb, M., Alhalabi, W.: Discriminant correlation analysis for feature level fusion with application to multimodal biometrics. In: 2016 IEEE International Conference on Acoustics, Speech and Signal Processing (ICASSP) (2016)

Obstructive Pulmonary Disease Prediction Through Heart Structure Analysis

Umaima Rahman[1], Parthasarathi Bhattacharyya[2], and Sudipto Saha[3($\boxtimes$)]

[1] Department of Computer Science and Engineering, University of Calcutta, Kolkata, India
umaimarahman21@gmail.com
[2] Institute of Pulmocare and Research, Kolkata, India
[3] Division of Bioinformatics, Bose Institute, Kolkata, India
ssaha4@gmail.com

Abstract. There are more than 200 million cases of lung diseases worldwide. Most commonly they include Obstructive Pulmonary Disease (OPD) like Chronic OPD (COPD), asthma and bronchiectasis. This paper aims to study the causal relationship between the shape of the heart and presence of obstructive pulmonary disease by analyzing HRCT scans which are sensitive and informative as it provides with multiple slices of a patient's internal structure. A mathematical model to predict disease gives confidence to the radiologists for correct and early diagnosis of the disease. Real life HRCT scans along with the disease information were obtained from the Institute of Pulmocare and Research (IPCR). Using Image Processing techniques we finally obtained the right and left atrium of the heart as individual gray-scale images from the HRCT scans; which were then converted into a gray-scale matrix and finally into a vector. We generated our data-set consisting of 40 patients. For patient diagnosed with Obstructive Pulmonary Disease we assigned the label as $+1$ and for those who have other disease as -1. Different machine learning algorithms such as kNN, SVM, Random Forest and Naive Bayes were applied to the dataset to find the algorithm with highest accuracy and maximum area under the ROC plot.

Keywords: Biomedical imaging · Image Processing · Machine Learning

1 Introduction

Large number of diseases that affect the worldwide population are lung-related. Therefore, research in the field of Pulmonology has great importance in public health studies and focuses mainly on asthma, bronchiectasis and Chronic Obstructive Pulmonary Disease (COPD). The World Health Organization (WHO) estimates that there are 300 million people who suffer from asthma,

© Springer Nature Singapore Pte Ltd. 2020
S. M. Thampi et al. (Eds.): SIRS 2019, CCIS 1209, pp. 106–117, 2020.
https://doi.org/10.1007/978-981-15-4828-4_10

and that this disease causes around 250 thousand deaths per year worldwide [1]. In addition, WHO estimates that 210 million people have COPD. This disease caused the death of over 300 thousand people in 2005 [2]. Recent studies reveal that COPD is present in the 20 to 45 year-old age bracket, although it is characterized as an over-50-year-old disease. Accordingly, WHO estimates by 2030 COPD will be the third cause of mortality worldwide. For the public health system, the early and correct diagnosis of any pulmonary disease is mandatory for timely treatment and prevents further death. From a clinical standpoint, diagnosis aid tools and systems are of great importance for the specialist and hence for the people's health [2].

Commonly used diagnosis methods for lung diseases are radiological. A chest X-Ray helps in visualization of the lungs. However, chest X-Rays present blurred images of the lungs and precise observations are not possible with them. In such cases computed tomography (CT) scans are used, which is a more sophisticated and powerful X-Ray that gives a 360° image of the internal organs, spine and the vertebrae. A more sensitive version of CT scan, HRCT (High Resolution CT) scan is used to study the morphological changes associated with certain disease.

Larrey-Ruiz et al. in [3] present an efficient image-driven method for the automatic segmentation of the heart from CT scans. The methodology relies on image processing techniques such as multi-thresholding based on statistical local and global features, mathematical morphology, or image filtering, and it also exploits the available prior knowledge about the cardiac structures involved. The development of such a segmentation system comprises of two major tasks: initially, a pre-processing stage in which the region of interest (ROI) is delimited and the statistical parameters are computed; and next, the segmentation procedure itself, which makes use of the data obtained during the previous stage [3].

HRCT[1] scanning shows cross sections (slices) through the heart and lungs. For one patient there are approximately 60 slices of HRCT scan. An expert in this field cannot give more than 30 s to each slice. Moreover there are many nuances present in a HRCT scan slice which are hidden from the naked eye. In that case even a close inspection from an expert may not be enough. With a dearth of experts in this field it is absolutely imperative to find a method that can capture as much information from the HRCT scan images and draw quick, efficient and reliable conclusions automatically.

The challenge is to use the HRCT scans and the expertise of a doctor to develop a model that can predict if a patient suffers from obstructive pulmonary disease. However, we are not using the information present in the lungs rather our focus will be to extract the information of the shape of the heart components. In order to draw a causal relationship between the shape of the heart and Obstructive Pulmonary Disease. One of the reasons for this approach is that it takes much less time to examine a heart present in an HRCT scan compared to the lungs.

---

[1] High-resolution computerized tomography.

2   Methodology

Given an HRCT slice of a patient, we propose to extract the portions of the heart which contains the left and right atrium. To the vector obtained from these images we apply different machine learning algorithms to understand which model of which algorithm performs better in terms of accuracy for classification of obstructive pulmonary disease against the background disease.

Our approach is to focus on the heart in order to classify the input to different categories of lung diseases. We do this for multiple reasons:

– It takes much less time to closely examine the heart than the lungs.
– A closer look tells us that the heart and lungs are very closely associated for blood circulation. As a result we try to exploit the proximity between the heart and lungs to study the impact of heart on lung conditions.
– Aberrations in the volume of the heart chambers can possibly point towards certain abnormalities in the lungs and vice versa.
– Study the pressure in the two chambers of the heart the Right Atrium (RA) and the Left Atrium (LA) by measuring their volume (which corresponds to area in 2D).
– Increased pressure in the RA suggests a rise in the Pulmonary artery pressure which goes to the lungs, similarly increased pressure in the pulmonary vein suggests a rise in the pressure in the LA of the heart.

3   Data Collection and Pre-processing

HRCT scans of patients is collected from IPCR[2], which are available in DICOM[3] format. A DICOM file consists of a header and image data sets, all packed into a single file [4]. As a result the DICOM files have to be converted to other Image formats such as JPEG, TIFF or PNG for easier visualization and faster analysis through image processing techniques.

Exploiting the mediastinal window[4] using one of the many tags present in the DICOM header, we only had to process 3000 image slices instead of 13000, for each of the 50 patients, which reduced the conversion time considerably.

3.1   Contrast Enhancement

Conversion from DICOM to PNG file format, results in loss of contrast. To correct this, histogram equalization was performed on the obtained PNG image.

---

[2] Institute of Pulmocare & Research, Kolkata, India.
[3] Digital Imaging and Communications in Medicine.
[4] Part of the chest that lies between the sternum and the spinal column.

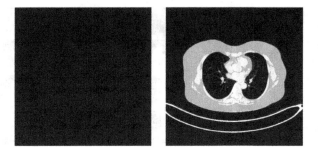

Fig. 1. Images (a) before and (b) after histogram equalization

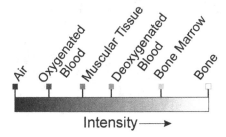

Fig. 2. Grayscale intensity level values of different parts of an HRCT scan

3.2   Multi-thresholding on CT Scan Image

Different parts of a CT scan image have different intensity levels. Exploiting this property, we can extract different portions of the heart by applying different intensity thresholds, which we compute, as per our requirements. This process, we call multi-thresholding[5] (Figs. 1 and 2).

3.3   Automatic Selection of the Region of Interest (ROI)

To select only the heart from the image we need to define a Region of Interest (ROI). However, for every patient the size and shape of the heart and lung varies significantly. Moreover the device used to obtain the HRCT scan varies. As a result we cannot fix an ROI, instead the ROI should be selected with respect to each slice of the HRCT scan. To do so, we find the automatic Region of Interest using the algorithm present in [3] with minor parametric tweaks. The algorithm produces a corresponding ROI which only selects the Region of the heart from the entire HRCT scan.

---

[5] It will help us obtain the different components belonging to the heart only.

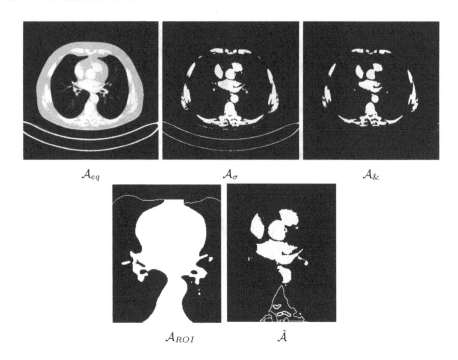

$\mathcal{A}_{eq}$ $\qquad\qquad$ $\mathcal{A}_{\sigma}$ $\qquad\qquad$ $\mathcal{A}_{\&}$

$\mathcal{A}_{ROI}$ $\qquad\qquad$ $\hat{\mathcal{A}}$

Fig. 3. Images corresponding to different steps of heart component extraction

---

Algorithm 1. Heart Component Extraction

1: $\mathcal{A}_{eq} \leftarrow \texttt{HistogramEqualize}(\mathcal{A})$
2: $\mathcal{A}_{\sigma} \leftarrow \texttt{SigmaEqualize}(\mathcal{A}, \mu_{sup}(k) + \sigma(k))$
3: $\mathcal{A}_{\&} \leftarrow \mathcal{A}_{eq}\&\mathcal{A}_{\sigma}$
4: $\mathcal{A}_{ROI} \leftarrow \texttt{ROI}(\mathcal{A}_{eq})$
5: $\hat{\mathcal{A}} \leftarrow \mathcal{A}_{ROI}\&\mathcal{A}_{\&}$

3.4 Algorithm to Extract the Components of Heart from the Background

Step 3 produces an image in which the different sections of the heart are very clearly visible. Step 4 removes the portions of vertebrae from the image.

3.5 Obtaining the Right Atrium and Left Atrium

We can study the ratio (RA:DA) of the area of right atrium (RA) to the descending aorta (DA), and then computing the matrix of the intensity values of finally generating the vectors of the contours of the RA and DA. This approach, however, fails, when the contours of the RA and DA are not clearly distinguishable from other components (Fig. 3).

In this case, we opt for another approach, where we first divide the image obtained after the implementation of the algorithm 1 mentioned above, into four equal parts. This division is supported by the rationale that:

– the RA will always be present in the top left corner of the image,
– the DA is in the bottom right corner of the image and
– the Left Atrium (LA) is in the top right corner of the image.

4   Dataset Generation

Each of the four images obtained after dividing the final image is of dimension 100×125. We select the image that contains the right atrium. This 100×125

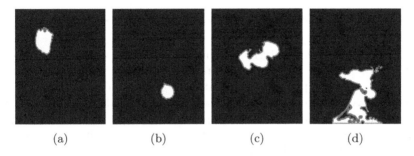

(a)   (b)   (c)   (d)

Fig. 4. Area of the right atrium (RA) and descending aorta (DA) contours in fig. (a) and (b), Indistinguishable components in fig. (c) and (d)

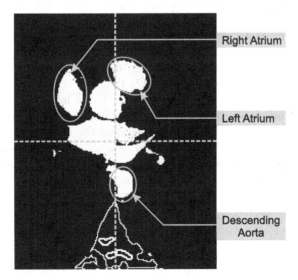

Fig. 5. Four equal sections of the image containing the heart for clearly distinguishable heart components

intensity matrix is converted to a vector of length 12500, which forms 12500 columns in the datasets. There are 40 rows, each corresponding to a patient (Figs. 4 and 5).

Another column, named `label` is added to the dataset, which is used for supervised learning to classify between Obstructive Pulmonary Disease (OPD) and Background (non-OPD) Disease. The OPD is assigned a label of $+1$, whereas, the background is assigned a label of -1. OPD consists of COPD and asthma while background diseases include TB, chronic cough and normal cough. This makes our dataset of size: 40 rows $\times$ 12501 columns.

On close inspection, it was found that the first 2000 columns contain a zero level intensity for each patient, so these columns were omitted, so that they do not act as noise for the machine learning algorithms. The resultant dataset has a size of 40 rows $\times$ 10501 columns.

5 Machine Learning

Machine Learning, a branch of Artificial Intelligence, relates the problem of learning from data samples to the general concept of inference [4–6]. Every learning process consists of two phases: (i) estimation of unknown dependencies in a system from a given dataset and (ii) use of estimated dependencies to predict new outputs of the system. ML has also been proven an interesting area in biomedical research with many applications, where an acceptable generalization is obtained by searching through an n-dimensional space for a given set of biological samples, using different techniques and algorithms [7].

Spathis et al. [8] attempted to choose a representative portion used in the literature comprising of multiple categories of classifiers including linear, non-linear, kernel-based, trees and probabilistic algorithms for the diagnosis of asthma and chronic obstructive pulmonary disease. For similar types of problems Decision Trees were used by Metting et al. [9] Mohktar et al. [10], Prosperi et al. [11] and Prasad et al. [12] and Kernel-based methods such as SVMs were used by Dexheimer et al. [13]. In addition to this Random Forest was also examined by Leidy et al. [14]as well as Prosperi et al. [11]. We have used the method of k-fold cross validation [15], i.e., we created a bunch of train/test splits, calculated the testing accuracy for each, and averaged the results together. We have deployed machine learning algorithms such as kNN, SVM, Random Forest and Naive Bayes on our dataset for different values of $k \in \{5, 7, 10\}$, i.e., 5-cross validation, 7-cross validation, 10-cross validation as well as Jack-Knife Validation. The Python library 'Sklearn' [16] was used for performing the ML classifications.

6 Results

In this section, we show a comparison among the performance of different machine learning algorithms for the different sets of data. We present the

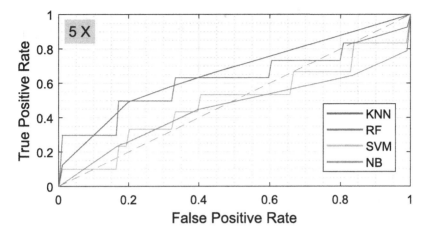

Fig. 6. ROC plot, 5 cross validation

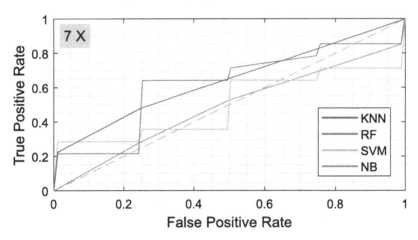

Fig. 7. ROC plot, 7 cross validation

threshold-dependent plots and threshold-independent ROC[6] plots, which serve as standard performance metric of the algorithms.

6.1 Testing Set

K-fold cross validation and Jack-Knife validation is done on the Testing set.

---

[6] Receiver Operating Characteristic, is a graphical plot that illustrates the diagnostic capability of a binary classifier system. The ROC curve is created by plotting the True Positive Rate against the False Positive Rate.

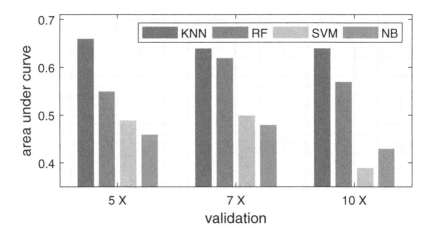

Fig. 8. Area under the curve of ROC plots

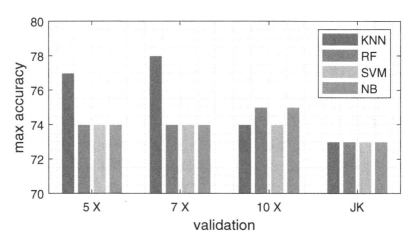

Fig. 9. Classification accuracy

6.2 Blind Set

From a total of 40 observations in our dataset, we kept aside a total 6 observations as the blind set. 2 observations with label +1 and 4 observations with label −1. The blind set is disjoint form the training/test set in order to avoid biased prediction (Figs. 6 and 7).

6.3 Best Model on the Blind Set

The confusion matrix generated for the kNN algorithm using 5-,7-,10-cross gives a 100% accuracy whereas it gives an accuracy of 83% using the Jack Knife validation techniques.

7   Discussion

Based upon the threshold-dependent plots and the Receiver operating Characteristic plots shown in the Results section we can say that the k-Nearest Neighbour
Classifier (k = 1) performs best on our data-set giving a 77% accuracy and an
area under the curve (AUC) of 0.66 for the ROC plots, using 5-cross validation technique on the training/testing data; and giving a 78% accuracy and an
area under the curve (AUC) of 0.64 for the ROC plots, using 7-cross validation
technique on the training/testing data. For, kNN, the blind set used for the validation of the model gives an accuracy of 100% with a 100% sensitivity using
5-cross, 7-cross and 10-cross validation techniques. Our data-set contains over
10,000 columns, making our data-set complex. The reason why K nearest neighbors performs well in our case can be because this algorithm classifies new cases
based on similarity measures, based upon the cases it stored earlier (Figs. 8, 9
and 10).

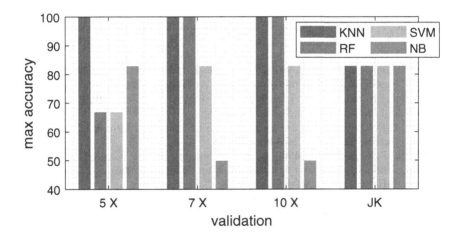

Fig. 10. Classification accuracy on Blind Set

Random Forest Classifier has also performed well giving an accuracy of 74%
and an area under the curve (AUC) = 0.62 for the ROC plots, using 7-cross
validation technique on the training/testing data.

Since the data-set contains a vector of image pixels, where the vector length
is greater than 10,000; the data-set is a very complex one and Naive Bayes being
a simple and crude algorithm could not process the complexity of the data and
performed poorly.

8   Conclusion

Based upon our results we can conclude that k-Nearest Neighbor (k = 1) performs
better for complex data. Random Forest Classifier has also performed well giving

an accuracy of 74%. Naive Bayes performed the worst comparatively and Support Vector Machines was a significant improvement over Naive Bayes.

9   Future Scope

In future, we plan to expand this problem to a multi-class classification problem, which will enable us to predict across a range of pulmonary diseases such as COPD, asthma, tuberculosis, ILD, DPLD, chronic cough, etc. We can also use Deep Neural Networks (DNNs) for image-based classification, owing to the complex nature of the available dataset. However, application of DNN demands a significantly larger sample size which will be possible with the availability of more patient data in the near future.

References

1. Campos, H.d.S., Lemos, A.C.M., et al.: A asma e a dpoc na visão do pneumologista (2009)
2. Mannino, D.M., Thorn, D., Swensen, A., Holguin, F.: Prevalence and outcomes of diabetes, hypertension and cardiovascular disease in COPD. Eur. Respir. J. **32**(4), 962–969 (2008)
3. Larrey-Ruiz, J., Morales-Sánchez, J., Bastida-Jumilla, M.C., Menchón-Lara, R.M., Verdú-Monedero, R., Sancho-Gómez, J.L.: Automatic image-based segmentation of the heart from CT scans. EURASIP J. Image Video Process. **2014**(1), 52 (2014). https://doi.org/10.1186/1687-5281-2014-52
4. Bishop, C.M.: Pattern Recognition and Machine Learning. Springer, Boston (2006). https://doi.org/10.1007/978-1-4615-7566-5
5. Mitchell, T.M.: The discipline of machine learning, vol. 9. Carnegie Mellon University, School of Computer Science, Machine Learning (2006)
6. Witten, I.H., Frank, E., Hall, M.A., Pal, C.J.: Data Mining: Practical Machine Learning Tools and Techniques. Morgan Kaufmann, Burlington (2016)
7. Niknejad, A., Petrovic, D.: Introduction to computational intelligence techniques and areas of their applications in medicine. In: Medical Applications of Artificial Intelligence, vol. 51 (2013)
8. Spathis, D., Vlamos, P.: Diagnosing asthma and chronic obstructive pulmonary disease with machine learning. Health Inf. J. (2017). https://journals.sagepub.com/doi/10.1177/1460458217723169#_i7, 1460458217723169
9. Metting, E.I., et al.: Development of a diagnostic decision tree for obstructive pulmonary diseases based on real-life data. ERJ Open Res. **2**(1), 00077–2015 (2016)
10. Mohktar, M.S., et al.: Predicting the risk of exacerbation in patients with chronic obstructive pulmonary disease using home telehealth measurement data. Artif. Intell. Med. **63**(1), 51–59 (2015)
11. Prosperi, M.C., Marinho, S., Simpson, A., Custovic, A., Buchan, I.E.: Predicting phenotypes of asthma and eczema with machine learning. BMC Med. Genomics **7**(1), S7 (2014). https://doi.org/10.1186/1755-8794-7-S1-S7
12. Prasad, B.D.C.N., Prasad, P.E.S.N.K., Sagar, Y.: A comparative study of machine learning algorithms as expert systems in medical diagnosis (asthma). In: Meghanathan, N., Kaushik, B.K., Nagamalai, D. (eds.) CCSIT 2011. CCIS, vol. 131, pp. 570–576. Springer, Heidelberg (2011). https://doi.org/10.1007/978-3-642-17857-3_56

13. Dexheimer, J.W., Brown, L.E., Leegon, J., Dominik, A., et al.: Comparing decision support methodologies for identifying asthma exacerbations. In: Medinfo 2007: Proceedings of the 12th World Congress on Health (Medical) Informatics; Building Sustainable Health Systems, p. 880, IOS Press (2007)
14. Leidy, N.K., et al.: Insight into best variables for COPD case identification: a random forests analysis. Chronic Obstr. Pulm. Dis. J. COPD Found. **3**(1), 406 (2016)
15. Kohavi , R., et al.: A study of cross-validation and bootstrap for accuracy estimation and model selection. In: Ijcai, vol. 14, pp. 1137–1145, Montreal, Canada (1995)
16. Pedregosa, F., et al.: Scikit-learn: Machine learning in python. J. Mach. Learn. Res. **12**, 2825–2830 (2011)

Speaker Specific Formant Dynamics of Vowels

Sharada Vikram Chougule[(⊠)]

Finolex Academy of Management and Technology, Ratnagiri, Maharashtra, India
shardavchougule@gmail.com

Abstract. Automatic Speech and Speaker Recognition technology has growing demands in variety of voice operated devices. Although the input for all such systems is speech signal, the features useful for each application/task are different. Of the different speech sounds, vowel sounds spectrally well-defined and well represented by formants. Formants which represent resonances of vocal tract are the result of physiology of individual's speech production mechanism as well as nature of speech (words) being spoken. In this way formants are features of speech as well as of speaker. In this paper significance of formants for speech and speaker recognition is explored through experimental analysis. Formant tracking and estimation is done using adaptive formant filter bank and single pole formant based filter. Twelve vowel sounds represented in ARPABET (Advanced Research Project Agency bet) form are used to estimate the first four formants. The analysis based on extracting and emphasizing speaker specific clues indicates that higher formants carry more speaker specific information than first (lower) formant.

Keywords: Automatic Speech and Speaker Recognition · Vowels · Formants

1 Introduction

Automatic Speech and Speaker Recognition technology has growing demands in variety applications such as automated customer services, healthcare applications, mobile banking, voice operated devices, education etc. Although the input for all such systems is speech signal, the features useful for each application/task are different. For speech recognition the purpose is to emphasize '*what*' is being spoken ignoring who is speaking, whereas in speaker recognition the emphasis is on '*who*' is speaking irrespective of contents of speech. Thus speech signal carries characteristics of both 'speech' as well as 'speaker'.

Human speech is result of physiology of vocal apparatus and articulatory features. During speaking, the short spurts of air are taken in lungs and released in a controlled manner through trachea (windpipe) towards vocal cords. The lungs and associated muscles act as source of air for exciting the vocal mechanism [1]. Voiced speech sounds are produced because of periodic vibration of tensed vocal cords. Unvoiced sounds are the result of relaxed state of vocal cords and constriction of vocal tract. Thus, the manner in which different speech sounds are produced is dependent on vocal structure and position of various articulatory elements (e.g. tongue, teeth, lips etc.).

Human vocal tract is comprised of larynx to the lips (or nose), which takes different length, shapes and cross-sections while speaking. One purpose of vocal tract is 'shape'

© Springer Nature Singapore Pte Ltd. 2020
S. M. Thampi et al. (Eds.): SIRS 2019, CCIS 1209, pp. 118–129, 2020.
https://doi.org/10.1007/978-981-15-4828-4_11

or 'modulate' the source making perceptually distinct speech sounds [2]. Vowel are generally long induration and can be distinguished better through spectrum (frequency domain analysis). Vowels are characterized by which represent distinct peaks of acoustic energy. Thus, formants or formant frequencies are resonances of vocal tract represented as peaks in the spectrum of vocal tract response, which are well distinguished during vowel sounds. In spite of specific properties of different vowels, there is much variability of vowel properties among the speakers because of articulatory differences [3].

The basic purpose in this work is to explore the characteristics of vowel sounds representing both speech as well as speaker in a distinct manner. Studies in literature [6–8] used short time power spectrum, Linear Prediction (LP) spectrum, mel-spectrum (holomorphic analysis) to extract typically first four to five formant frequencies. Similar spectral and cepstral features like Linear Prediction Cepstral Features (LPCC) and Mel Frequency Cepstral Coefficients (LPCC) are widely used as features for speaker recognition which mainly relate to vocal tract parameters and human perception of speech sounds [9–11].

In this paper, speaker specific formant dynamics of vowel sounds is studied for Twelve vowel sounds represented in ARPABET (Advanced Research Project Agency bet) form generated by [13]. Formant tracking and estimation is done using adaptive formant filter bank and single pole formant based filter. Section 2 discusses the methodology used for formant tracking and estimation. Results and discussion are given in Sect. 3 and conclusion in discussed in Sect. 4.

2 Methodology for Formant Tracking and Estimation

In this work, Linear Prediction Coefficients (LPC) are most commonly used method to represent the vocal tract parameters (formants) through its spectrum. Vowel parameters are useful for analysis and discrimination of speech patterns for diseases such as neurodegenerative disorder [12]. The disadvantage of using simple LPC to model the speech sounds is that the spectral and temporal characteristics of speech signal varies because of physiological, behavioral characteristics as well as by transducer/channel effects. Such variabilities present a challenge to speech as well as speaker recognition. To minimize the variabilities in LP spectrum and to better track and estimate the formant frequencies, methodology based on work in [14] and [15] is used. The steps in formant estimation algorithm are as follows:

i.   Pre-emphasis
ii. Formant tracking filters
iii. Voicing detector and spectral estimation
iv. Decision maker

2.1 Pre-Emphasis

Speech is a quasi-stationary signal, in which high frequency components are having lower energy than low frequency components. Also high frequency speech components get easily distorted with noise because of low SNR. Pre-emphasis improves the SNR and helps to boost the energy of high frequency speech signals. Pre-emphasis filter is first order Butterworth IIR High pass filter. The real valued pre-emphasized speech signal is converted into analytic signal using Hilbert transformer in order to design complex valued filters.

2.2 Formant Tracking Filters

As speech is quasi-stationary signal and having transients in harmonic frequencies especially during voiced speech, fixed filter bank will not be reasonable. Therefore an adaptive filter bank proposed in [13] and [14] used to track narrowband speech signal. Also in practice, there is leakage or aliasing of neighboring formants, which affects the estimate of desired formant location and creates variability in true estimate. Prior to estimating the formant frequencies, the speech signal passed through a set of adaptive band pass filters, also known as formant filters. The band pass filters are basically complex all-zero filters designed such that complex zeros are placed at the formant location other than single formant which is to be estimated. For example to estimate 4th formant (F4), the band pass filters have complex-zeros at the location of first (F1), second (F2) and third formants (F3) respectively. These filters are designed to have normalized gain and zero phase lag at the filter center frequency [13].

The zeros of all zero filters are given by:

$$z_k = r_z\, e^{\frac{j2\pi F_k}{F_s}} \tag{1}$$

Here r_z indicates distance of zero from origin and angle $\theta = \frac{j2\pi F_k}{F_s}$ decides location of zero in z-plane.

Thus, the transfer function of kth all-zero formant filter for $k = 1,2,3,4$ is given by [13]:

$$H(z, n) = k_k(n, z) \prod_{l=1, l\neq k}^{4} 1 - r_z e^{-j2\pi Fl(n)} z^{-1} \tag{2}$$

Here $r_z = 0.98$. The location of zeros at formant frequencies ensures minimum response of formant filters except for the kth formant. Unity gain and zero phase lag response at estimated formant frequency at k^{th} component is obtained by the term $k_k(n, z)$.

$$k_k(n) = \frac{1}{\prod\limits_{\substack{l=1 \\ l \neq k}}^{4} 1 - r_z * e^{-j2\pi f_l(n) - f_k(n)}} \tag{3}$$

A single-pole (IIR) filter is cascaded after each all zero FIR filter to estimate peak in the respective filter output. The transfer function of the first order peak tracking IIR filter is given by:

$$H_{DTk}(z) = \frac{1 - b_k}{1 - b_k\, e^{-j2\pi Fl(n)} z^{-1}} \tag{4}$$

where $b_k = 0.9$ is pole radius, $(1 - b_k)$ is DC gain and F_l is l^{th} formant frequency.

The initial set of estimated formant frequencies is updated over time using previous formant frequency estimate and complete set of filters is named as formant tracking filters. The analytic speech signal is filtered by these four formants tracking filters.

2.3 Voicing Detector and Spectral Estimation

Voicing detector based on zero crossing rate (ZCR) [16], [17] is used to detect the voiced frame of speech. Analytic speech signal is sampled at 8 kHz sampling frequency and framed using 20 ms Hamming window (160 samples). The LPC provides good model for voiced regions of speech signal in which all-pole model of LPC provides a good approximation to the vocal tract spectral envelope. Speech sample at a time s(n) can be approximated as linear combination of the past p speech samples such that [1]:

$$s(n) = \sum_{i=1}^{p} a_i s(n - i) + G * u(n) \tag{5}$$

where a_i are weights of samples, G is gain of excitation and u(n) is normalized excitation, leading to p^{th} order all-pole system as:

$$H(z) = \frac{S(z)}{GU(z)} = \frac{1}{1 - \sum_{i=1}^{p} a_i z^{-i}} = \frac{1}{A(z)} \tag{6}$$

Here a single linear predictive coefficient (LPC) of the windowed frame is calculated for each band using the autocorrelation method [2, 3] for each band. The LPCs are only calculated from the bands if the entire previous 20-ms window of the speech signal is voiced. Moving average value of LPC estimated current formant frequency over the frame is calculated based on decision of voicing detector. Temporal changes in the envelope structure are often derived from estimated formant tracks [18]. Formant based vowel perceptual space classification is possible using LPC spectrum [19]. The correlation of LP analysis with human speech production mechanism is also being used to identify an individual (speaker identification) irrespective of content of speech [20].

3 Results and Discussion

Speech samples from 10 speakers (five women, five men) are used for experimental analysis. Formant estimation and analysis is carried out on twelve vowel sounds represented in ARPABET form using the method discussed Sect. 2. Table 1 gives the ARPABET

Table 1. Vowel codes used for formant analysis

ae- "had"	ah- "hod"	aw- "hawed"	eh- "head"	ei- "haid"	er- "heard"
ih- "hid"	iy- "heed"	oa- "hoat"	oo- "hood"	uh- "hud"	uw- "who'd"

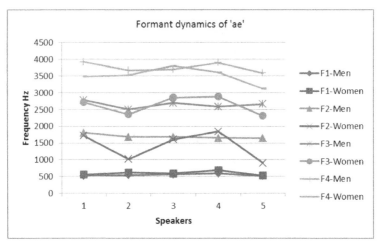

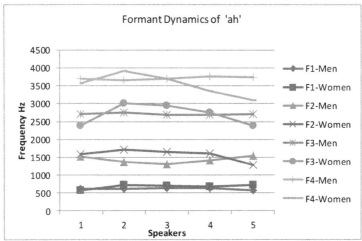

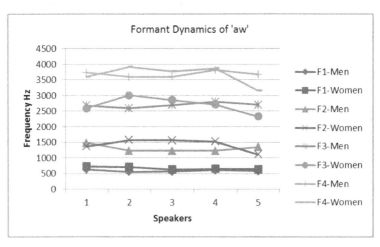

Fig. 1. Plots of Formant dynamics of vowels among Men and Women Speech samples

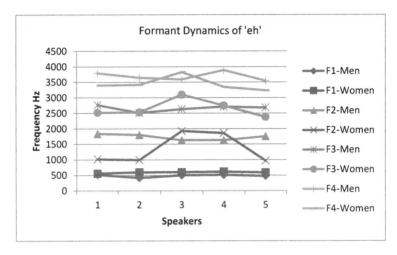

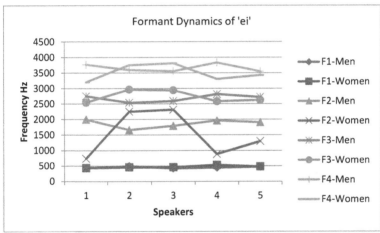

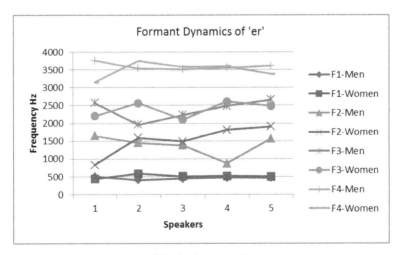

Fig. 1. (*continued*)

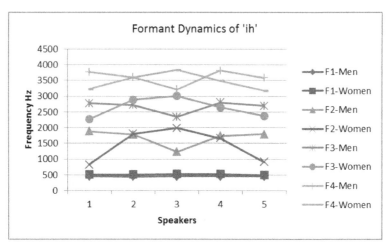

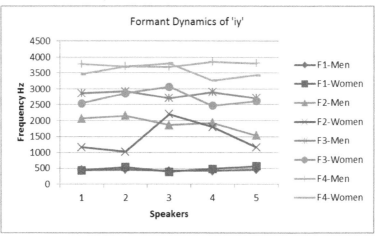

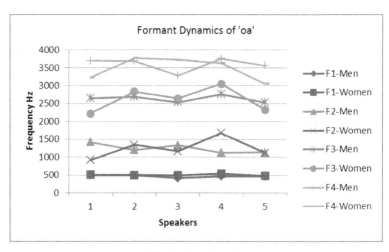

Fig. 1. (*continued*)

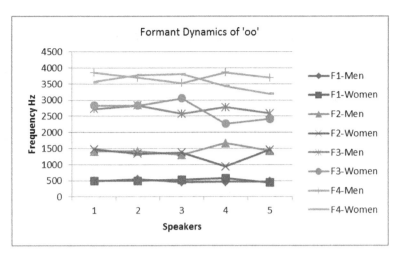

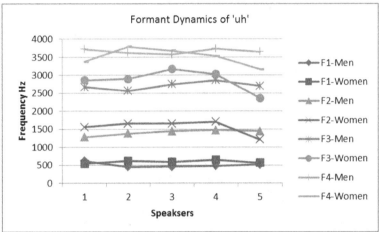

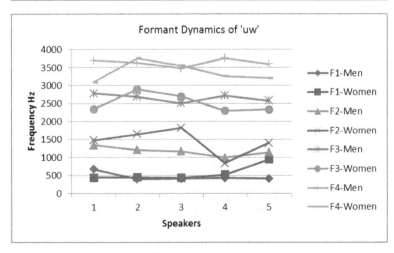

Fig. 1. *(continued)*

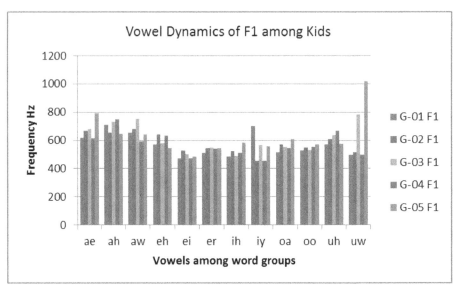

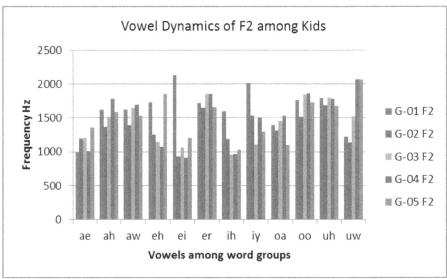

Fig. 2. Plots of Formant dynamics of vowels among Kids (Girls) Speech Samples

symbols of 12 vowels and vowel code used for each group [13]. The plats shows the estimated formants (F1 to F4) for ten of speakers for each vowel code. For comparison of characteristics of male and female speakers, formant plots of speech of both men and women are shown in same plot.

Figure 1 shows the plots of first four formants estimated for 05 male and 05 female speakers for 12 code (word) groups of vowel data. It was observed from the plots that, there is less deviation of first formant (F1) among the speakers for all vowels. Also, the

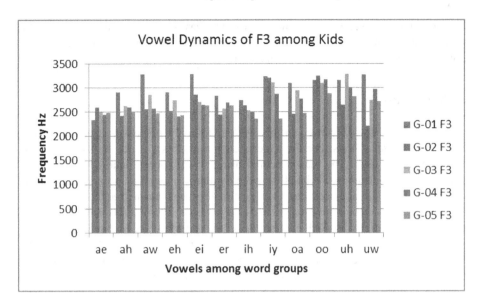

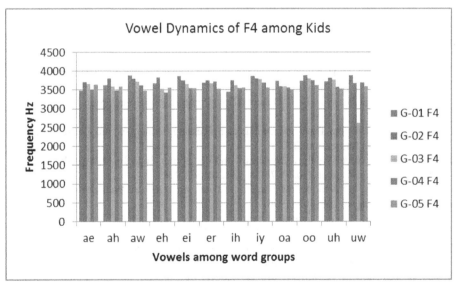

Fig. 2. (*continued*)

frequency of first formant for female speakers is much greater than that of male speakers. In general, female speaker formants are higher than male speakers. There is substantial variation in second formant except vowels codes ah, aw and uh. Higher formants of all vowels are incomparable for different speakers, i.e. there is speaker specific variation of higher formants.

Figure 2 shows the plots of formant dynamics (F1 to F4) of speech samples of five kids (girls). As compared to results in Fig. 1, there is more variation of first formant frequency of kids speaker for all vowel types, whereas first formant is observed to be nearly constant among men and women speakers. Also average formant frequencies among children is higher that of men and women.

4 Conclusion

The purpose of the work carried in this work is to analyze speaker specific formant dynamics of vowel sounds. For the experimental work a set words having of twelve vowel sounds are used. The reason behind using vowel speech data is, vowels carry most significant information related to speech patterns as well as properties of vocal mechanism of individual. In order to get true estimate of the formant frequencies, formant filters and voicing detector are used prior to LPC analysis. The first formant consistently carries the information of speech phoneme, over the speakers. The higher formants vary with respect to speaker for each vowel. The deviation is larger for third and fourth formant (F3 and F4). This variability in higher formants can be used as key point for speaker or voice recognition. Thus, similarity in estimated formants for a particular vowel can be used as feature for speech recognition, whereas variation or differences in estimated formants for same vowel sounds as a feature for speaker recognition.

References

1. Rabiner, L., Juang, B.-H.: Fundamentals of Speech Recognition. Prentice-Hall, Upper Saddle River (1993)
2. Deller, J., Hansen, J., Proakis, J.: Discrete-Time Processing of Speech Signals. IEEE Press, New York (2000)
3. Quatieri, T.F.: Discrete-Time Speech Signal Processing: Principles and Practice. Pearson Education, Third Impression (2007)
4. Flanagan, J.L.: Speech Analysis, Synthesis, and Perception, 2nd edn. Springer-Verlag, New York (1972). https://doi.org/10.1007/978-3-662-01562-9
5. Welling, L., Ney, H.: A model for efficient formant estimation. In: Proceedings of the IEEE ICASSP, Atlanta, pp. 797–800 (1996)
6. Welling, L., Ney, H.: Formant estimation for speech recognition. IEEE Trans. Speech Audio Process. 6(1), 36–48 (1998)
7. Holmes, J.N., Holmes, W.J.: The use of formants as acoustic features for automatic speech recognition. In: Proceedings IOA, vol. 18, part 9, pp. 275–282, November 1996
8. Rao, P., Das Barman, A.: Speech formant frequency estimation: evaluating a nonstationary analysis method. Sign. Process. 80, 1655–1667 (2000)
9. Anusuya, M.A., Katti, S.K.: Speech recognition by machine: a review. Int. J. Comput. Sci. Inf. Secur. (IJCSIS) 6(3), 501–531 (2009)
10. Kinnunen, T., Li, H.: An overview of text independent speaker recognition: from features to supervectors. J. Speech Commun. 52(1), 12–40 (2010)
11. Reynolds, D.A.: An over view of automatic speaker recognition technology. In: IEEE International Conference on Acoustics, Speech, and Signal Processing (ICASSP), vol. 4 (2002). MIT Lincoln Laboratory, Lexington, MA USA
12. Mirarchi, D., Vizza, P.: Signal analysis for voice evaluation in Parkinson's disease. In: 2017 IEEE International Conference on Healthcare Informatics (2017)

13. Vowel Data: James M. Hillenbrand, Speech Pathology and Audiology, Western Michigan University
14. Rao, A., Kumaresan, R.: On decomposing speech into modulated components. IEEE Trans. Speech Audio Process. **8**(3), 240–254 (2000)
15. Mustafa, K., Bruce, I.C.: Robust formant tracking for continuous speech with speaker variability. IEEE Trans. Audio Speech Lang. Process. **14**(2), 435–444 (2006)
16. Rabiner, L.R., Schafer, R.W.: Digital Processing of Speech Signal, 1st edn. (2003)
17. Yang, X., Tan, B., Ding, J., Zhang, J., Gong, J.: Comparative study on voice activity detection algorithm. In: IEEE International Conference on Electrical and Control Engineering (2010)
18. Craciu, A., Paulus, J., Sevkin, G., Backstrom, T.: Modeling formant dynamics in speech spectral envelopes. In: 25th European Signal Processing Conference (EUSIPCO) (2017)
19. Dey, S., Ashraful Alam, Md.: Formant based Bangla vowel perceptual space classification using support vector machine and k-nearest neighbor method. In: 21st International Conference of Computer and Information Technology (ICCIT) (2018)
20. Almaadeed, N., Aggoun, A., Amira, A.: Text-independent speaker identification using vowel formants. J. Sign. Process. Syst. **82**, 345–356 (2016)

User Recognition Using Cognitive Psychology Based Behavior Modeling in Online Social Networks

A. Saleema[✉] and Sabu M. Thampi

Indian Institute of Information Technology and Management-Kerala,
Kazhakkoottam, India
{saleema.res17,sabu.thampi}@iiitmk.ac.in

Abstract. Computer-mediated social contexts, specifically Social Media has been identified as an appropriate platform to assess the human personality as well as behavior as it involves massive interactions and communications. Social Behavioral Biometrics is a very recently introduced research area that attempts to extract unique behavioral features from the social interactions of individuals, which are powerful enough to be used as biometric identifiers. This research investigates the feasibility of generating biometric templates from user profiles, which satisfies the basic requirements of a biometric identifier namely distinctiveness and stability. This work proposes a user identification as well as user behavior modeling approach using some novel concepts on the grounds of certain unassailable theories in Human Psychology. The authors have performed an empirical study on Online Social Network data by extracting certain psychometric properties of temperament from user profiles in order to generate unique and stable user behavior patterns. Based on the Cognitive Affective System theory of Human Personality which has been proved indubitable years ago in offline social contexts, we have created Situation-Behavior profiles of a sample set of Twitter users and analyzed that whether the theory befits the OSN data. The Big Five behaviors of users are measured by content analysis and are analyzed in different situations like Political, Personal, Entertainment, Sports and Technical. The results of the empirical study reveal that the Situation-Behavior profiles are pertinent enough to be used as biometric patterns with high stability and distinguishability.

1 Introduction

Nowadays, Social Media has become a standard way of life and thereby the identity of human beings is extending beyond the real world to an equivalent virtual world. User profiles in typical online social networks such as Facebook, LinkedIn, Twitter, Instagram, etc assigns them a virtual identity through which they perform massive interactions and communications. As evidenced by the ComScore report, approximately 1 of every 5 min spent online by internet users is for social networking sites [1]. Among these, Twitter-the rapidly evolving microblogging

© Springer Nature Singapore Pte Ltd. 2020
S. M. Thampi et al. (Eds.): SIRS 2019, CCIS 1209, pp. 130–149, 2020.
https://doi.org/10.1007/978-981-15-4828-4_12

platform launched in 2006, is the most propitious one utilized for research due to its tight limits and restrictions in posting.

In the real world, a person can be identified using their Physiological as well as Behavioral traits which are categorized under the two classes of Biometrics. Physiological Biometrics make use of the static traits of the human body that are not subjected to change over aging, which includes the fingerprint, retina, iris, finger vein, face, palm, etc. Behavioral Biometrics, on the other hand, corresponds to the unique individual behavioral patterns which span from the walking style, typing style, speech, handwriting, etc to the most recent ones using social interactions and communications.

As far as computer-mediated social contexts are concerned, person authentication relies upon the username and password method, which doesn't guarantee adequate security, especially when we log into various social networking sites from different locations such as libraries or cafes. Also, the username password method is associated with security questions, which are quite painstaking to the users especially because of the need for memorizing the circumstantial questions and their answers. Most importantly, password-based authentication do not ensure a strong identity check, as it can be stolen/hacked. That means the security has highly relied upon the strength and confidentiality of the passwords.

The majority of people nowadays use social networks through mobile phones or other smart device applications, where login credentials are not verified every time. So in case of a device steal/loss or sometimes in cases of death, the social media profiles of the person can be misoperated by others. Therefore, the need for new static as well as continuous authentication techniques in computer-mediated social contexts is being critical to ensuring authenticity, in such an era of unfettered social media usage.

Social Behavioral Biometrics(SBB) is such an authentication approach that is defined as the identification of an actor/person/avatar based on their social interactions and communication in different social settings [2]. Social settings for studying SBB is described as the environments and their properties, over which the communication and interaction between actors take place. SBB is a most recently introduced area that seeks immediate research attention. The scope of social behavioral biometrics resides in continuous authentication in cyberspace as well as online social networking platforms. In today's scenario where millions of people are being connected to the OSNs in a day to day basis, the compromise of one user's account could pose malicious security threats to many of the user's acquaintances. While users of SNSs are authenticated continuously based on social behavioral biometric features, if the trust level of the user drops significantly, the system can take necessary actions such as forbidding the user from using the service, generating alarms, etc. This application would facilitate anomaly, fraud, or intrusion detection as well. The social behavioral features extracted from social networking sites contain predominant information about the user and there comes the possibility of automatically generating security questions which would increase the security to the users' account as well as lessen the strain of setting security questions and answers. The scope and relevance of the Social Behavioral Biometric(SBB) feature extend to the integration of SBB with Multimodal/Multifactor authentication systems. The combination of SBB with other behavioral biometric methods opens a new door of research since it improves the uniqueness and enhances identification or verification rates of users in the virtual world. SBB features can be used for author identification in cases of law enforcement and forensic applications of the investigations of cybercrimes. The scope of social behavioral biometrics is not confined to person authentication but spans to risk analysis, situation analysis, anomaly detection, access control, customer profiling, etc [2].

The massive data on online social networks have already been explored by researchers in the past decade for new application fields such as sentiment analysis, stock market, election results prediction, education tools, opinion mining, etc. A multitude of works has been carried out utilizing the concepts of personality to detect the factors influencing the use of social networks, the addictive tendencies towards social networking sites, the way how people behave online, etc. Most of the research work investigated how the five factors of personality are correlated to the social networking sites' duration and frequency of usage, sharing of information, the number of friends, etc. Also, using the big five model as a basis, researchers have identified how the different traits are related to different types of users namely listeners, influencers, popular, etc. Beyond these, personality-based research using online social network data is proliferating in diverse directions such as the development of friend suggestion methods, targeted marketing, automatic website morphing using cognitive styles, authorship attribution, etc.

Even Though the human behavioral patterns have been studied for individual behavior analysis over the past decade, only a few researchers considered the potential of OSN data as social behavioral biometric templates/features. The previous works on SBB explored frequency-based and knowledge-based social behavioral biometric features, extracted from online social networking sites. More specifically, they have created distinct biometric profiles of users by utilizing the reply/retweet networks, hashtag networks, URL, friendship network, etc as features. The major challenges in creating such biometric profiles include the difficulty in extracting features from huge random datasets, selection of small consistent features for identification and adaptation to the change in behavior over time. Although these methods could achieve reasonable recognition rates, these biometric profiles were not sufficiently unique to identify a person with high precision. However, in combination with other modalities, identification systems with high accuracy and precision were obtained. Also, the spatial and temporal information collected from social networking sites while combined with these SBB profiles resulted in high recognition accuracy. So, there is a big motive in developing social behavioral biometric algorithms to efficiently authenticate individuals from the online social network data, that provide high uniqueness and accuracy.

We have performed an extensive study on psychology and its applications in online social networks and discovered that there is substantial scope in exploring cognitive psychology as well as human psychological theories of Disposition in order to generate SBB features. Since the traditional cognitive psychological personality theories based on temperament have been well investigated and proved to be irrefutable, the use of these aspects in the context of online social networks will be highly admissible with appreciable reliability and accuracy, as the conversations/interactions in OSN possess much resemblance to the real-world situations. Our paper gives a direction to move a first step towards utilizing these aspects for person authentication, providing high stability and uniqueness.

Besides person identification, a good deal of applications is expedient by integrating the social network analysis with cognitive psychology. According to our analysis, the utilization of the social media platform for person identification along with context awareness and cognitive psychology will actualize the future generation intelligent automated biometric systems, that are expected to mimic the human brain. Figure 1 is an illustration of such an intelligent system that will automate the process of user recognition, user change detection, behavior analysis, rumor detection, targeting and recommendation systems, etc.

Our research work aims to collaborate with cognitive psychology and online social network analysis in order to extract person distinctive properties. The notion behind this collaboration is the clear and solid descriptions of the Dispositions and Temperament aspects of personality in human psychology. The term personality itself is defined in psychology as a combination of characteristics and qualities that form an individual's distinctive character. Temperament refers to the qualities distinguishing a person while Disposition implies the customary

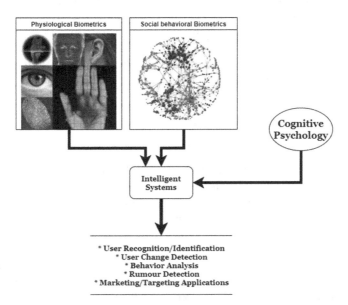

Fig. 1. The depiction of a next-generation Intelligent System that automatically recognizes users, profile changes, rumors, etc.

moods and attitude towards the life around one. Our research work employs the conception of personality in terms of behavioral dispositions or traits that predispose individuals to engage in relevant behaviors.

The key contributions of this research work are as follows:

1. Our work is the first to collaborate the concept of cognitive psychology with online social network analysis in order to check the feasibility of generating biometric templates.
2. We have performed an empirical study on social network data based on the Cognitive Affective System Theory of Personality to extract Situation-Behavior Profiles of users. Two hypotheses related to the properties of stability and distinctiveness are formulated and tested on the OSN data.
3. This work proposes a behavior modeling technique that makes use of emotion extraction based content analysis and mapping the emotions to the PleasureArousal-Dominance temperament space. Further, it employs the evaluation of Big Five traits from the PAD space.
4. We have compiled a Twitter user dataset with user tweets from two separate periods and annotated with situations such as Personal, Political, Entertainment, Technical, etc to suit our research problem.

The organization of the remaining sections in the paper are as follows: Sect. 2 examines some of the relevant works in the area of social behavioral biometrics, personality prediction based research works in social networks, modeling of psychological theories in social media and some of the computational models of personality from psychology. In Sect. 3, we state our hypotheses, maps the CAPS theory to the social networks and elaborate our proposed approach for modeling human behavior and creating the Behavioral Signatures. In Sect. 4, we brief the experimental setup in which we outline the details of dataset collection, summarize the findings based on the preliminary analysis of the dataset and discuss the results of our empirical study. Section 5 concludes the work and specifies the limitations and future works.

2   Related Works

This section reviews some research works on social behavioral biometrics, related works on behavioral modeling using psychological theories in social media, some relevant works on behavior and personality predictions on social media and some significant computation models of personality from the literature. The concept of SBB was introduced in [2] and it defines SBB as the identification of an actor based on the communications and social interactions in different social settings. The interaction between actors, the communication manner, etc is considered while generating SBB. It was the first study on social behavioral biometrics which deals with human social interactions from the aspect of identifying a person. The authors have created reply networks, retweet networks, URL networks, etc and applied as the features for person identification.

In [3], the authors have attempted to fuse the social behavioral data with other physiological biometrics using confidence-based decision fusion to develop a multimodal approach, being the first work of decision fusion for multimodal biometrics using social network data. Later in 2017, they have attempted to recognize users in a computer-mediated social context [4]. They have created a new framework by identifying several social biometric features, proposed a new matching method and evaluated three basic biometric properties namely accuracy, uniqueness and stability. The above-discussed ones are purely based on the extraction of social media profile features but not related to psychological concepts. Now we discuss some of the approaches which model the traditional theories of human psychology in OSNs, but not in an authentication perspective.

Not many research works have been done to model social media using theories of human psychology and that's why it remains a fertile research field. Significant work was done by Tyshchuk [5] in which the well-known Theory of Planned Behavior is applied. According to the theory of planned behavior [6], an individuals intention to perform a behavior is driven by factors like subjective norms, attitude towards the behavior and perceived behavioral control. In this work, the authors have identified features from the social media profiles that can be well mapped to the factors of the theory and applied TPB to predict their behavioral intents. They have concentrated on the behavior that happens in response to a trigger, specifically an event. A set of elements are selected as features and they were measured by using natural language processing and social network analysis techniques. Another significant work was done by Tripathi [7] which used the Big Five Model as the basis.

Bachrach et al. [8] have presented a work based on examining the correlation between user's personality and the features of online profiles such as the density, size, etc. of friendship networks, number of events, group memberships, number of tags, etc. They could show significant relationships between user personality and the SNS features and showed how multivariate regression is utilized for the prediction of the personality traits of an individual user, given their Facebook profile. Significant work to understand the information sharing behavior using socio-cognitive theory was done in [9].

Bandura's socio-cognitive theory [10] is a well-accepted theory in psychology and is an extension of his own social learning theory [11]. The socio-cognitive theory can be stated as the fact that people do not learn solely by their success or failure, but it is depended upon the replication of the action of others. Bandura has applied SCT to analyze the influence and diffusion of new behaviors through Social networks in [12]. Most of the papers in the literature are focused on personality predictions and behavior predictions which can be applied to the recommendation likes systems.

A considerable amount of research works have been found in the literature which relates to the study of behavior and personality prediction in social media, identifying the psychological connection between personality and social media usage, etc. The majority of the reported works explored the Big Five model of personality which is a strong consensus on the basic traits of human beings,

Table 1. Related Works on Behavior and Personality predictions in social media

Author and Year	Purpose	Social Network Type	Features used
Ross et al. 2009 [13]	Study the influence of personality and competency factors related to facebook usage	Facebook	Attitude, Online sociability
Hamburger et al. 2010 [14]	Examines the connection between personality and faceboook usage	Facebook	No: of friends The user bio Group, admin
Wilson et al. 2010 [15]	Identify psychosocial characters of young people's use of social networks	Myspace, Facebook	Time spend, Attitude
Quercia et al. 2011 [16]	Predict user personality and find the similarity and differences between different types of users	Twitter	Followers, Following, Listed counts
Golbeck et al. 2011 [17]	Predicting Big Five traits from profile information	Twitter	Tweet analysis
Buettner 2016 [18]	Product recommendation system from personality analysis	Xing	No: of contacts, Groups, Page views, Work experience
Pornsakulvanich 2017 [19]	To find the contributing factors to online social support satisfaction and frequency	Facebook, Instagram, Line	Usage duration, Attitude, Social influence, Friends etc
Tedesse et al. 2018 [20]	Predicting the personality traits	Facebook	Content analysis, networksize, betweenness, density, brokerage, transitivity

Table 2. Computational models of personality

Model type	Author and year
Neural Network Model	Pozanski and Thagard 2005 [21]
Neural Network Model	Quek and Moskowitz 2007 [22]
Knowledge and Appraisal Model	Cervone [23]
Personality Systems Interaction Model	Kuhl 2000 [24]
Cognitive Affective System Theory	Mishel and Schoda 1995 [25]

initially evidenced by Fiske in 1949. Table 1 summarizes some of the recent relevant works in this area, all of which have used the Big Five model as the basis.

Now we would like to appraise some of the computation models of personality, that can be explored in the future, on the grounds of social network data. Table 2 delineates several well regarded computational models of personality which will benefit in the study of applying psychological aspects to social media.

From the above-reviewed literature, it is evident that not a single work made an attempt to model any human psychological theories to extract person distinctive properties with a view of authentication.

3   Proposed Approach

We have performed an empirical study on online social network profiles on the grounds of the Cognitive Affective Personality System Theory. Also, we have used the concept of the Pleasure-Arousal-Dominance temperament model and the big five personality traits from psychology. The proposed approach is outlined in three subsections namely the background theory, formulation of hypotheses and the methodology of behavior extraction.

3.1   Background Theory and Formulation of Hypotheses

3.1.1   Cognitive Affective Personality System

The theory accounts for the differences in individuals in predictable patterns of variability across situations. ie If situation A, then she behaves X, but if situation B, then she behaves Y. Also, it states that the average levels of behavior are the behavioral signatures of the same fundamental personality system. The reason for this is illustrated by a framework termed as the Cognitive Affective Personality System.

The Situation-Behavior Profiles (if-then profiles) are proved to be intra-individually stable by an empirical study. Also, these profiles are considered as behavioral signatures of a person, which means that these are unique for each person. Since the properties of uniqueness and stability are satisfied in their studies according to the theory, we have considered the possibility of generating

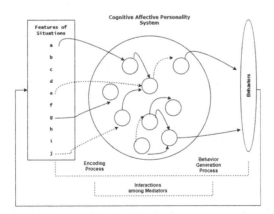

Fig. 2. The simplified illustration of the cognitions and effects as mediating processes in a Cognitive Affective Personality System.

Table 3. Cognitive affective units in the personality system

Encodings: Categories (constructs) for the self, people, events, and situations (external and internal)

Expectancies and Beliefs: About the social world, about outcomes for behavior in particular situations, about self- efficacy

Affects: Feelings, emotions, and affective responses (including physiological reactions)

Goals and Values: Desirable outcomes and affective states; aversive outcomes and affective states; goals, values, and life projects

Competencies and Self-regulatory Plans: Potential behaviors and scripts that one can do and plans for organizing action and affecting outcomes and one's own behavior and internal states

---

Note: From [25] in 1995

a biometric template from these Situation-Behavior profiles, on the background of social networks.

In order to explain how and why these behavioral signatures are exhibited, the authors incorporated the role of situations, events or contexts into the conception of personality. Figure 2 shows a combined view of the personality system. Each individual is characterized in terms of two things as follows

(a) The cognitions and affects.
(b) The distinctive way of organization of the interrelations among cognitions and effects and psychological features of situations.

When certain situations are encountered by an individual, some particular situational features are activated and it inturn activates a set of cognitions and affects which are characteristic in nature. So, as the figure indicates, within an individual a characteristic set of interrelationships among the cognitive and affective units further activates certain units and finally activates some behavior. The types of cognitions and affects are described as shown in Table 3 and the structure of CAPS is depicted in Fig. 2.

3.1.2   Formulation of Hypothesis

Based on the conception of behavioral signatures of disposition, we have formulated two hypotheses on the grounds of online social networks.

Hypothesis (H1): People in online social networks show considerable variability in similar behaviors say extroversion, agreeableness, conscientiousness, emotional stability and sophistication across various situations like political, entertainment, technical, personal, sports, etc

Hypothesis (H2): People tend to show high consistency in those behaviors within the same classes of situations and their Situation-Behavior profiles are stable and distinctive enough to be utilized as a biometric template.   □

Traditional researchers in human personality commented that the basic coherence in the underlying personality (Disposition) results in cross-situational consistency. But later on, as research embellished, researchers unveiled the role of situations and theorized about the low cross-situational consistency in behavior. Mishel and Shoda have later proved that individuals' behavior shows less cross-situational consistency and the behavior patterns across situations are intra-individually stable. They have provided empirical evidence by analyzing the if....then...profiles of a set of individuals. The if...then..profiles are termed as Situation-Behavior profiles and are considered as signatures of Personality. In their study, the situations were defined in nominal terms as places and activities in the setting, ie arithmetic tests, dining halls, woodworking activities, school playgrounds, etc. They have selected some dimensions of behavior such as verbal aggression, friendliness, withdrawal, prosocial behavior, etc in a study conducted in a residential camp setting. For each person, the Situation-Behavior profiles are examined and assessed the stability and distinctiveness.

In our study, we have mapped the entire entities from the real world to a virtual cyber world. The conviction behind this mapping is that the interactions, interrelationships, communications, etc in online social networks possess much resemblance with the real face to face conversations. Therefore we have reformulated the hypotheses in the context of the online social networks as in hypotheses 1 and 2.

Hypothesis 1 corresponds to the behavior variability across situations more specifically the if-then relations, which are to be proved intra-individually stable. Hypothesis 2 corresponds to the stability and distinctive property, ie the if..then..profiles are stable as well as unique for each individual.

3.2   Mapping of CAPS to the Context of Online Social Networks

The real-life situations like a family party, group meeting, worksite, etc, when mapped to the online social network becomes political, technical, entertainment, personal, etc according to the posts. Especially on twitter, there will be hashtags or URLs associated with the majority of the posts. In the real-world, the behaviors taken for the study were friendliness, verbal aggressiveness, anxious, being silent, etc, but in OSN we have considered the big five traits proposed by J.M.Digman. Table 4 shows the mapping of the situations and behaviors from the real world to the online social network.

In the study conducted with students in the residential camps, they have identified the situational features as the valence of the interaction which can be either positive or negative, the type of person(whether adult counselor or child peer), etc. Here we have identified the situational features like whose tweet, the no: of replies, it has obtained so far, the no: of retweets, whether it is associated with a hashtag or a URL, which all people replied for it, etc. The organization of the cognitions and affects shown inside the circle is the same for the OSN

Table 4. Mapping of Situations and Behaviors from real world to Online Social Network

Real life contexts	OSN contexts	Real life behaviors	Behaviors in OSN
Family Party	Political	Verbal Agressiveness	Extroversion
Group Meeting	Technical	Friendliness	Conscientiousness
Working Site	Personal	Anxious	Emotional Stability
Working in an urgent project	Disasters	Tease and make fun of	Agreeableness
Individual Assignmen	Entertainment	Surf web	Sohistication
Trying to get a date	Business	Be Silent	
	Sports	Explore environment	

context also, as it purely represents the underlying cognitive structure of the personality system.

The whole system of CAPS in the context of OSN is shown in Fig. 3 and can be described as follows. The figure shows how behavior signatures are produced by a personality system characterized by the available cognitive-affective units, organized in a distinct way of inter-relationships. When a situation is encountered, say a tweet related to a political party, some situational features say whose tweet, which all friends replied will stimulate certain cognitions and affects inside that person. Since the way of organization is different for different individuals, the stimulations or activations of the units will vary. Within any individual, a rich set of relationships exists between the units and according to them, they will activate or deactivate further units and ultimately behavior is generated. The situational features that causes stimulation of these units also will be different for different individuals. For example, some people will reply if their friends are

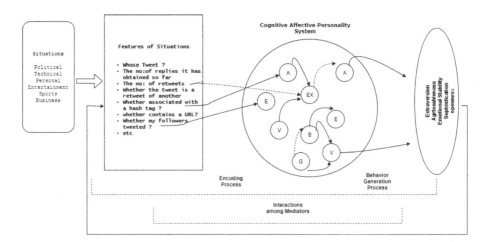

Fig. 3. The illustration of how behavioral signatures are generated, with the help of CAPS theory in the Online Social Network context

replied, some others will retweet or reply if there is a presence of a hashtag or a URL since these things represent clear shreds of evidence of the subject in hand. The mediating units are activated in response to some situational features and are deactivated with respect to some situations. ie The relation may be positive (represented by the solid line) or negative (represented by the dashed line).

3.3 Methodology

The Big Five Behaviors in OSN can be measured in various ways. In [7], two methods, content-based analysis, and linkage data-based analysis are discussed for evaluating the behavior. The huge amount of contents such as text, image, video, etc are used in content analysis while the network data which includes the relationship among the entities are used for linkage data-based analysis. Factors like no: of friends, whether profile photo is displayed, whether user biodata is displayed etc helps in determining which category of behavior the users belong to. Several deep learning models are available for detecting the big five traits directly from the text. We have performed a content analysis by predicting the emotion probabilities from the tweet text instead of the big five traits since we were interested to analyze the emotional dispositions of individuals. The proposed process of extraction of big five behavior from tweets is as explained below.

Initially, the tweets in two separate periods of the sample set of users are collected and labeled with situations such as technical, entertainment, political, personal, sports, etc. The tweets are preprocessed by removing the URLs, Mentions, Hashtags, Twitter-reserved words, punctuations, single letter words, blank spaces, stopwords, profane words, and numbers.

The processed tweets are then given to a trained Recurrent Neural Network model presented in [26] for predicting emotions. Most of the emotion recognition models for twitter use lexicons and bag-of-word models. We have taken the Recurrent Neural Network model for consideration since it proved to be more efficient than the state-of-the-art.

The Ekman's emotions are then mapped to the Pleasure-Arousal-Dominance (PAD) temperament space using a transformation matrix. Several mapping techniques have been proposed in the literature for transforming the Ekman's emotions to the PAD space, among which the one illustrated in [27] seems the most appealing. The mapping matrix presented presented in [27] is as follows
PAD[Happy, sad, Angry, Scared, Disgusted, Surprised, 1]=

$$
\begin{bmatrix}
0.46 & -0.30 & -0.29 & -0.19 & -0.14 & 0.24 & 0.52 \\
0.07 & -0.11 & 0.19 & 0.14 & -0.08 & 0.15 & 0.53 \\
0.19 & -0.18 & -0.02 & -0.10 & -0.02 & 0.08 & 0.50
\end{bmatrix}
\tag{1}
$$

The mapping of Ekman's emotions to the Pleasure-Arousal-Dominance temperament space can be defined mathematically as the matrix multiplication of the mapping matrix in Eq. 1, with the emotion vector obtained from the pre-trained model.

We have chosen the PAD temperament space since these are characteristic emotional predispositions that are stable over the lifetime of an individual. The three measures Pleasure, Arousal and Dominance are proved to be orthogonal. The trait Pleasure refers to the relative predominance of positive versus negative affective states across a life sample of situations. The scale Arousability is the measure of how easily a person is aroused and returned by certain stimuli. The third one is the Dominance-Submissiveness scale that refers to a person's characteristic feelings of influence and control over his life circumstances versus feelings of being controlled and influenced by others or events.

Then, as the last step, we have computed the big five Traits of behavior using the PAD space with the linear mapping equations proposed in [28], as shown below.

$$Extraversion = 0.23P + 0.12A + 0.82D \tag{2}$$

$$Agreeableness = 0.83P + 0.19A - 0.21D \tag{3}$$

$$Conscientiousness = 0.32P + 0.30D \tag{4}$$

$$Emotional Stability = 0.57P - 0.65A \tag{5}$$

$$Sophistication = 0.33A + 0.67D \tag{6}$$

The transformation from PAD space to the big five space can be expressed as $f(P, A, D) : R^3 \rightarrow R^5$, where R^3 represents the PAD space and R^5 Five space. If V is the PAD vector and W is the Big Five Traits vector, this linear transformation can be represented using matrix multiplication as Five space.

$$\overrightarrow{W} = A\overrightarrow{V} \tag{7}$$

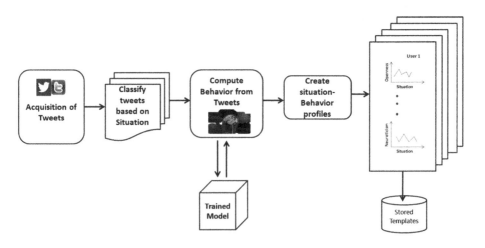

Fig. 4. The illustration of extracting the Situation-Behavior Profiles

$$whereA = \begin{bmatrix} 0.23 & 0.12 & 0.82 \\ 0.83 & 0.19 & -0.21 \\ 0.32 & 0 & 0.30 \\ 0.57 & -0.65 & 0 \\ 0 & 0.33 & 0.67 \end{bmatrix} \tag{8}$$

The whole process of Situation-Behavior Profile extraction is depicted in Fig. 4.

4 Experimental Results

This section provides the experimental results of our evaluation of the Hypothesis 1 and 2 on a sample set of twitter users. All the experiments including the implementation of our methodology for Situation-Behavior Profile extraction are carried out in a 3.50 GHz Intel(R)Xenon(R) CPU with 64GB RAM. Python 3.7 is used for the implementation of emotion and behavior extraction.

A dataset of 50 sample users from two non-overlapping periods(sessions) of the time was collected for the analysis in the following manner. Initially, a user with a high engagement timeline is selected as a seed. From his acquaintances, another 2 prolific users were selected and continued this manner until we got 50 prolific users.

The dataset was collected using 'vicinitas', which is an online tool that helps track and analyze real-time and historical tweets. For each user, we have collected information such as the tweet Identifier, the tweet text, screen name, the

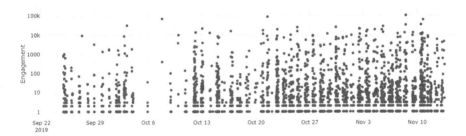

Fig. 5. The engagement timeline of a particular user during a particular session, analyzed using vicinitas tool

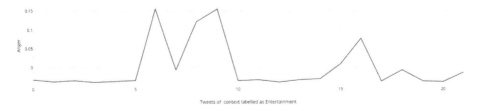

Fig. 6. Graph plotting the standard deviation of the emotion 'Anger' across the tweets of same context 'Entertainment' of User #1

Table 5. Sample tweets annotated with Situations

Tweet	Situation
Heading to Hamburg to meet old friends and new ideas	Entertainment
I hope that's the case. But Android seemed to have confirmed it	Technical
Seriously impressed with Deepak Chahar's bowling, he can strike a few with the bat too	Sports
Delhi's air strays deeper into 'severe' zone, likely to dip to 'severe+' by Thursday	General
Again Joh** baffles his opponents. The video is a piece of clever political communication	Political
Top films for me Super30	Entertainment
I need to work harder?	Personal
Captain Rohit. I shudder to think had he not been leading today	Sports

date and time of the tweet, Tweet type, the no: of retweets, etc. The users we have selected are those having an average of 25 tweets per day. We could analyze the engagement timeline day-wise and hour-wise, the percentage of tweet types, the percentage of rich media, etc using this tool. The engagement timeline of a particular user at a particular session is shown in Fig. 5. The tweet type represents whether the tweet is a reply, retweet, mention or a normal tweet. We have done pre-processing of the dataset by removing elements such as URL, mentions,

Fig. 7. Graph plotting the standard deviation of the emotion 'Joy' across the tweets of same context 'Entertainment' of User #1

Fig. 8. Graph plotting the standard deviation of the emotion 'Disgust' across the tweets of same context 'Personal' of User #1

hashtags, twitter reserved words, punctuation, single letter words, blank spaces, stopwords, and numbers, as these will not contain any necessary information about the users' characteristics. Also, we have filtered the tweet lists by removing the retweets since it may be written by another user and will not benefit our content analysis.

After filtering and pre-processing, we have annotated the tweets of each user with the 'situation' field, which represents which context the tweet can be related to. This was a time-consuming task, which required a lot of manual effort and considered as the major limitation of this study. However, automatic annotation of the context/situation of a post can be made practical in the near future, utilizing the well advanced natural language processing and artificial intelligence techniques, which will ease the work of all the researchers related to OSN. A sample set of tweets with the annotated situations are shown in Table 5. The number of tweets for each situation varied for different users. So the probability of each tweet to belong to a particular situation differed for every user. The interest or enthusiasm for each user is different for different situations and therefore some people may not respond at all in particular situations, say sports, politics, etc.

We have evaluated hypothesis 1, by plotting each of the big five behaviors with respect to various situations like entertainment, technical, personal, sports, etc and found to be true for each of the users in our dataset. The individuals have shown ample variations in the conditional probability of exhibiting behaviors across various situations. These plots are termed as Situation-Behavior profiles or if..then.. relations and it confirms the intra-individual behavior variability across situations. The Situation-Behavior profiles of two randomly selected users are shown in Fig. 11.

We have analyzed Ekman's emotions extracted using the method prescribed in [26] for the whole tweets and arrived at the finding that during a particular context, the specific behavior of a person remains more or less similar. An

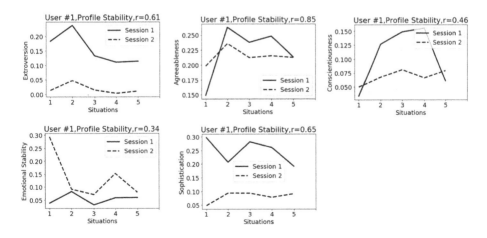

Fig. 9. Situation-behavior Profiles of User #1 at Sessions 1 and 2

example illustration of the emotion 'Anger' extracted from the tweets of the same situation 'Entertainment' during session 1 is shown in Fig. 6. On the vertical axis, the standard deviation of emotion 'Anger' with respect to the normative mean(zero) is shown with respect to the tweets related to entertainment in the horizontal axis. This clearly shows that the values are always nearer to zero, with an overall standard deviation of 0.063. This indisputably indicates the high intra-situational consistency of individuals in expressing emotions. Figure 7 shows the standard deviation of another emotion joy in the same situation entertainment for the same user. Its overall standard deviation is 0.26. The graph plotting Disgust against the tweets of 'Personal' situation is shown in Fig. 8.

About 65% of the If..then Profile shows the stability coefficient greater than 0.5 and when the Agreeableness-Situation profiles alone are considered, around 89% of individuals show stability coefficients greater than 0.65. This strongly recommends Agreeableness as the most stable behavior.

Figure 9 shows the Situation-Behavior Profiles of user #1 for each of the behaviors extroversion, agreeableness, conscientiousness, emotional stability, and sophistication. We have plotted and analyzed the If..then relations of all users in our dataset in two separate sessions producing two profiles for each individual. The profile at session 1, having a duration of 4 months is plotted as a solid line, where the dashed line represents the profile of the same person at session 2 of another 4 months duration. The time period of sessions 1 and 2 are non-overlapping. We have evaluated the stability of the profile using the correlation coefficient, r and found out that the profile at two times shows impressive stability. Here in Fig. 9, the person shows high stability in agreeableness, extroversion, and sophistication while the behaviors conscientiousness, emotional stability is less stable. The user #29 shown in Fig. 10 has exhibited high stability in extroversion, agreeableness, and emotional stability. The traits conscientiousness and sophistication are less stable with r = 0.30 and 0.37 respectively. We have

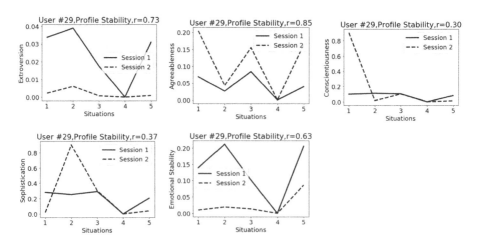

Fig. 10. Situation-behavior Profiles of User #29 at Sessions 1 and 2

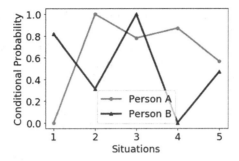

Fig. 11. Intra-individual pattern of Agreeableness of two users plotted across different situations.

analyzed the profile stability of each individual and arrived at the conclusion that the behavior 'Agreeableness' is the one which gives the highest stability across the two time periods and the profile Agreeableness-Situation is the most suitable one to be considered as a behavioral signature that characterizes a person.

Figure 11 shows the intra-individual behavior variability of two persons with the behavior agreeableness plotted over the different situations. The conditional probability of agreeableness is plotted across the situations, since agreeableness was identified as the most intra-individually stable one among the big five, according to our experiments using the profile data.

The Figs. 9, 10 and 11 clearly shows the profile stability and distinguishability property, which are in support of the hypothesis 2. We have examined the stability and distinguishability properties of all the users in our dataset and found complying with hypothesis 2.

5   Conclusion and Future Works

This paper presents a novel approach to social behavioral biometrics by utilizing cognitive psychology in order to develop behavioral signatures of individuals from online social network profiles. To the best of our knowledge, this work is the first to make use of the conceptions of personality dispositions/temperament in a biometric perspective. The notion behind this is the idiosyncrasies and uniqueness in each individual's personality, behind which the cognitive organizations inside a person plays a role. The paper describes how and why the personality and behaviors differ with the help of the Cognitive Affective Personality Systems theory in psychology. We have provided clear descriptions of how the theory can be applied to the online social network with proper mappings and explanations. We have tested two hypotheses related to the stability and distinguishability requirements of biometrics, on a self-compiled Twitter dataset and are proved to be true. The Situation-Behavior profiles of a set of individuals are created and arrived at a conclusion that the profiles are showing considerably good stability and distinguishability to be used as behavioral signatures. Among the five behaviors, the Agreeableness-Situation profile is identified as the most stable and recommended profile.

The annotation of the situation for each individual tweet was a tedious task, which required a lot of manual effort and it is considered as a major limitation of our work. In our work, we have calculated the behavior metrics by content analysis, specifically the text and context of the tweets. In the future, we can utilize a lot of other information like user engagement pattern, temporal data, the network of interactions, etc that may produce better templates/signatures.

Acknowledgements. This research work was funded by the Kerala State Council for Science, Technology and Environment [KSCSTE/5623/2017-FSHP-ES].

References

1. Comscore report. https://wearesocial.com. Accessed 15 Nov 2019
2. Sultana, M., Paul, P.P., Gavrilova, M.: A concept of social behavioral biometrics: motivation, current developments, and future trends. In: 2014 International Conference on Cyberworlds, pp. 271–278. IEEE (2014)
3. Paul, P.P., Gavrilova, M.L., Alhajj, R.: Decision fusion for multimodal biometrics using social network analysis. IEEE Trans. Syst. Man Cybern. Syst. **44**(11), 1522–1533 (2014)
4. Sultana, M., Paul, P.P., Gavrilova, M.L.: User recognition from social behavior in computer-mediated social context. IEEE Trans. Hum. Mach. Syst. **47**(3), 356–367 (2017)
5. Tyshchuk, Y., Wallace, W.A.: Modeling human behavior on social media in response to significant events. IEEE Trans. Comput. Soc. Syst. **5**(2), 444–457 (2018)
6. Ajzen, I.: The theory of planned behavior. Organ. Behav. Hum. Dec. Process. **50**(2), 179–211 (1991)
7. Tripathi, A.K., Hossain, S., Singh, V.K., Atrey, P.K.: Personality prediction with social behavior by analyzing social media data-a survey. University of Winnipeg (2013)
8. Bachrach, Y., Kosinski, M., Graepel, T., Kohli, P., Stillwell, D.: Personality and patterns of Facebook usage. In: Proceedings of the 4th Annual ACM Web Science Conference, pp. 24–32. ACM (2012)
9. Yoon, H.J., Tourassi, G.: Analysis of online social networks to understand information sharing behaviors through social cognitive theory. In: Proceedings of the 2014 Biomedical Sciences and Engineering Conference, pp. 1–4. IEEE (2014)
10. Bandura, A.: Social cognitive theory. In: Annals of Child Development, vol. 6. Six Theories of Child Development (1989)
11. Bandura, A.: Social learning theory of aggression. J. Commun. **28**(3), 12–29 (1978)
12. Bandura, A.: Social cognitive theory of mass communication. In: Media Effects, pp. 110–140 Routledge (2009)
13. Ross, C., Orr, E.S., Sisic, M., Arseneault, J.M., Simmering, M.G., Orr, R.R.: Personality and motivations associated with facebook use. Comput. Hum. Behav. **25**(2), 578–586 (2009)
14. Amichai-Hamburger, Y., Vinitzky, G.: Social network use and personality. Comput. Hum. Behav. **26**(6), 1289–1295 (2010)

15. Wilson, K., Fornasier, S., White, K.M.: Psychological predictors of young Adults' use of social networking sites. Cyberpsychol. Behav. Soc. Network. **13**(2), 173–177 (2010)
16. Quercia, D., Kosinski, M., Stillwell, D., Crowcroft, J.: Our Twitter profiles, our selves: predicting personality with twitter. In: 2011 IEEE Third International Conference on Privacy, Security, Risk and Trust and 2011 IEEE Third International Conference on Social Computing, pp. 180–185. IEEE (2011)
17. Golbeck, J., Robles, C., Edmondson, M., Turner, K.: Predicting personality from Twitter. In: 2011 IEEE Third International Conference on Privacy, Security, Risk and Trust and 2011 IEEE Third International Conference on Social Computing, pp. 149–156. IEEE (2011)
18. Buettner, R.: Predicting user behavior in electronic markets based on personality-mining in large online social networks. Electron. Mark. **27**(3), 247–265 (2016). https://doi.org/10.1007/s12525-016-0228-z
19. Pornsakulvanich, V.: Personality, attitudes, social influences, and social networking site usage predicting online social support. Comput. Hum. Behav. **76**, 255–262 (2017)
20. Tadesse, M.M., Lin, H., Xu, B., Yang, L.: Personality predictions based on user behavior on the facebook social media platform. IEEE Access **6**, 61959–61969 (2018)
21. Thagardz, P.: Changing personalities: towards realistic virtual characters. J. Exp. Theor. Artif. Intell. **17**(3), 221–241 (2005)
22. Quek, M., Moskowitz, D.: Testing neural network models of personality. J. Res. Pers. **41**(3), 700–706 (2007)
23. Cervone, D., et al.: The architecture of personality. Studia Universitatis Babes-Bolyai-Psychologia-Paedagogia **49**(1), 3–44 (2004)
24. Kuhl, J.: A functional-design approach to motivation and self-regulation: the dynamics of personality systems interactions. In: Handbook of self-regulation, pp. 111–169. Elsevier (2000)
25. Mischel, W., Shoda, Y.: A cognitive-affective system theory of personality: reconceptualizing situations, dispositions, dynamics, and invariance in personality structure. Psychol. Rev. **102**(2), 246 (1995)
26. Colnericˆ, N., Demsar, J.: Emotion recognition on Twitter: comparative study and training a unison model. IEEE Trans. Affect. Comput., 1 (2018)
27. Landowska, A.: Towards new mappings between emotion representation models. Appl. Sci. **8**(2), 274 (2018)
28. Mehrabian, A.: Analysis of the big-five personality factors in terms of the pad temperament model. Aust. J. Psychol. **48**(2), 86–92 (1996)

Indian Semi-Acted Facial Expression (iSAFE) Dataset for Human Emotions Recognition

Shivendra Singh and Shajulin Benedict[(✉)]

Indian Institute of Information Technology Kottayam, Kottayam, Kerala, India
shivendra15@alumni.iiitkottayam.ac.in, shajulin@iiitkottayam.ac.in
http://www.iiitkottayam.ac.in/shajulin.php

Abstract. Human emotion recognition is an imperative step to handle human computer interactions. It supports several machine learning based applications, including IoT cloud societal applications such as smart driving or smart living applications or medical applications. In fact, the dataset relating to human emotions remains as a crucial pre-requisite for designing efficient machine learning algorithms or applications. The traditionally available datasets are not specific to the Indian context, which lead to an arduous task for designing efficient region-specific applications. In this paper, we propose a new dataset that reveals the human emotions that are specific to India. The proposed dataset was developed at the IoT Cloud Research Laboratory of IIIT-Kottayam – the dataset contains 395 clips of 44 volunteers between 17 to 22 years of age; face expressions were captured when volunteers were asked to watch a few stimulant videos; the facial expressions were self annotated by the volunteers and they were cross annotated by annotators. In addition, the developed dataset was analyzed using ResNet34 neural network and the baseline of the dataset was provided for future research and developments in the human computer interaction domain.

Keywords: Human computer interaction · Affective computing · Human emotions · Facial expression recognition

1 Introduction

Perceptual processing of facial expressions plays a major role while expressing emotions. Emotion exercises a powerful influence on human action [5]. A better emotion recognition, obviously, brings better expression to life in animated movies and *animojis*. In human computer interaction, therefore, deciding the course of action heavily depends on emotions. In fact, the identification of emotion is applied in several domains such as computer vision, artificial Intelligence and emotional science; it could help in medical science [17], it could automate the selection of the best suited music for user's current mood and it could also provide a better music visualization than animated patterns [4].

S. M. Thampi et al. (Eds.): SIRS 2019, CCIS 1209, pp. 150–162, 2020.
https://doi.org/10.1007/978-981-15-4828-4_13

Ekman and Friesen in his research work presented six basic emotions such as *Anger, Disgust, Fear, Happiness, Sadness, Surprise* that are universally present among humans irrespective of cultural differences [6]. But, due to the differences in the physical appearance of people from different race there is a need for region specific databases. The authors, in their work, had identified six basic emotions based on their study on the isolated culture of people from the Fori tribe in Papua New Guinea in 1972. The tribe members were able to identify these six emotions on the pictures.

Precisely, a well-annotated media content of facial expressions is needed for training and testing machine learning algorithms which could be applied in recognition systems or IoT cloud applications. This paper focuses on developing a dataset that highlights the emotions in the Indian scenario. In this dataset, we have collected temporal and spatial expressions that are extracted from the recorded video when stimulant video was watched by volunteers. Temporal information is utilized to identify the emotions and capture more information when compared to the spacial information. We have disclosed an exploratory survey on the available databases. In addition, we have analyzed the importance of the proposed dataset by conducting experiments at the IoT Cloud research laboratory of our premise.

The rest of the paper is organized as follows: Sect. 2 reviewed the available databases in line to human emotion recognition; Sect. 3 described the details on the proposed database and its attributes; Sect. 5 explained about the experimental setup and the analysis report on the database; and, finally, Sect. 6 reports on the conclusion of the paper.

2   Literature Survey

Machine learning and data science domains, including IoT cloud domain [15, 16], in general, demand high quality datasets for accurate modeling or accurate predictions. Research works were carried out in the past to create human emotion datasets. For instances, a large label based dataset was developed through croud sourcing by [2] for representation learning in the facial emotion recognition domain. However, the images available in the dataset were only 48 * 48 in dimension. Similarly, in 1998, a Japanese database [12] of 213 images was developed with 7 different emotions by 10 posers.

Recently, Cohn-Kanade's AU-coded facial expression database [10] has become quite popular among researchers. It is a French database which contains 486 clips by 97 posers. Later, they have released it's extension known as CK+ database which contains 593 clips of 123 posers. The database has also provided a baseline for future research and for pursuing comparative studies.

In the meantime, Affect Net has emerged as one of the widely known dataset among researchers in recent years. This database contains over 1 million images [13]. The authors of Affect Net have collected images by querying three major search engines using 1250 keywords relating to emotions in six different languages. Out of 1 million images, 40% of images are manually annotated and they are annotated by applying machine learning based training model on the manually annotated images.

A few researchers have attempted to create an online-based database – Daniel et al. [11] recorded clips of emotions over internet. The participants were subjected to watch commercial products and the emotions were captured through webcams. This particular database contains 242 facial videos. In addition, there were efforts in the past to capture emotions using physiological signals [9] and audio signals [3].

Region specific datasets are crucial for several machine learning based applications in order to achieve quality results. For instance, Happy et al. [7] have presented an Indian dataset named as Indian Spontaneous Expression Dataset (ISED) – ISED bestowed lots of emphasis on the spontaneity of emotions; it contained near frontal face clips of recorded videos of 50 participants (29 male, 21 female) when exposed to stimulant videos; ISED utilized clips rather than a single image, which helped to capture temporal as well as spatial sequence of changes in behaviour, while capturing emotions. ISED dataset is more relevant to our work. However, the ISED dataset has only four emotions.

Our proposed iSAFE dataset, in this paper, provides temporal information and contains the required 6 basic emotions for facial recognitions. The dataset is generated and stored in databases which could be collected over internet via. http protocol.

3   iSAFE - Dataset Creation Approach

This section describes the approach of creating iSAFE – i.e., the protocols involved, stimulant video utilized, and the volunteers involved (see Fig. 1).

In order to capture high quality videos, proper lighting conditions were arranged in the recording room. Hearing devices were provided so that volunteers could capture intense details of stimulant clips. The stimulant clips were played on a laptop, which was kept in front of the volunteer, and the camera was placed behind it in order to capture the emotions.

3.1   Camera

Nikon D5000 camera with auto ISO and exposure settings was utilized for recording videos. The video was recorded at high resolutions. For instance, in our experiments, we have utilized a 1920×1080 resolution with the rate of 60 frames per second.

3.2   Protocols

Emotions were captured using the above mentioned camera while setting up an open environment rather than a closed laboratory. Before exposing stimulant videos to the volunteers, some general conversations were done by the experimenter who knows the subject well in order to make volunteers to feel comfortable with the exercises. In fact, the volunteers were not monitored or interrupted while watching the stimulant clips.

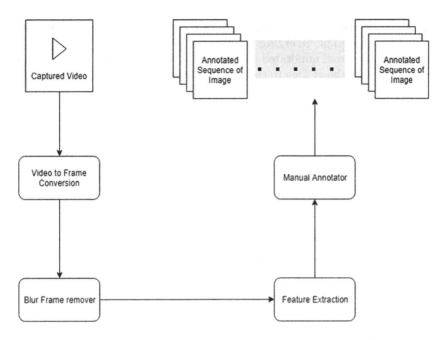

Fig. 1. Data pre-processing and annotation steps for iSAFE

3.3 Stimulant Video

Videos for inducing emotions were carefully selected and tested before the experiments were conducted by an experienced person. They were downloaded from various sources on internet and emotion inducing parts were trimmed and merged together. However it was experienced that stimulant videos were not much effective in stimulating sadness and anger as they were so for happiness, disgust, fear and surprise.

3.4 Volunteers

All volunteers were between 17 to 22 years of age. Out of 44 volunteers, 25 volunteers were male and 19 were female. They were educated about the six basic emotions before recording the emotions. At the end of the recording, consent were obtained from them in order to utilize the videos for the research and education purposes. To do so, they were asked to sign in the "Data Usage Agreement" (DUA) document. All the volunteers were fit and healthy while recording the emotions. However, no formal medical tests were conducted to the volunteers for the experiments.

4 iSAFE - Data Pre-processing and Annotations

In this section, we have described the data pre-processing stages and the annotation approaches carried out during the process of creating iSAFE database.

All the clips are split from the recorded video of volunteers while watching a few stimulant clips. The emerging facial expressions are self annotated by the volunteers and are cross annotated by an annotator. Both of the annotations are entered into the database. Due to the presence of sequence of images of emotions, the movement of facial muscles provides better classifications over time. A brief description of the iSAFE database is provided in Table 1.

Table 1. Brief description of iSAFE database

Number of clips	395
Number of volunteers	44
Age	17–22 years
Emotion labels	Happy
	Sad
	Surprise
	Disgust
	Fear
	Anger
	Uncertain
	No-Emotion
Number of annotator	2 + 1 (self-annotation)
Race and ethnicity	Indo-Aryan and Dravidian (Asian)

The sequence of emotions was carefully annotated in iSAFE database. In fact, the frames, at the start of the captured video, would not provide sufficient emotions when compared to the last frames. In addition, the feature extractions might lead to inaccuracy due to the presence of objects such as hand, spectacles, and so forth. In short, the sequence of extracted emotions was carefully annotated in the iSAFE database (as shown in Fig. 2).

4.1   Conversion of Recorded Video to Frames

The recorded videos of volunteers are trimmed out to collect the regions where the presence of expressions are detected. Later, these trimmed video clips are converted to images by taking each frame out of the video clips. Python codes were written in order to convert video clips to frames.

4.2   Removal of Blurry Frames

To ensure the quality of database, blurred and shaky frames were removed from the database. For removing the blur in an image, there exists a few techniques [14]. Out of those techniques, fast fourier transform and laplacian of image performs well according to the literature. Hence, we have utilized the variance of

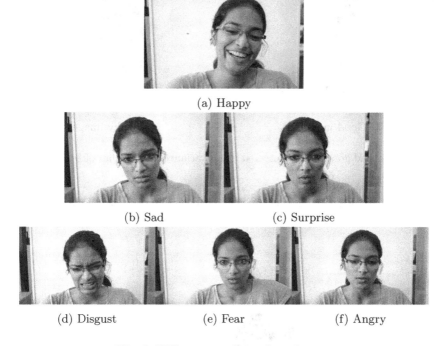

(a) Happy

(b) Sad   (c) Surprise

(d) Disgust   (e) Fear   (f) Angry

Fig. 2. Different emotions of a volunteer

laplacian of image to find out the blurred frame as it provides single floating point value to represent the blur in a given image. If the variance of image reaches below a certain threshold, the image is considered to be shaky or blurry; otherwise, the image is graded as a good image (see Fig. 3).

Laplacian approach focuses on the regions of a picture containing the rapid change in the intensity and sometimes it is utilized for edge detections. Our assumption here is that if an image has a high variance, then there is a wide spread of responses. If there is very low variance, then there is a tiny spread of responses, indicating that there are very less edges in the image. This is due to the fact that the more an image is blurred, the lesser the edges are present.

4.3   Feature Extraction

With visual data, the extracted features of images are also provided with the database. Features from the images are extracted using openface [1]. The features that were extracted from the images include a subset of facial Action Units (AUs), intensity of AUs, 45 Facial landmarks in 2D and 3D, head pose and Eye gaze movement data and so forth.

The volunteers involved in creating the iSAFE database are aware that they would be filmed for extracting required features. However, they are not instructed

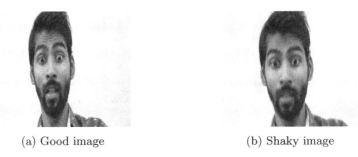

(a) Good image   (b) Shaky image

Fig. 3. Blurred and good image detected using variance of laplacian of images method

to act. Hence, iSAFE database is considered to be a Semi-Acted Feature Extraction based Indian dataset.

Fig. 4. GUI for manual annotations

The entire dataset is available for future reference and research works. Interested candidates could sign an End User License Agreement (EULA) and obtain the dataset through the github resource links of our lab.

4.4   Manual Annotation

Manual annotation of the database is done by a trained annotator and by the volunteers itself. The trained annotator, in our approach, is the psychologist of our premise who is trained in assessing the human emotions.

In order to identify the emotions experienced by the volunteer, they were asked to self-annotate these data. The self-annotation was carried out by an user-interface portal as depicted in Fig. 4.

As the assessment of emotion plays an important role in the success of creating database, the annotations were carried out very carefully. In addition, the annotation score of all the annotators are available in the database.

5   iSAFE Analysis and Baseline Creation

We have also analysed the database and proposed a baseline for the classification of images into various emotions. In general, convolutional neural networks (CNN) are the state of the art algorithms for image classification. We have utilized ResNet34 (Residual Networks with 34 layers) convolutional neural network for the evaluations.

In succinct, ResNet34 explores the bigger parameter space of images in a deep mode in order to solve the vanishing gradiant problems. To do so, convolution and pooling steps are iteratively targeted on the input images [8].

For the baseline in iSAFE, we have removed some images from each sub-sessions. And, multiple images from the same session had provided a good augmentation with some small variation in images (see Fig. 6).

Fig. 5. Masked image

(a) Frame 1 of same sub-(b) Frame 10 of same sub-
session   session

Fig. 6. Different frame from same sub-session (Augmented data)

5.1   Training, Testing and Validation Dataset

Three annotations are provided for each sub-session which includes self annotation, annotation from trained professional and annotation from trained unprofessional. The agreement between self annotation and trained professional is shown in Fig. 7 and agreement between trained unprofessional and trained professional is shown in Fig. 8. The agreements were represented as confusion matrices.

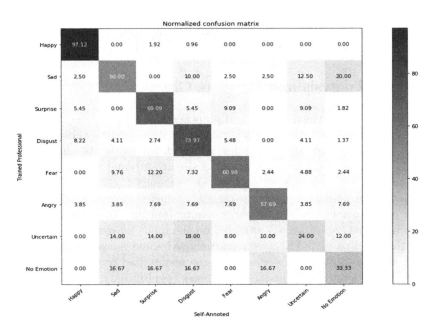

Fig. 7. Agreement between self annotation and trained professional

From the analysis we can find out that there is xx% agreement between self annotation and professional while xx% agreement between unprofessional and professional. Notably, annotating happy emotion observed 94.23% agreement between trained and un-trained professional; 97.12% agreement between trained and self-annotated emotions. Uncertain class of annotation was not agreed upon effectively – 40% agreement between trained and untrained professional, and 24% agreement between trained and self-annotated professionals.

Distribution of processed images used for training was skewed towards happy emotions so that we could make the distribution in a even manner. In addition, we have dropped some of the augmented frames of happy emotions in order to reduce the bias. Uncertain class was creating confusion while training models so we had dropped that class too.

Initially, we started training on images without cropping out faces. However, it ended up with very poor results – i.e., we observed around 30% accuracy for the training models. Subsequently, we have cropped out faces from the image

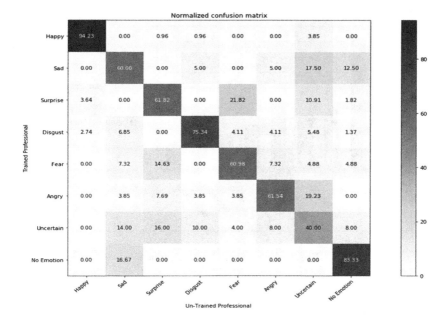

Fig. 8. Agreement between trained and untrained professional

and masked the unnecessary area with black masks as shown in Fig. 5. By this approach, we could improve the accuracy and drop the anticipated losses.

The processed dataset for evaluation is divided randomly into 60%, 30%, and 10% for training, validation and testing respectively.

5.2   Evaluation Metrics

For multi class classification, the predicted class labels should perfectly match with actual class labels. Accuracy score gives a closer insight towards the performance of model. It performs well on class balanced dataset. For binary classification, accuracy score can be defined as

$$AccuracyScore = \frac{TP + TN}{TP + TN + FP + FN} \tag{1}$$

where, TP equals True Positives; TN equals True Negatives; FP equals False Positives; and, FN equals False Negatives.

We are using the *AccuracyScore* for evaluating the proposed model. Figures 9 and 10 illustrate the skewness of data with class imbalance dataset and without class imbalance datasets.

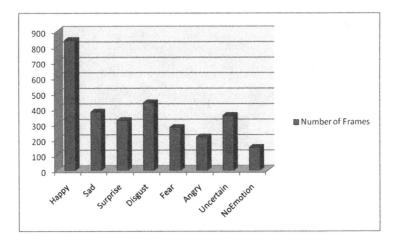

Fig. 9. Data with class imbalance

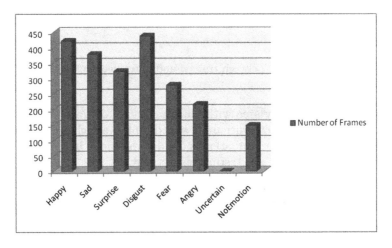

Fig. 10. Data without class imbalance

5.3   ResNet Analysis Results

Our Model took around 50 min on an Nvidia K80 Tesla GPU with 12 GB memory and reached to 98.32 % of accuracy on validation dataset; and, 92% on test dataset.

6   Conclusions

Emotion recognition using facial expressions by computer needs a quality facial expression databases with reliable annotations. The proposed iSAFE fulfills a number of aspects of the desired requirements. It has mild to strong facial expressions, and it's quality build can help researchers to develop algorithms for the recognition of human emotions in practical situations – especially, in emerging applications such as IIoT applications. Several protocols and strategies are adopted for the creation of the database which have been briefly described in the paper. Stimulant clips were presented to keep the subject engaged and to induce spontaneous emotions. Due to the presence of large number of volunteers and the variation of expression across time, several expressions of the same participants are included in iSAFE. The recorded videos were further processed and trimmed into small videos and further converted to sequence of images. Blurry frames are also removed from the sequence. The video clips of the database are annotated carefully by annotators and by the volunteers itself.

In data science, the amount of data provides direct advantage in availing concrete results and leverages to train model better. iSAFE is still very small in size; it requires more recordings for appending to the database. Some evaluation of database needs to be carried out in order to provide reference evaluation results for researchers. And, an increase in the number of annotations will help the database to become more reliable.

Acknowledgement. The authors thank IIIT Kottayam officials for granting space and support in order to carry out this research work at IoT Cloud research lab of IIIT Kottayam.

References

1. Baltrusaitis, T., Zadeh, A., Lim, Y.C., Morency, L.: OpenFace 2.0: facial behavior analysis toolkit. In: 2018 13th IEEE International Conference on Automatic Face and Gesture Recognition (FG 2018) (FG), pp. 59–66, May 2018
2. Barsoum, E., Zhang, C., Ferrer, C.C., Zhang, Z.: Training deep networks for facial expression recognition with crowd-sourced label distribution. In: ACM International Conference on Multimodal Interaction (ICMI) (2016)
3. Busso, C., et al.: IEMOCAP: interactive emotional dyadic motion capture database. Lang. Resour. Eval. **42**(4), 335 (2008)
4. Chen, C.-H., Weng, M.-F., Jeng, S.-K., Chuang, Y.-Y.: Emotion-based music visualization using photos. In: Satoh, S., Nack, F., Etoh, M. (eds.) MMM 2008. LNCS, vol. 4903, pp. 358–368. Springer, Heidelberg (2008). https://doi.org/10.1007/978-3-540-77409-9_34
5. Dolan, R.J.: Emotion, cognition, and behavior. Science **298**(5596), 1191–1194 (2002)
6. Ekman, P.: Universals and cultural differences in facial expressions of emotion, pp. 207–283 (1971)
7. Happy, S.L., Patnaik, P., Routray, A., Guha, R.: The Indian spontaneous expression database for emotion recognition. IEEE Trans. Affect. Comput. **8**(1), 131–142 (2017)
8. He, K., Zhang, X., Ren, S., Sun, J.: Deep residual learning for image recognition. CoRR, abs/1512.03385 (2015)
9. Koelstra, S., et al.: DEAP: a database for emotion analysis; using physiological signals. IEEE Trans. Affect. Comput. **3**(1), 18–31 (2012)

10. Lucey, P., Cohn, J.F., Kanade, T., Saragih, J., Ambadar, Z., Matthews, I.: The extended Cohn-Kanade dataset (CK+): a complete dataset for action unit and emotion-specified expression. In: 2010 IEEE Computer Society Conference on Computer Vision and Pattern Recognition - Workshops, pp. 94–101, June 2010
11. McDuff, D., Kaliouby, R., Senechal, T., Amr, M., Cohn, J.F., Picard, R.: Affectiva-MIT facial expression dataset (AM-FED): naturalistic and spontaneous facial expressions collected "In-the-Wild". In: 2013 IEEE Conference on Computer Vision and Pattern Recognition Workshops, pp. 881–888, June 2013
12. Kamachi, M., Gyoba, J., Lyons, M.J., Akemastu, S.: Coding facial expressions with gabor wavelets, pp. 200–205 (1998)
13. Mollahosseini, A., Hassani, B., Mahoor, M.H.: AffectNet: a database for facial expression, valence, and arousal computing in the wild. CoRR, abs/1708.03985 (2017)
14. Pech-Pacheco, J.L., Cristobal, G., Chamorro-Martinez, J., Fernandez-Valdivia, J.: Diatom autofocusing in brightfield microscopy: a comparative study. In: Proceedings 15th International Conference on Pattern Recognition, ICPR 2000, vol. 3, pp. 314–317, September 2000
15. Gowtham, N., Benedict, S., Giri, D., Sreelakshmi, N.: Real time water quality analysis framework using monitoring and prediction mechanisms. In: IEEE CiCT2018, pp. 1–6 (2018). https://doi.org/10.1109/INFOCOMTECH.2018.8722381
16. Ajith, S., Kumar, S., Benedict, S.: Application of natural language processing and IoTCloud in smart homes. In: Proceedings of IEEE-ICCT2019 (2019)
17. Tivatansakul, S., Ohkura, M., Puangpontip, S., Achalakul, T.: Emotional healthcare system: emotion detection by facial expressions using Japanese database. In: 2014 6th Computer Science and Electronic Engineering Conference (CEEC), pp. 41–46, September 2014

Deep Learning Approach for Intelligent Named Entity Recognition of Cyber Security

K. Simran[1(✉)], S. Sriram[1], R. Vinayakumar[1,2], and K. P. Soman[1]

[1] Center for Computational Engineering and Networking,
Amrita School of Engineering, Amrita Vishwa Vidyapeetham, Coimbatore, India
simiketha19@gmail.com, sri27395ram@gmail.com
[2] Division of Biomedical Informatics,
Cincinnati Children's Hospital Medical Centre, Cincinnati, OH, USA
Vinayakumar.Ravi@cchmc.org, vinayakumarr77@gmail.com

Abstract. In recent years, the amount of Cyber Security data generated in the form of unstructured texts, for example, social media resources, blogs, articles, and so on has exceptionally increased. Named Entity Recognition (NER) is an initial step towards converting this unstructured data into structured data which can be used by a lot of applications. The existing methods on NER for Cyber Security data are based on rules and linguistic characteristics. A Deep Learning (DL) based approach embedded with Conditional Random Fields (CRFs) is proposed in this paper. Several DL architectures are evaluated to find the most optimal architecture. The combination of Bidirectional Gated Recurrent Unit (Bi-GRU), Convolutional Neural Network (CNN), and CRF performed better compared to various other DL frameworks on a publicly available benchmark dataset. This may be due to the reason that the bidirectional structures preserve the features related to the future and previous words in a sequence.

Keywords: Information extraction · Named Entity Recognition · Cyber Security · Deep Learning

1 Introduction

Technology has evolved tremendously in the previous couple of decades making diverse online web sources (news, forums, blogs, online databases, etc.) available for common use. Newfound information regularly seems first on these kinds of online web sources in unstructured text format. Mostly, it requires a ton of time, sometimes years, for this information to properly classify into a structured format. However, cases related to Cyber Security does not have that kind of time, so timely extraction of this information plays an imperative job in numerous applications. Information modeling of cyber-attacks can be used by many individuals such as auditors, analysts, researchers, etc. [1]. At the core, knowledge

© Springer Nature Singapore Pte Ltd. 2020
S. M. Thampi et al. (Eds.): SIRS 2019, CCIS 1209, pp. 163–172, 2020.
https://doi.org/10.1007/978-981-15-4828-4_14

extraction task is recognition of named entities of a domain, such as products, versions, etc. The best accuracy given by the current Named Entity Recognition (NER) devices, in this area of Cyber Security, depend on feature engineering. However, feature engineering has many issues, for example, depending on look-ups or dictionaries to recognize known entities [2]. With profoundly dynamic fields such as Cyber Security, it is difficult to fabricate as well as harder to keep up the word references and in turn, consumes a lot of time for creating these NER tools. The limitation of these tools is that they are domain-specific and cannot accomplish good accuracy when applied to any other domain. These issues can be overcome by using Part of Speech (POS) tagging. POS tagging is a task of identifying the words as nouns, adjectives, verbs, adverbs, and so on (Distributed Denial of Service (DDoS) is mostly taken as a noun). POS tagging can also diminish the time delay in this domain.

Conditional Random Fields (CRFs) emerged in recent years as the most successful method for entity extraction. CRFs are best for tagging and sequential data as they can predict sequences of labels for a set of input samples, taking the context of the input data into account. Mikolov et al. [7] proposed a method called word2vec which converts every word in a corpus into a low dimensional dense vector representation. These vectors can reflect the semantic relationship between the words which was not possible in classical one-hot vectors. For example, the difference between the vectors representing the words 'king' and 'queen' is similar to the difference between the vectors representing the words 'man' and 'woman'. These relationships result in the clustering of semantically similar words in the vector space.

Another leap forward in the ongoing years is the Deep Learning (DL) field. DL is an enhanced classical neural network model which naturally learns non-linear combinational features on its own whereas classical methods, for example, CRFs can just learn linear combinations of the defined features. This reduces the user work of tedious feature engineering [1]. The expansion in information accessibility, increase in hardware processing power particularly GPUs, different activation functions, etc. has made DL exceptionally practical and feasible. Recent days, the applications of DL have been leverage towards various applications in the field of Cyber Security [3–6].

The hybrid model of Long Short Term Memory (LSTM) and CRF architecture proposed by Lampal et al. [8] for NER is a blend of LSTM, CRFs, and word2vec model. As this architecture is domain and entity type agnostic, it very well can be connected to any area as long as the input is an annotated corpus of the same format as the CoNLL-2000 dataset. This architecture is applied to the domain of Cyber Security NER in [9]. As the corpus of Cyber Security is not widely available, they utilized the corpus created in the work done by Bridges et al. [1] for training the model. The author compared the performance of LSTM-CRF with CRFSuite and LSTM-CRF beat CRFSuite with two percent accuracy.

The major contributions are:

1. This work proposes a deep learning based framework for NER in Cyber Security.
2. Various deep learning architectures are evaluated on well known and benchmark data set.
3. The importance of non-linear text representation is discussed in the context of Cyber Security NER.

The remaining paper is arranged as follows. Section 2 documents the related works followed by background related to deep learning architectures in Sect. 3. Section 4 describes the details of the data set used and Sect. 5 provides a description of proposed architecture. Section 6 reports the experiments, results, and observation whereas Sect. 7 will conclude this paper with remakes on future work of research.

2   Related Works

A lot of efforts are made for automatically labeling Cyber Security entities in the past years. Concepts in Common Vulnerability Enumeration (CVE-3) were annotated to instantiate a security ontology in [10]. Reuters [11] likewise did with the assistance of his web journals and OpenCalais. Mulwad et al. [12] searched the web to identify security relevant text and then train an SVM classifier to identify the descriptions of potential vulnerability. Afterward utilizing Open-Calais alongside the taxonomy of Wikipedia to identify and classify vulnerability entities. In spite of the fact that these sources rely upon standard entity recognition software, they are not prepared to recognize domain specific ideas. This is because of the general nature of their training data [13,14]. Joshi et al. [15] noted comparative discoveries where NERD system [16], Open-Calais and Stanford Named Entity Recognizer were all commonly helpless to recognize Cyber Security domain entities. This is due to the fact that these tools do not utilize domain-specific corpus. Stanford NER system with hand-labeled domain specific training corpus was created in [15] and delivered better outcomes for the greater part of their domain-specific entities.

Joshi et al. [15] additionally performed hand annotating for a small corpus. This small corpus was then encouraged to Stanford NER "off-the-shelf" layout for train a CRF entity extractor [13]. Their work has recognized the equivalent Cyber Security issue yet they didn't address a more broad issue of automatically the labeling process.

Semi-supervised methods are additionally utilizing entity extraction rather than supervised methods as they function admirably with very small training data also. [17] have used semi-supervised methods for entity extraction and have got beneficial outcomes. A bootstrapping algorithm created by McNeil et al. [18] depicts a prototypical execution to extract data about vulnerabilities and exploits. [19] utilizes known databases for seeding bootstrapping strategies. [20] developed an exact strategy to automatically label text from different data sources and then provide public access to a data set which is annotated with

Cyber Security entities. [21] proposed an algorithm for extracting security entities and their relationships using semi-supervised NLP with a small amount of input data. The precision of 0.82 was achieved using small corpus. Either rule-based or machine learning [22] approaches are utilized for NER and regularly the two are mixed [23]. Rule-based is a blend of lookups and rules for pattern matching that is hand-coded by an area expert. These rules use the contextual information of the entity to decide if candidate entities are substantial or not. Machine learning based NER approaches utilize an assortment of models, for example, Maximum Entropy Models [24], Hidden Markov Models (HMMs) [25], CRFs [26], Perceptrons [27], Support Vector Machines (SVMs) [28], or neural networks [2]. The best NER approaches incorporate those dependent on CRFs. CRFs address the NER issue utilizing an arrangement labeling model. In this model, the label of an entity is demonstrated as reliant on the labels of the previous and following entities in a predefined window.

As deep neural networks address many short comes of classical statistical technique such as feature engineering, they are considered as a potential alternative [29]. As feature are learned automatically by the neural network, they decline the measure of human endeavors required in various applications. Significant enhancements in accuracy in neural network feature with respect to hand-engineered features are shown by empirical outcomes over a wide arrangement of domains. Recurrent Neural Networks (RNNs) property of having long time memory as they process input with variable length results in outstanding accomplishment with several NLP tasks [30]. In [31], LSTM property to permit the learning between arbitrary long-distance dependencies enhances its performance over RNN.

In [32], LSTM-CRF technique was applied to the issue of NER in the Cyber Security domain utilizing the corpus made accessible by Joshi et al. [15]. Given an annotated corpus of a decent size, it was observed that this technique outflanks the state of the art methods given an annotated corpus of a decent size.

3   Background

3.1   Convolutional Neural Network (CNN)

CNN is very famous network as it uses less learnable parameters. This network consists of three types of layers which are convolution layers, pooling layers, and fully connected layers. In the convolution layer of this neural network, convolution operation is performed using several filters which slide through the input and learns the features of the input data. The pooling layer is utilized to decrease the size of the previous layer. Min pooling, max pooling, and average pooling are the different types of pooling layers available. The network is connected with a dense layer at the end so that the features can be mapped to the classification.

3.2   Recurrent Neural Network (RNN)

RNN is a very important deep learning architecture which can handle sequential information whereas feedforward neural networks fail at doing this. That's why

RNN is utilized for applications related to sequential data, for example, speech data as it can learn multi-variation like PoS tagging [33] using the hidden units present in the recurrent cells. The issue of exploding or vanishing gradients is the main problem with RNN.

3.3 Long Short-Term Memory (LSTM)

LSTM was introduced to solve the problem of vanishing gradient in RNN. It selectively remembers the patterns for a long duration of time. The error that will be backpropagated through layers is preserved which is a big help to LSTM. Because of the three gates in LSTM, it has the ability to handle long term as well as short term information whereas RNN can handle only short term data. The cell takes decisions about when to allow read, write, etc. operations as well as what to store and erase.

3.4 Gated Recurrent Unit (GRU)

GRU is an extended version of standard RNN and are also considered as a minor variation from LSTM. An LSTM without an output gate is basically called as GRU. It has update gate and reset gate which chooses what data should be passed through them. LSTM can easily perform unbounded counting, so in this way, it is stronger than GRU. That's why the LSTM is able to learn simple languages whereas GRU fails at this.

3.5 Bidirectional Architectures

To understand the context better and resolve ambiguities in the text, bidirectional recurrent structures are utilized which learns the information from the previous and future time stamps. The two types of connections are one going forward in time and the other going backward in time. These connections help in learning the previous and future representations respectively. The module in this bidirectional recurrent structure could be an RNN, LSTM or GRU.

4 Description of the Data Set

Auto-labeled Cyber Security domain text corpus provided by Bridges et al. [1] comprising of around 40 entity types was used in this work. All the corpora is stored in a single JSON file and high-level JSON element are used to represent each corpus. Each word in these corpora is auto-annotated with an entity type. The files were additionally changed over into CoNLL2000 format in order to feed the data into the model. Every word was annotated in a separate line and separation between the three corpora were removed. When the preprocessing is completed, each line in the data set consists of a word as well as its entity type. The data set was divided into three subsets that is training, validation, and testing consisting of 70%, 10%, and 20% data respectively.

5   Proposed Architecture

The proposed architecture for Cyber Security named entity recognition is shown in Fig. 1. The architecture has three different sections, namely input, feature generation, and output. In the input layer, the text data is converted into numerical features using Keras embedding. These numerical features are high dimensional in nature so they are passed into a bidirectional GRU layer. Bidirectional GRU layer refines the vectors and feeds them to a CNN layer. CNN network generates more optimal features and feeds it to a CRF layer for enhanced learning. This final layer uses the Viterbi algorithm to generate the most possible tag for the input word.

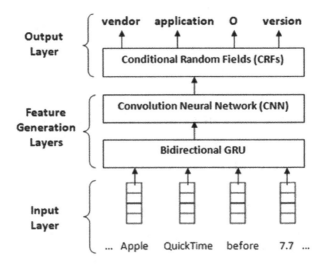

Fig. 1. Proposed architecture.

6   Experiments, Results, and Observations

TensorFlow[1] with Keras[2] were used to implement all the deep learning architectures. All the models are trained on GPU enabled TensorFlow. Various statistical measures are utilized in order to evaluate the performance of the proposed classical machine learning and deep learning models.

Initially, to identify the optimal architecture for auto-labeled Cyber Security domain text corpus we proposed an architecture which has been explained in Sect. 5 and we compared the performance of the proposed method with the

---

[1] https://www.tensorflow.org/.
[2] https://keras.io/.

architectures already published. Various deep learning architectures were implemented and their performance was compared with Bidirectional GRU+CNN-CRF. Bidirectional GRU+CNN-CRF outperformed all other architectures. LSTM is computationally expensive whereas GRU is computationally inexpensive which can lead to the same performance in some cases and some cases the performance of the GRU is better than LSTM.

The training dataset was used to train the models and these trained models were tested using the testing data set. LSTM, RNN, GRU, Bidirectional GRU contains 128 hidden units. Embedding dimension of 128, learning rate of 0.005, and Stochastic Gradient Descent (SGD) learning method are used in all the experiments. The models were compared in terms of precision, recall, and f1-score and the detailed results are reported in Table 1. The performance metrics in terms of precision, recall, and f1-score shows that Bidirectional GRU+CNN-CRF outperformed all other architectures.

Table 1. Average weighted performance metrics for all entity types.

Model	Precision	Recall	F1-score
LSTM-CRF [9]	85.3	94.1	89.5
CRF [9]	82.4	83.3	82.8
CNN-CRF	83.1	93.9	88.2
RNN-CRF	83.5	85.6	84.5
GRU-CRF	86.5	95.7	90.9
Bidirectional GRU-CRF	88.7	95.4	91.9
Bidirectional GRU+CNN-CRF	**90.8**	**96.2**	**93.4**

7   Conclusion and Future Work

This work evaluates the efficacy of various deep learning frameworks along with CRF model for NER Cyber Security text data and Bidirectional GRU+CNN-CRF model performed better than other deep learning architecture. The best part about the proposed architecture is that it does not require any feature engineering. This work can be extended for Relation Extraction (RE). RE basically attempts to find occurrences of relations among entities. Product vulnerability description can be better understood by using RE. For instance, RE can be used to identify and differentiate the product that is being attacked and the product that is meant to attack by using the information extracted from the product vulnerability description.

Acknowledgment. This research was supported in part by Paramount Computer Systems and Lakhshya Cyber Security Labs. We are grateful to NVIDIA India, for the GPU hardware support to research grant. We are also grateful to Computational Engineering and Networking (CEN) department for encouraging the research.

References

1. Bridges, R.A., Jones, C.L., Iannacone, M.D., Testa, K.M., Goodall, J.R.: Automatic labeling for entity extraction in cyber security. arXiv preprint arXiv:1308.4941 (2013)
2. Collobert, R., Weston, J., Bottou, L., Karlen, M., Kavukcuoglu, K., Kuksa, P.: Natural language processing (almost) from scratch. J. Mach. Learn. Res. **12**, 2493–2537 (2011)
3. Vinayakumar, R., Soman, K.P., Poornachandran, P.: Detecting malicious domain names using deep learning approaches at scale. J. Intell. Fuzzy Syst. **34**(3), 1355–1367 (2018)
4. Vinayakumar, R., Soman, K.P., Poornachandran, P.: Evaluating deep learning approaches to characterize and classify malicious URL's. J. Intell. Fuzzy Syst. **34**(3), 1333–1343 (2018)
5. Vinayakumar, R., Alazab, M., Jolfaei, A., Soman, K.P., Poornachandran, P.: Ransomware triage using deep learning: Twitter as a case study. In: 2019 Cybersecurity and Cyberforensics Conference (CCC), pp. 67–73. IEEE, May 2019
6. Vinayakumar, R., Soman, K.P., Poornachandran, P., Akarsh, S.: Application of deep learning architectures for cyber security. In: Hassanien, A., Elhoseny, M. (eds.) Cybersecurity and Secure Information Systems. ASTSA, pp. 125–160. Springer, Cham (2019). https://doi.org/10.1007/978-3-030-16837-7_7
7. Mikolov, T., Sutskever, I., Chen, K., Corrado, G.S., Dean, J.: Distributed representations of words and phrases and their compositionality. In: Advances in Neural Information Processing Systems, pp. 3111–3119 (2013)
8. Lample, G., Ballesteros, M., Subramanian, S., Kawakami, K., Dyer, C.: Neural architectures for named entity recognition. arXiv preprint arXiv:1603.01360 (2016)
9. Gasmi, H., Bouras, A., Laval, J.: LSTM recurrent neural networks for cybersecurity named entity recognition. In: ICSEA 2018, vol. 11 (2018)
10. More, S., Matthews, M., Joshi, A., Finin, T.: A knowledge-based approach to intrusion detection modeling. In: 2012 IEEE Symposium on Semantic Computing and Security Security and Privacy Workshops (SPW), pp. 75–81. IEEE (2012)
11. Reuters, T.: OpenCalais (2009)
12. Mulwad, V., Li, W., Joshi, A., Finin, T., Viswanathan, K.: Extracting information about security vulnerabilities from web text. In: Proceedings of the 2011 IEEE/WIC/ACM International Conferences on Web Intelligence and Intelligent Agent Technology, ser. WI-IAT 2011, vol. 03, pp. 257–260. IEEE Computer Society, Washington, DC (2011). https://doi.org/10.1109/WI-IAT.2011.26
13. Finkel, J.R., Grenager, T., Manning, C.: Incorporating non-local information into information extraction systems by Gibbs sampling. In: Proceedings of the 43rd Annual Meeting on Association for Computational Linguistics, ser. ACL 2005, pp. 363–370. Association for Computational Linguistics, Stroudsburg (2005). https://doi.org/10.3115/1219840.1219885
14. Tjong Kim Sang, E.F., De Meulder, F.: Introduction to the CoNLL 2003 shared task: language-independent named entity recognition. In: Proceedings of the Seventh Conference on Natural Language Learning at HLT-NAACL 2003, vol. 4, pp. 142–147. Association for Computational Linguistics (2003)
15. Joshi, A., Lal, R., Finin, T., Joshi, A.: Extracting cybersecurity related linked data from text. In: Proceedings of the 7th IEEE International Conference on Semantic Computing. IEEE Computer Society Press (2013)

16. Rizzo, G., Troncy, R.: NERD: a framework for unifying named entity recognition and disambiguation extraction tools. In: Proceedings of the Demonstrations at the 13th Conference of the European Chapter of the Association for Computational Linguistics, pp. 73–76. Association for Computational Linguistics (2012)

17. Huang, R., Riloff, E.: Bootstrapped training of event extraction classifiers. In: Proceedings of the 13th Conference of the European Chapter of the Association for Computational Linguistics, pp. 286–295. Association for Computational Linguistics (2012)

18. McNeil, N., Bridges, R.A., Iannacone, M.D., Czejdo, B., Perez, N., Goodall, J.R.: PACE: pattern accurate computationally efficient bootstrapping for timely discovery of cyber-security concepts. In: 2013 11th International Conference on Machine Learning and Applications (ICMLA). IEEE (2013)

19. Geng, J., Yang, J.: AUTOBIB: automatic extraction of bibliographic information on the web. In: Proceedings of the International Database Engineering and Applications Symposium, ser. IDEAS 2004, pp. 193–204. IEEE Computer Society, Washington, DC (2004). https://doi.org/10.1109/IDEAS.2004.14

20. Bridges, R.A., et al.: Automatic labeling for entity extraction in cyber security. arXiv preprint arXiv:1308.4941 (2013)

21. Jones, C.L., et al.: Towards a relation extraction framework for cyber-security concepts. In: Proceedings of the 10th Annual Cyber and Information Security Research Conference. ACM (2015)

22. Cimiano, P., Handschuh, S., Staab, S.: Towards the self annotating web. In: Proceedings of the 13th International Conference on World Wide Web, pp. 462–471 (2004)

23. Pantel, P., Pennacchiotti, M.: Automatically harvesting and ontologizing semantic relations. In: Proceedings of the 2008 Conference on Ontology Learning and Population: Bridging the Gap Between Text and Knowledge, Amsterdam, The Netherlands, pp. 171–195 (2008)

24. Chieu, H.L., Ng, H.T.: Named entity recognition: a maximum entropy approach using global information. In: Proceedings of the 19th International Conference on Computational Linguistics, vol. 1, pp. 190–196 (2002)

25. McCallum, A., Freitag, D., Pereira, F.C.N.: Maximum entropy Markov models for information extraction and segmentation. In: ICML, pp. 591–598 (2000)

26. McCallum, A., Li, W.: Early results for named entity recognition with conditional random fields, feature induction and web-enhanced lexicons. In: Proceedings of the Seventh Conference on Natural Language Learning at HLT-NAACL 2003, vol. 4, pp. 188–191 (2003)

27. Carreras, X., Màrquez, L., Padró, L.: Learning a perceptron-based named entity Chunker via online recognition feedback. In: Proceedings of the Seventh Conference on Natural Language Learning at HLT-NAACL 2003, vol. 4, pp. 156–159 (2003)

28. Isozaki, H., Kazawa, H.: Efficient support vector classifiers for named entity recognition. In: Proceedings of the 19th International Conference on Computational Linguistics, vol. 1, pp. 1–7 (2002)

29. Goldberg, Y.: A primer on neural network models for natural language processing. J. Artif. Intell. Res. (JAIR) **57**, 345–420 (2016)

30. Graves, A., Mohamed, A., Hinton, G.: Speech recognition with deep recurrent neural networks. In: IEEE International Conference on Acoustics, Speech and Signal Processing (ICASSP), pp. 6645–6649 (2013)

31. Gers, F.A., Schmidhuber, J.A., Cummins, F.A.: Learning to forget: continual prediction with LSTM. Neural Comput. **12**, 2451–2471 (2000)

32. Hochreiter, S., Schmidhuber, J.: Long short-term memory. Neural Comput. **9**(8), 1735–1780 (1997)
33. Collobert, R., Weston, J., Bottou, L., Karlen, M., Kavukcuoglu, K., Kuksa, P.: Natural language processing (almost) from scratch. J. Mach. Learn. Res. **12**(Aug), 2493–2537 (2011)

A Chronic Psychiatric Disorder Detection Using Ensemble Classification

V. J. Jithin, G. Manoj Reddy, R. Anand, and S. Lalitha$^{(\boxtimes)}$

Department of Electronics and Communication Engineering,
Amrita School of Engineering, Bengaluru, Amrita Vishwa Vidyapeetham,
Bengaluru, India
jithinvj06@gmail.com, gmanojreddy1996@gmail.com, anand.ramandran@gmail.com,
s_lalitha@blr.amrita.edu

Abstract. The objective of this work is to detect depression that is a more chronic psychiatric disorder found in humans using speech samples. This work is first of its kind in depression detection using Audio Visual and Emotional Challenge 2011 (AVEC 2011) and Computational and Paralinguistics challenge 2016 (ComParE 2016) feature sets. A novel method of ensemble classification using simple machine learning algorithms of Instance-Based classifier with parameter K (IBK), Stochastic Gradient Descent (SGD) and Random Forest is proposed for the projected task with gender dependent and independent systems. Experimental results demonstrate the superiority of ComParE 2016 over AVEC 2011 in determining the psychological state of an individual. Feature selection method is applied to reduce feature vector size, maintaining the accuracy of depression detection with that obtained using large size feature sets. The ensemble-based classification provide better accuracy performance than the individual classifiers.

Keywords: AVEC 2011 · ComParE 2016 · Ensemble voting classifier · IBK · Random Forest · SGD

1 Introduction

World Health Organization discerned that depression is a psychiatric aberration that leads to low spirit, loss of pleasure or interest, low energy levels, poor concentration and bad appetite [1]. Anxiety and depressive disorders are highly prevalent around the world. Different ways of depression detection are proposed in the past time. Many of the depression detection technique uses psychological and behavioral signals for characterizing depression. These include facial expression, body gestures, eye movement, speech, etc. [2]. Amongst these, speech is considered to be an effective method to detect depression since it is simple and non-invasive. Conventional method of depression detection using speech involve feature extraction followed by classification. The prosodic and acoustic features of the speech signals are extracted and fed into various classifiers viz Support

© Springer Nature Singapore Pte Ltd. 2020
S. M. Thampi et al. (Eds.): SIRS 2019, CCIS 1209, pp. 173–185, 2020.
https://doi.org/10.1007/978-981-15-4828-4_15

Vector Machines (SVM), Hierarchical Fuzzy Signature (HFS), Gaussian Mixture Models (GMM), Multilayer Perceptron (MLP), etc. [3]. These classifiers predict the speech samples as depressed or non-depressed. Literature reveals various attempts by research community in depression detection. Moore et al. achieved a classification accuracy from 87% to 94% using glottal waveform features for differentiating a patient group and control group [4]. Ooi used a major voting classifier on a parallel ensemble. Depression in adolescents was predicted using a multichannel weighted classifier, giving an accuracy of 73% [5]. Teager Energy Operator (TEO) and glottal features provide an accuracy of 69% for predicting major depression in adolescents in the work proposed by Ooi [6]. Ozdas et al. distinguished participants as suicidal, major depressed and healthy by investigating their glottal flow spectrum and vocal jitter using spectral slope getting an accuracy of 90% [7]. Alghowinem used different classifiers: Support Vector Machines (SVM), Hierarchical Fuzzy Signature (HFS), Gaussian Mixture Models (GMM), and Multilayer Perceptron (MLP) and found a hybrid combination of SVM and GMM giving better results [3]. In a proposed work by Yingthawornsuk [8], two speaking styles were used: a text-reading in and a clinical interview. On analyzing, it was found that depressed subjects possessed a slow rate of speech, monotonous delivery and a low range of fundamental frequency. Hence, they formulated small changes in characteristics of speech (e.g. rate of speaking, loudness, differences in pitch etc.) as indicators. Stress is one of the major factor that leads to psychological disorder. Narayanan et al. in their work [9] used the Gammatone frequency cepstral coefficients and multiclass SVM for stress detection. The non-linear TEO features classified using Probabilistic Neural Network (PNN) and the Multi-Layer Perceptron Neural Network (MLPNN) also achieved better performances. Lalitha et al. in [10] used Mel and Bark scale dependent perceptual features to detect emotion. The proposed model outperformed other models in terms of recognition accuracy. K Nearest Neighbor (KNN) and Simple Logistic Classifier (SL) provided an accuracy of 84.7% for affective state recognition in the work proposed by Murali et al. [11]. The Mel Frequency Cepstral Coefficients (MFCC), Perceptual Linear Predictive Cepstrum (PLPC), and Mel Frequency Perceptual Linear Prediction Cepstrum (MFPLPC) are found to play a prominent role for SER as proposed by Lalitha [12]. From the survey, it is evident that the performance of the individual classifiers proved to provide less accuracy performance comparatively. It is observed that there are setbacks in advanced warning techniques to acknowledge the depressed person. Due to less combinations and ensemble pruning methods, the performance of the systems are less efficient. This resulted in less accuracy performance between 60% to 70%. The ComParE feature set have been used in the automatic recognition of eating conditions in speech, that is whether the people are eating while speaking. The average recalls obtained using ComParE features witnessed above 90% [13]. The AVEC features set have been used in building artificial listener agents that can captivate a person in emotionally colored conversation [14]. This work contributes a method to detect the psychiatric disorder of an individual. The focus of this work is on investigation of feature sets i.e ComParE 2016 over

AVEC 2011 and classifier combinations for the detection of psychiatric disorder. A search to find the appropriate feature set and classifier is proposed here. The work also focuses on comparing the performance of the individual classifiers like IBK, SGD, Random Forest and an ensemble of these classifiers using the appropriate feature sets in determining the psychiatric state.

This work is organized in five sections. Section 2, explains the proposed work for depression detection and performance metrics used. In Sect. 3, experiment and analysis is given. Section 4 has a comparative analysis. Finally, Sect. 5 discusses the conclusion and future scope.

2   Proposed Work for Depression Detection Using Ensemble Classification Method

The proposed model for depression detection using ensemble classification consists of five stages - speech database, pre-processing, feature extraction, feature selection and classification. The flow chart of the proposed model is depicted in Fig. 1.

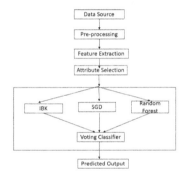

Fig. 1. Flow chart of proposed work

2.1   Speech Database

The database used in this work is DAIC-WOZ depression database. This database includes questionnaire responses and audio recordings, incorporated from a larger collection, the Distress Analysis Interview Corpus (DAIC). It comprises of 189 sessions of interactions of duration 7 to 33 min each. Speech samples were recorded from 53 speakers, of which 31 male and 22 female speakers were used [15].

2.2 Pre-processing

The audio of the interviewer and interviewee are separated initially. Audio file corresponding to each speaker (interviewee) are divided into segments of 10 s. This is performed with help of an audio tool Audacity [16]. A total of 1763 audio sub-samples were segmented from the samples of 53 interviewees. Features for each segment are extracted individually. Waveform of the depressed and non-depressed interviewee is depicted in Figs. 2 and 3 respectively. It can be observed that non-depressed voice sample has no large variations in amplitude levels while the depressed voice sample has variations in intensity by a large extent.

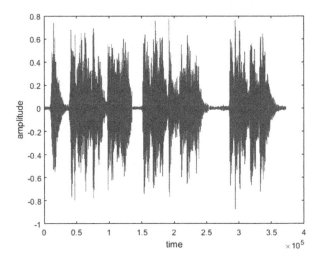

Fig. 2. Non depressed speech sample

2.3 Feature Extraction

The ComParE 2016 [17] and AVEC 2011 [18] features are extracted from the audio samples using openSMILE [19]. openSMILE is an open source software for automatic extraction of features. The software is likewise fit for perceiving the attributes of a given speech or music fragment. The Computational and Paralinguistics challengE 2016 (ComParE 2016) feature set consists of 6373 features. Energy, spectral, cepstral (mfcc), and voice related low level descriptors are few among those. These features come mainly under three groups. First group comprises of 4 energy LLDs, 55 spectral LLDs and its delta coefficients. Second group consists of 6 voice related LLDs and delta coefficients. A lot of 54 functionals is connected to LLDs of first group and 46 functionals to its delta coefficients,

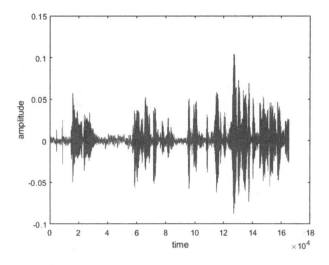

Fig. 3. Depressed speech sample

bringing about $59*(54+46) = 5900$ features. A lot of 39 functionals is connected to second group and its delta coefficients, bringing about $6*(39+39) = 468$ features. The third group comprises of five static descriptors, registered for voiced portions. Along these lines the final set comprises of $5900 + 468 + 5 = 6373$ features [13]. An insight of the ComParE 2016 feature set is mentioned in Table 1.

Table 1. Insight of ComParE 2016 features

Sl. no.	Group	Functionals on LLDs	Functionals on delta coefficients	ComParE features
1	4(energy) + 54(spectral) = 59 LLDs	54	46	5900
2	6(voice related) LLDs	39	39	468
3	5(temporal static descriptors)	–	–	5
4	Total Features			6373

The Audio Visual and Emotional Challenge 2011 (AVEC 2011) feature set comprises of 1941 highlights. It incorporates 25 spectral and energy related low-level descriptors, 6 voicing related LLDs, 25-delta energy/spectral LLD, 6 delta coefficients of the voice related LLDs and 10 voiced/unvoiced durational highlights [14]. An insight of AVEC 2011 feature set is mentioned in Table 2. These features are extracted for the entire 1763 audio samples and further fed into the classifiers.

Further the feature selection method is employed to select the relevant features.

Table 2. Insight of AVEC 2011 features

Sl. no.	Group	Functionals on LLDs	Functionals on delta coefficients	AVEC 2011 features
1	25 LLDs(energy and spectral)	42	23	1625
2	6(voicing related) LLDs	32	19	306
3	10(voiced/unvoiced durational features)	–	–	10
4	Total Features			1941

2.4 Feature Selection

Among the features obtained after feature extraction, some features contribute more to the prediction output than certain other features. These features needed to be selected and this helps reduce the complexity of the classification. Thus to select those relevant features, attribute selection is used. The feature selection is carried out using WEKA [20] in two parts, Attribute Evaluator and Search method. In this work, cfssubsetEval Attribute evaluator along with Best First Search method is used. cfssubsetEval is a correlation based feature selection subset evaluator. Different combinations of attributes are looked into and the best attributes are chosen. The Best First [21] search technique looks through the space of attribute subsets by slope climbing enlarged with a backtracking facility. Setting the quantity of continuous non-improving hubs controls the dimension of backtracking done. Best First strategy begins with either an unfilled set of attributes and inquiry forward or begin with full set of attributes and hunt in reverse, or begin anytime and look in both directions. After performing the attribute selection, classification of the samples into depressed and non-depressed are performed.

2.5 Classification

Classification is a significant data mining strategy with wide applications. It is used to organize everything in a set of data into one of predefined set of classes or groups. The classification is performed using SGD, IBK, Random Forest and an ensemble voting classifier in Weka.

SGD utilizes stochastic slope drop to learn models. Stochastic Gradient Descent is a steady calculation that can be connected to bigger datasets. It replaces every single missing values and transforms nominal attributes into binary ones. It replaces each and every missing values and changes nominal features into binary ones. It additionally standardizes all characteristics, so the coefficients in the yield depend on the standardized information. This ensures better prediction [22].

IBK is a case based classifier which finds the preparation case nearest in the Euclidean separation to the given test tests and predicts the comparable class as this preparation occurrence. It's a k-closest neighbor classifier which decides the quantity of closest neighbors naturally utilizing leave-one-out cross-validation or can be indicated in the object editor. A kind of different search algorithms can be utilized to quicken the assignment of finding the closest neighbors. A linear search is the default yet further choices fuse search trees, for example, KD-trees and ball trees. Predictions from more than one neighbor can be weighted by their distance from the test occasion and after that changing over the distance into a weight [23].

Random Forest is a combination of tree predictors and an ensemble learning method for classification. A multitude of decision trees are used for the classification. Each individual tree performs the class prediction from which the final class of the Random Forest classifier is predicted using a voting classifier of the individual classifiers. Random Forest provides a powerful and effective prediction solution for many practical solutions [24].

Ensemble Voting Classifier is a meta classifier for combining different classifiers for classification via majority voting. Each of the classifiers selected will predict a class label for a test sample. The label which is predicted the most will be selected as output of the voting classifier. An outfit of SGD, IBK, and Random Forest have been ensembled to build up the proposed model.

2.6 Performance Metric

In this work, performance of the depression detection system is evaluated by determining accuracy. The weighted average of recall of the depressed and non-depressed classes gives the accuracy of the system.

$$Accuracy = \frac{TruePositive}{FalseNegative + TruePositive}$$

True positives are information focuses delegated positive by the model that really are positive (which means they are correct), and false negatives are information focuses the model distinguishes as negative that are really positive (off base) [25].

3 Experiment and Analysis

Experimentation is performed to analyze the performance analysis of proposed method using ensemble classification for depression detection on the DAIC-WOZ speech dataset. Speech feature sets derived from ComParE 2016 and AVEC 2011 are referred to as feature set 1 and feature set 2 respectively. ComParE 2016 feature set consists of 6373 features and AVEC 2011 feature set consists of 1941 features respectively. SGD, IBK and Random Forest classifier will comprise the individual classifiers while combination of all these three result in ensemble classifier.

3.1 Proposed Approach for Gender Dependent System (Experiment 1)

Comprehensive test was performed on male and female speakers independently to analyze the performance of ensemble classifier. Two data sets from DAIC-WOZ database containing 1084 voice samples from 31 male speakers and 703 voice samples from 22 female speakers respectively were tested independently with ensemble and individual classifiers for both feature sets. Comparison of accuracy performance for male and female speakers are shown in Tables 3 and 4 respectively.

Table 3. Accuracy (%) analysis for male speakers using feature-set 1 and feature set 2

Classifier	Accuracy using feature set 1	Accuracy using feature set 2
SGD	92.4	73.6
IBK	91	79.1
Random Forest	91.4	79.9
Ensemble classifier	95.6	80.6

Table 4. Accuracy (%) analysis for female speakers using feature-set 1 and feature set 2

Classifier	Accuracy using feature set 1	Accuracy using feature set 2
SGD	95.4	82.7
IBK	92.9	85
Random Forest	92.2	84.1
Ensemble classifier	96.0	85.4

From Table 3 for male speakers, depression is well detected using ensemble method with a 3% rise from the result provided by the individual classification using feature set 1 from SGD. Table 4 depicts with female speakers an improvement of 0.6% is observed with ensemble classifier against SGD using feature set 1. Thus, ensemble classifier provided best results for both male and female speakers. As for gender dissimilarity, the classification accuracies for female speakers for both feature sets were higher than that of the male speakers since females are more expressive. The next experiment involved to analyze the proposed method on depression detection using gender independent system.

3.2 Proposed Approach for Gender Independent System (Experiment 2)

In the experiment 2, the accuracy analysis of the system for entire data set using both the feature sets is compared in Table 5. The system performance using feature set 1 is considerably higher and error is narrowed to 10% for individual classifiers and to a 5% margin for ensemble classifier. So, it is evident from experiments 1 and 2 the features from the ComParE 2016 feature set provides much better assistance while predicting the depression state of the person than that of AVEC 2011 feature set. So, in further experiments classification will be performed only using feature set 1. The narrow deprivation in the performance of ensemble classifier when compared to the results of experiment 1 conveys that system performance is more effective when classification is carried out for different genders separately.

Table 5. Accuracy (%) analysis for different classifiers using feature set 1 and feature set 2.

Classifier	Accuracy using feature set 1	Accuracy using feature set 2
SGD	91.3	71.0
IBK	89.4	78
Random Forest	91.7	76.8
Ensemble classifier	94.3	78.5

Also, in both the experiments it is observed that feature set 1 that constitute ComParE 2016 features are more effective for depression detection using speech. Henceforth, the work is carried on using feature set 1. Owing to the huge size of feature set 1, the following experiments involve application of feature selection and analyze the proposed system performance on gender dependent and gender independent context.

3.3 Proposed Approach for Gender Dependent and Independent System Using Feature Selection (Experiment 3)

In experiment 3, attribute selection classifier was used that employs the strategy of best search method to find the most relevant features from 6373 features of the feature set 1 for gender dependent scenario with male and female speakers as well as gender independent case. It was found that 130 features were selected. These features responded better with depression state. Feature selection was imparted on data sets for male and female separately. The accuracy obtained by the system with and without feature selection are interpreted in the Table 6. It can be observed that the ensemble method dominated the best individual classifier results around 3% and 0.6% for male and female speakers respectively while approximately 2% rise achieved with gender independent samples. That though there is a narrow deprivation in systems performance using feature selection, significant performance was achieved using compact feature set. Using feature selection systems performance for female speakers is slightly higher than male speakers. With feature selection application, although around 1% reduction in accuracy is observed against the results of experiment 1 of Tables 3 and 4 of gender dependence while 1.5% accuracy dip attained with gender independence of Table 5, but size of the feature vector is drastically reduced from 6373 to 130 features.

Table 6. Accuracy (%) analysis for different classifiers using feature set 1 using feature selection.

Classifier	Male speakers	Female speakers	Gender independent
SGD	91.4	94.1	89.5
IBK	81.0	90.9	87.8
Random Forest	90.4	91.1	89.2
Ensemble classifier	94.2	94.7	92.7

Next, a comparative analysis is performed with the proposed methodology and existing state of art systems of depression detection using speech. Although, existing state of art methods employ different databases, features and classification, to indicate the robustness of the proposed method for depression detection the comparison is performed.

4   Comparative Analysis

A comparative analysis of the accuracies using different methods and classifiers are summarized in Fig. 4. The authors Song, Shen and Valstar [26] used SVM and CNN classifiers for the DAIC-WOZ depression database to achieve an accuracy of 85% and 86.6% respectively. Deep learned spectral features were used for the classification. Whereas Paula Lopez and Laura Docio-Fernandez [27] used a text embedding method called GloVe for depression detection. Automatic Speech Recognition on this database using GloVe method gave an accuracy of 82.9%. The comparison of the existing work with the proposed work indicates the advantage of using ensemble voting classifier. Ensemble voting classifier with and without feature selection achieved better results than the existing work.

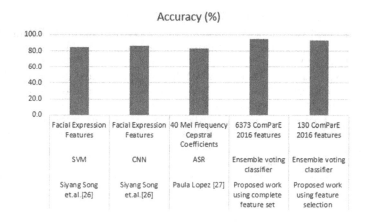

Fig. 4. Existing systems and proposed model for depression detection

5   Conclusion and Future Scope

Depression detection from speech is performed for gender dependent and independent systems wherein ComParE 2016 features and AVEC 2011 features are extracted using OpenSMILE open source toolkit. Segmentation of the speech samples of equal length on DAIC-WOZ speech database was implemented. ComParE 2016 features and AVEC 2011 features were extracted from each segment. Classification was performed using IBK, SGD, Random Forest classifiers and an ensemble voting classifier. The performance is evaluated using the accuracy obtained from each of these classifiers. The ensemble voting classifier gives better prediction results than the individual classifiers, for both the ComParE feature set as well as AVEC feature set. For speaker dependent and independent samples with and without feature selection, the performance of the ensemble classifier was superior. One of the limitation of depressive state detection is the difficulty in collecting real world speech database. In future, the method can be repeated with other database using different feature-sets. Also, the combination of ComParE 2016 and AVEC 2011 features with other relevant features can be used for depression detection. Since for speaker independent scenario, the accuracy is degraded, deep learning can be employed to obtain higher accuracy.

References

1. World Health Organization: Depression: a global public health concern (2012)
2. Scherer, S., Stratou, G., Mahmoud, M., Bober, J.: Automatic behavior descriptors for psychological disorder analysis. Image Vis. Comput. **32**, 648–658 (2014)
3. Alghowinem, S., Goecke, R., Wagner, M., Epps, J.: A comparative study of different classifiers for detecting depression from spontaneous speech. In: IEEE International Conference on Acoustics, Speech and Signal Processing (ICASSP), pp. 8022–8026. IEEE (2013)

4. Moore, E.I., Clements, M., Peifer, J., Weisser, L.: Investigating the role of glottal features in classifying clinical depression. In: Proceedings of the 25th Annual International Conference of the IEEE Engineering in Medicine and Biology Society, vol. 3, pp. 9–12 (2003)

5. Ooi, K.E.B., Lech, M., Allen, N.B.: Multichannel weighted speech classification system for prediction of major depression in adolescents. Biomed. Sig. Process. Control **14**(1), 228–239 (2014)

6. Ooi, K.E.B., Low, L.S.A., Lech, M., Allen, N.: Early prediction of major depression in adolescents using glottal wave characteristics and Teager Energy parameters. In: Proceedings of the IEEE International Conference on Acoustics, Speech and Signal Processing, ICASSP, vol. 980, pp. 4613–4616 (2012)

7. Ozdas, A., Shiavi, R.G., Silverman, S.E., Silverman, M.K., Wilkes, D.M.: Investigation of vocal jitter and glottal flow spectrum as possible cues for depression and near-term suicidal risk. IEEE Trans. Biomed. Eng. **51**(9), 1530–1540 (2004)

8. Yingthawornsuk, T., Keskinpala, H.K., Wilkes, D.M., Shiavi, R.G., Salomon, R.M.: Direct acoustic feature using iterative EM algorithm and spectral energy for classifying suicidal speech. Perform. Eval. 766–769 (2007)

9. Narayanan, V., Lalitha, S., Gupta, D.: An epitomization of stress recognition from speech signal. Int. J. Eng. Technol. (UAE) **7**, 61–68 (2018)

10. Lalitha, S., Tripathi, S.: Emotion detection using perceptual based speech features. In: 2016 IEEE Annual India Conference (INDICON), Bangalore, pp. 1–5 (2016). https://doi.org/10.1109/INDICON.2016.7839028

11. Murali Krishna, P., Pradeep Reddy, R., Narayanan, V., Lalitha, S., Gupta, D.: Affective state recognition using audio cues. J. Intell. Fuzzy Syst. **3**, 2147–2154 (2019)

12. Lalitha, S., Tripathi, S., Gupta, D.: Enhanced speech emotion detection using deep neural networks. Int. J. Speech Technol. **22**(3), 497–510 (2018). https://doi.org/10.1007/s10772-018-09572-8

13. Hantke, S., Weninger, F., Kurle, R., Ringeval, F., Batliner, A.: I hear you eat and speak: automatic recognition of eating condition and food type, use-cases, and impact on ASR performance (2016)

14. McKeown, G., Valstar, M., Cowie, R., Pantic, M., Schroder, M.: The SEMAINE database: annotated multimodal records of emotionally colored conversations between a person and a limited agent. IEEE Trans. Affect. Comput. **3**(1), 5–17 (2012)

15. Gratch, J., et al.: The Distress Analysis Interview Corpus of human and computer interviews. In: Proceedings of Language Resources and Evaluation Conference (LREC) (2014)

16. www.audacityteam.org

17. Schuller, B., et al.: Affective and behavioural computing: lessons learnt from the first computational paralinguistics challenge

18. Schuller, B., Valstar, M., Eyben, F., McKeown, G., Cowie, R., Pantic, M.: AVEC 2011–the first international audio/visual emotion challenge. In: D'Mello, S., Graesser, A., Schuller, B., Martin, J.-C. (eds.) ACII 2011. LNCS, vol. 6975, pp. 415–424. Springer, Heidelberg (2011). https://doi.org/10.1007/978-3-642-24571-8_53

19. https://www.audeering.com/opensmile/

20. WEKA 3 - Data Mining with open source machine learning software. https://www.cs.waikato.ac.nz/ml/weka/downloads.html

21. Narayanan, V., Lalitha, S., Gupta, D.: Stress recognition using auditory features for psychotherapy in Indian context. In: ICCSP, pp. 0426–0432 (2018)

22. Bottou, L.: Large-scale machine learning with stochastic gradient descent. In: Lechevallier, Y., Saporta, G. (eds.) Proceedings of COMPSTAT 2010. Springer, Heidelberg (2010). https://doi.org/10.1007/978-3-7908-2604-3_16
23. Aha, D., Kibler, D., Albert, M.: Instance-based learning algorithms. Mach. Learn. **6**, 37–66 (1991). https://doi.org/10.1007/BF00153759
24. Pal, M.: Random forest classifier for remote sensing classification. Int. J. Remote Sens. **26**(1), 217–222 (2005)
25. https://towardsdatascience.com/understanding-confusion-matrix-a9ad42dcfd62
26. Song, S., Shen, L., Valstar, M.: Human behaviour-based automatic depression analysis using hand-crafted statistics and deep learned spectral features. In: 2018 13th IEEE International Conference on Automatic Face and Gesture Recognition (2018)
27. Lopez Otero, P., Docio-Fernandez, L., Abad, A., García-Mateo, C.: Depression Detection Using Automatic Transcriptions of De-Identified Speech, pp. 3157–3161 (2017). https://doi.org/10.21437/Interspeech.2017-1201

Voice Controlled Media Player: A Use Case to Demonstrate an On-premise Speech Command Recognition System

Arunesh Kumar Singh$^{(\boxtimes)}$, Snehlata Sinha, Bhushan Jagyasi, C. C. Abinaya, Shashank Mishra, Ravi Mylavarapu, Harika Maripalli, Pallavi Gawade, and Gopali Contractor

Accenture, Advanced Technology Centers in India, Mumbai, India
{arunesh.k.singh,snehlata.sinha,bhushan.jagyasi,abinaya.cc,
shashank.v.mishra,r.mylavarapu,harika.maripalli,pallavi.s.gawade,
gopali.contractor}@accenture.com

Abstract. Voice User Interface will play an important role in the next generation of applications and IoT devices. In this paper, we propose a pipeline for Streamed Speech Command Recognition (SSCR) system by training Convolutional Neural Network (CNN) based model using the speech command dataset. The dataset was generated using our own crowdsourcing application and the data was further augmented for increasing the scale of it. The proposed approach results in a small memory footprint model, which makes the on-premise deployment feasible. The trained model is then integrated with a unique use case of Voice Controlled Media Player (VCMP) which demonstrates the on-premise capability of the proposed SSCR system. An exhaustive testing of the live application resulted in some valuable insights.

1 Introduction

Over 53 million people in the U.S. already own at least one smart speaker and it has increased 78% year-over-year as compared to 2017 [1]. Nowadays, with an enormous growth of gadgets, speech processing has become an important feature to enhance the human machine interaction. Speech processing is also used abundantly in chatbots for providing answers to queries asked by the customers [2]. This requires a large vocabulary speech recognition along with a robust language model. However, in several use cases, we need to recognize only words belonging to a certain limited vocabulary. For instance, voice enabled switch boards, smart home appliances, voice controlled car audio-systems, handling mechanical machines with speech command interface etc. Here, full-blown speech recognition system becomes an expensive affair requiring a lot of computing and communication bandwidth. Hence for these use cases, Speech Command Recognition (SCR) proves to be an efficient and reliable solution [2–4]. An ideal SCR system must be able to detect limited speech commands, with small footprint, high accuracy and

© Springer Nature Singapore Pte Ltd. 2020
S. M. Thampi et al. (Eds.): SIRS 2019, CCIS 1209, pp. 186–197, 2020.
https://doi.org/10.1007/978-981-15-4828-4_16

minimal delay. It should also support multiple languages and dialects, should be noise-robust and speaker-independent.

In this paper, we propose a CNN based Streamed Speech Command Recognition (SSCR) system which works on single command as well as on stream of commands. Further, we enhance the dataset through speech augmentation techniques, as discussed in [5], by adding background noise, speed variation and time shifting. To demonstrate the utility of SSCR, we have developed end-to-end Voice Controlled Media Player (VCMP) controlled by 7 pre-defined commands. Rigorous quality assurance (QA) test was further carried out to verify its performance. We present comparative analysis of the two models (A) CNN model trained on Noise-free dataset (Noise-Unaware model) (B) CNN model trained on Noisy dataset (Noise-Aware model). Our key contributions are summarized below:

1. Convolutional Neural Network (CNN) model trained on Tensorflow dataset [6] for 10 commands like *"yes"*, *"no"*, *"up"*, *"down"* etc.
   - Increased robustness of model by using data augmentation techniques.
2. Voice Controlled Media Player (VCMP) with 7 commands *"Play"*, *"Pause"*, *"Stop"*, *"Previous"*, *"Next"*, *"Volume-Increase"* and *"Volume-Decrease"*.
   - Data collection using our own crowd sourcing application
   - Increased robustness of the model by using data augmentation techniques
   - End to end implementation for streamed command recognition
3. Exhaustive evaluation of live functionality of SSCR model integrated with VCMP. Here accuracies are compared for offline and online by using utterances with varying speed, noise level and loudness.

2   Literature Review

In Speech command recognition system, the data collection, pre-processing, feature extraction and model development plays a major role. In this section, we review some of the important work done earlier in these areas.

Tensorflow speech command recognition dataset [6,7] is a widely used limited vocabulary speech recognition dataset. This dataset has been contributed by many speakers with multiple utterances of 35 spoken commands comprising of total 65000 utterances. This dataset also includes different background noises which can be used for data augmentation. Speech augmentation techniques like adding background noise, speed variation, time shifting etc. have helped SCR systems to achieve more robustness in real world scenario [5].

Dave et al. [8] have performed comparative analysis on different feature extraction methods like Linear Predictive Coding (LPC), Perceptual Linear Predictor (PLP) and Mel-Frequency Cepstral Coefficient (MFCC) for audio data, out of which MFCC shows better results [9].

For keyword spotting and SCR, Convolution Neural Network (CNN) based approaches have shown better performance than Deep Neural Network (DNN) and Vanilla networks [10–12]. Recently, Fernández-Marqués et al. [13] have presented a compact binary architecture for several applications of keyword spotting.

In this paper, we have implemented the CNN based model architecture presented by [11] for SCR on the Tensorflow SCR dataset [6] and on the dataset collected by our team using our crowdsourcing application for a voice controlled media player use case.

3   Proposed Streamed Speech Command Recognition (SSCR) - Model Development

We propose an end-to-end Streamed Speech Command Recognition (SSCR) system which can continuously accept an input stream of speech signal and recognize the specific commands. The key features of this system is to detect the commands, on-premise and in a portable manner, to serve the applications such as controlling car audio system, home appliance control system and many more. To demonstrate this, we have implemented a Voice Controlled Media Player (VCMP) which can be operated by the utterance of one of the seven commands including Play, Pause, Stop, Previous, Next, Volume-Increase and Volume-Decrease. In this section, we present the proposed training pipeline (refer Fig. 1) for the development of the model for VCMP use case.

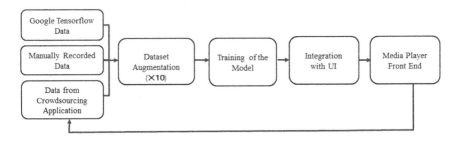

Fig. 1. Training pipeline for media player

3.1   Dataset

To control the media player with speech commands, experiments were performed on seven commands, special label "unknown" for the words other than these seven commands and "silence" when no speech is detected. For the above commands, we have collected the dataset through our crowdsourcing application. In our dataset, each utterance is stored as a 1.2 s (or less) WAVE format file, with the audio data encoded as linear 16-bit single-channel PCM values, at 16 KHz sampling rate. On an average there are 300 utterances for each class (details in Table 1). Further, data augmentation is done for improving the robustness of the model for the real-world scenarios. We carried out data augmentation by addition of noises and by varying the speed of the utterances.

Data Augmentation Through Addition of Noisy Speech Data. We have simulated noisy speech command data by mixing the speech command signals with the noise signals obtained from the crowd sourcing application as well as from the Tensorflow dataset [6]. We have randomly selected the utterances and have combined them with standard noises such as white noise, pink noise, outside environment, washing dishes, human generated cat voice, people talking, running tap, conversations etc, with one example shown in Fig. 2.

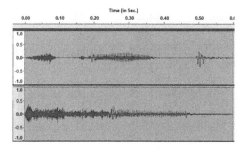

Fig. 2. Waveform of Stop Command top: before adding noise and bottom: after adding noise

Data Augmentation Through Speed Variation. In order to accommodate fast and slow accent in a real scenario, speed variation was done [5] to further scale-up the dataset. The utterances were randomly selected and then by using python package LibROSA their speed was modified by 30% i.e. speed variation was in the range of 0.7 to 1.3. Also, in order to maintain uniform length of all the audio files at 1.2 s, padding was performed for the audio files smaller than 1.2 s.

After augmentation, the dataset scaled up from 6062 utterances to 23,009 utterances as mentioned in Table 1. This dataset was then used for training and testing of the models by maintaining 90:10 ratio, respectively.

3.2 Feature Extraction

We have extracted Mel-Frequency Cepstral Coefficients (MFCC) with 40 coefficients from speech signal and used it as an input feature [8,9]. Figure 3 represents a sample MFCC for one of the commands. The window length is of 30 ms with the hop length of 10 ms without any padding. If Ns is number of samples in the speech signal, Nw is number of samples in a window, and Nh is number of samples in a hop then number of MFCC bins would be $N = 1 + (Ns - Nw/Nh)$. Hence, *one-second* audio with sampling rate of 16 KHz will result in 98 MFCC bins (N) with 40 coefficients each ($N = 1 + (16000 - 480)/160 = 98$).

Table 1. Training dataset collected using crowd sourcing application and after data augmentation. Unknown utterances are taken from Tensorflow and silence is collected through background noise.

Commands	Utterances	Augmented	Total
Play	314	2424	2738
Pause	289	2464	2753
Stop	318	2409	2727
Next	314	2326	2640
Previous	321	2425	2746
Volume-Increase	307	2441	2748
Volume-Decrease	292	2458	2750
Unknown	3434	NA	3434
Silence	473	NA	473
Total	6062	16947	23009

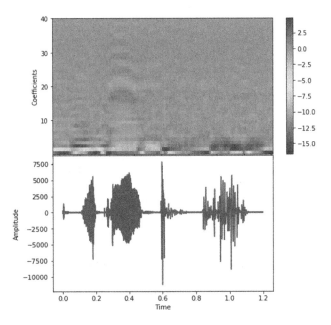

Fig. 3. MFCC features for *"Stop"* command and it's corresponding speech waveform

3.3   CNN Based Model Architecture and Training

In speech processing, Convolutional Neural Network (CNN) has been used successfully in keyword-spotting and speech command recognition [10]. The proposed CNN architecture of this network (Fig. 4) is similar to the architecture suggested by Sainath and Parada [11]. The network consists of input layer, two

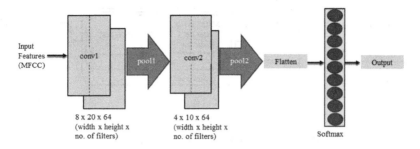

Fig. 4. CNN based model architecture for Voice Controlled Media Player (VCMP)

convolution layers which are followed by ReLU activation, dropout and max-pool. Finally, the output layer is a dense layer with softmax activation. The first convolutional layer has 64 filters with width and height of 8 and 20, respectively. The second filter also has 64 filters with width and height of 4 and 10, respectively. We have tried to keep number of training parameters minimal to keep the model size small. Training was carried out on augmented dataset. We have trained upto 50,000 steps in total, 30,000 steps with learning rate of 0.01 while another 20,000 with learning rate of 0.001, keeping batch-size of 32.

4 Streamed Speech Commands Recognition (SSCR) Model Deployment for Voice Command Media Player (VCMP) Use Case

In this section, we present the integration of the trained SSCR model with User Interface (UI) developed specifically for VCMP use case. The deployment pipeline is presented in Fig. 5 and screenshot of UI designed for VCMP is shown in Fig. 6.

UI is designed in Angular with Typescript language. User can operate the UI using defined voice commands and it performs actions with respect to it. To maintain continuous stream of speech data, we have used web socket from the front-end to back-end. Voice Activity Detection (VAD) is implemented using state-of-the-art WebrtcVAD [14,15] over web sockets to detect speech from continuous stream of commands spoken through the microphone. The trained model is Tensorflow for small-footprint and is packaged into REST API to work on HTTP. It is then integrated with VAD to work in real-time.

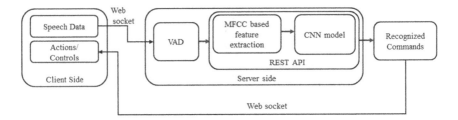

Fig. 5. Model deployment for Voice Controlled Media Player

Fig. 6. User interface of the media player

5   Results

In this section, we present an exhaustive evaluation of the CNN based speech command recognition for following scenarios:

1. Models developed for commands with existing Tensorflow dataset were evaluated. Here we propose the Noise-Aware model to make the models more robust to the noisy environment.
2. Live testing of accuracy for VCMP with test data from different distribution.

5.1   Speech Command Recognition for 10 Commands with and Without Noise

In order to cater to noisy environment, we compared two methods for training of Tensorflow speech commands dataset:

- Training was carried out with non-augmented data and resultant model is called Noise-Unaware model M_{NU}
- Training was carried out with Noisy data and the resultant model is called Noise-Aware model M_{NA}.

The noise was added from [6,7] in the dataset which resulted in training dataset to make sure models are robust to noisy environment. We have used

1000 examples of noisy and noise-free test dataset to evaluate performances of both models M_{NU} and M_{NA}. The consolidated test accuracies are presented in Table 2. The Noise-Unaware model M_{NU} works reasonably well (with 90% test accuracy) on non-noisy dataset, it results in a very poor performance (with 64.9% test accuracy) if the test data is noisy. On the other hand the Noise-Aware model M_{NA} results in a significantly better performance for both non-noisy (test accuracy of 94.4%) as well as noisy dataset (test accuracy of 86.9%). This leads to an insight that preprocessing for noise removal is not required on the noisy speech data if CNN model considers enough noisy data for the training, which results in a very robust Noise-Aware model M_{NA}.

Table 2. Accuracies of CNN based Noise-Aware model M_{NA} and Noise-Unaware model M_{NU} (Performed on Google's Speech Command Dataset)

Noise-Unaware model M_{NU}	Noise-Aware model M_{NA}
Non noisy dataset	
90.06%	94.40%
Noisy dataset	
64.90%	86.95%

5.2 Speech Command Recognition for Seven Commands of Voice Controlled Media Player

We now present the training and test accuracies for CNN based model trained by us for a new set of commands - *Play, Pause, Stop, Next, Previous, Volume-Increase and Volume-Decrease* for which dataset was not readily available. We hence used our own crowd sourcing application to collect the utterance for these commands, Table 1 summarises the dataset.

The resulting learning curves in the form of training accuracy and cost are represented in Figs. 7 and 8 respectively. The training accuracy and cost shows consistent learning with the increase in number of training steps.

The models were then tested on the 10% of the dataset reserved for testing. The confusion matrix for test results is given in Table 3. The diagonal values which indicates correct predictions in the confusion matrix are high, while the other values which indicates incorrect predictions are very low except in case of silence where we have used huge no-utterance for testing.

Table 4 summarizes accuracy, precision, recall and F1-score for VCMP. The training accuracy observed during the final steps were close to 96% and test accuracy close to training accuracy showing no over-fitting of the model. High precision of the model indicates low false positive rate while high recall indicates low false negative rate.

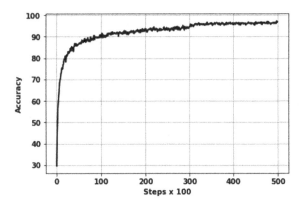

Fig. 7. Learning curve showing training accuracy of the model.

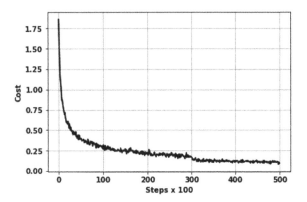

Fig. 8. Learning curve showing cost of training

5.3 Quality Assurance of Live Application with Streamed Speech Command Recognition

Speech command recognition Quality Assurance (QA) is a process in which testers ensures that the application correctly recognizes the uttered commands which results in the expected action. There are number of factors that influence the accuracy of speech recognition such as noise from the environment, the user's age, gender, user accent, speed and voice loudness. The trained models may hence incorrectly classify the sound uttered by the user. Eventually, it would perform actions that do not meet the business user's expectations.

We have performed rigorous live testing using the new distribution of data. A total of seven testers - three male and four female testers participated in this testing process. Each tester has uttered every command five times by varying speaking volume intensity, speaking rate and background noise. Volume-intensities considered were low, normal, high and speaking rate as slow, medium, fast. This entire test was conducted in two different environments, less noisy and noisy.

Table 3. Confusion matrix for test results (Performed on our own speech command dataset for voice controlled media player)

		Total Test Utterance	Predicted Hypothesis								
			Play	Pause	Stop	Next	Previous	Vol. Inc.	Vol. Dec.	Unknown	Silence
True Hypothesis	Play	307	292	1	5	5	2	1	0	0	1
	Pause	309	7	292	7	1	0	0	1	0	1
	Stop	308	3	1	300	1	1	0	0	1	1
	Next	286	1	1	3	277	0	2	1	1	0
	Previous	310	2	0	5	5	295	2	0	0	1
	Vol. Inc.	310	0	1	0	2	0	296	10	1	0
	Vol. Dec.	310	0	0	0	0	1	0	6	302	0	1
	Unknown	297	3	2	6	7	0	0	1	276	2
	Silence	1142	69	7	117	30	5	0	0	5	909

Table 4. Accuracy, Precision, Recall and F1-score of CNN model (Performed on our own speech command dataset for voice controlled media player)

Accuracy	Precision	Recall	F-measure
96.01%	93.82%	90.13%	91.94%

For less noisy environment, enclosed noise-free studio room was used whereas for noisy environment common office rooms, canteen with background music were used. Considering these combinations, total of 4410 utterances were used for this live testing experiment.

Figures 9 and 10 summarizes the test accuracies in all three dimensions (volume, speed and environment). We observe that the utterances spoken in a non-noisy environment with slow speed and high volume-intensity results in highest

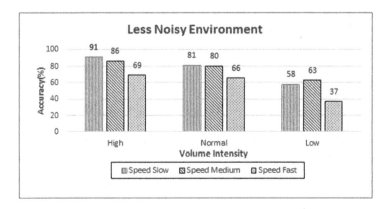

Fig. 9. Live testing accuracies in less noisy environment with varying volume intensity and speed

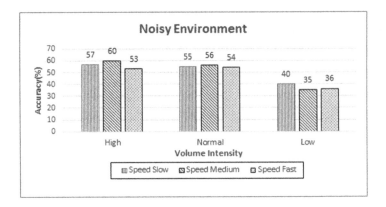

Fig. 10. Live testing accuracies in noisy environment with varying volume intensity and speed

accuracy of 91%. Accuracy is seen to drop significantly with the increase in speed of the utterances and with the reduction in volume-intensity of utterances in both less-noisy and noisy environment. This behaviour of deep learning models is hence seen to mimic the human cognition ability.

6   Conclusion and Future Work

This paper presents an implementation of speech command recognition for a voice controlled media player use case. An end to end approach was established from data collection using crowd sourcing, to data augmentation, training of models and integration with a live system. Without loss of generality, this work can be extended to other similar use cases which requires lighter models for on-premise speech command recognition. The exhaustive test results showcase accuracy variation observed across different dimensions like voice loudness, speed and ambient noise. To further gain in accuracy for noisy environment, our next step is to explore new network architectures, use transfer learning and improvise augmentation of the speech data.

References

1. The smart audio report. Report, NPR and Edition Research (2019)
2. Shum, H.-Y., He, X., Li, D.: From eliza to xiaoice: challenges and opportunities with social chatbots. Front. Inf. Technol. Electron. Eng. **19**(1), 10–26 (2018). https://doi.org/10.1631/FITEE.1700826
3. Kevin, O., Kasamani, B.: The application of real-time voice recognition to control critical mobile device operations. Int. J. Res. Stud. Sci. Eng. Technol. **2**(07), 174–184 (2015)
4. Schalkwyk, J., et al.: "Your word is my command": Google search by voice: a case study. In: Neustein, A. (ed.) Advances in Speech Recognition, pp. 61–90. Springer, Boston (2010). https://doi.org/10.1007/978-1-4419-5951-5_4

5. Ko, T., Peddinti, V., Povey, D., Khudanpur, S.: Audio augmentation for speech recognition. In: Sixteenth Annual Conference of the International Speech Communication Association (2015)
6. Tensorflow speech commands recognition dataset. http://download.tensorflow.org/data/speech_commands_v0.01.tar.gz. Accessed 02 Sept 2018
7. Warden, P.: Speech commands: a dataset for limited-vocabulary speech recognition. CoRR, abs/1804.03209 (2018)
8. Dave, N.: Feature extraction methods LPC, PLP and MFCC in speech recognition. Int. J. Adv. Res. Eng. Technol. **1**, 1–4 (2013)
9. Chakraborty, K., Talele, A., Upadhya, S.: Voice recognition using MFCC algorithm. Int. J. Innov. Res. Adv. Eng. (IJIRAE) **1**(10), 158–161 (2014). ISSN 2349–2163
10. Chen, G., Parada, C., Heigold, G.: Small-footprint keyword spotting using deep neural networks. In: 2014 IEEE International Conference on Acoustics, Speech and Signal Processing (ICASSP), pp. 4087–4091. IEEE (2014)
11. Sainath, T.N., Parada, C.: Convolutional neural networks for small-footprint keyword spotting. Google Inc. (2015)
12. Zhou, Z., Li, X.: Speech command recognition with convolutional neural network (2017). stanford.edu
13. Fernández-Marqués, J., Tseng, V.W.-S., Bhattachara, S., Lane, N.D.: Deterministic binary filters for keyword spotting applications. In: Proceedings of the 16th Annual International Conference on Mobile Systems, Applications, and Services, MobiSys 2018, New York, NY, USA, pp. 529–529. ACM (2018)
14. WebRTCVAD. https://webrtc.org. Accessed 28 Feb 2019
15. Tong, S., Chen, N., Qian, Y., Yu, K.: Evaluating VAD for automatic speech recognition. In: 2014 12th International Conference on Signal Processing (ICSP), pp. 2308–2314. IEEE (2014)

Efficient Traffic Signboard Recognition System Using Convolutional Networks

Siva Krishna P. Mothukuri$^{(\boxtimes)}$, R. Tejas, Soham Patil, V. Darshan, and Shashidhar G. Koolagudi

National Institute of Technology Karnataka, Surathkal, India
msivakrish@gmail.com, tejas1908@gmail.com, sohampatil798@gmail.com, darshan.blh@gmail.com, koolagudi@nitk.edu.in

Abstract. In this paper, a smart automatic traffic sign recognition system is proposed. This signboard recognition system plays a vital role in the automated driving system of transport vehicles. The model is built based on convolutional neural network. The German Traffic Sign Detection Benchmark (GTSDB), a standard open-source segmented image dataset with forty-three different signboard classes is considered for experimentation. Implementation of the system is highly focused on processing speed and classification accuracy. These aspects are concentrated, such that the built model is suitable for real-time automated driving systems. Similar experiments are carried in comparison with the pre-trained convolution models. The performance of the proposed model is better in the aspects of fast responsive time.

Keywords: Advanced Driver Assistance System · Convolutional neural networks · GTSRB · Traffic sign · Detection · Recognition

1 Introduction

Traffic signs [5] help drivers in vehicles think about the traffic principles and help the drivers for better and safe driving. Wider expansion in the number of vehicles on roads leading drivers to experience perils while driving and this may in like manner cause setbacks. Many mishaps are happening each year throughout the world. These setbacks are overwhelmingly an aftereffect of the driver's weakness to process all the visual data that is accessible while driving. Traffic signs, which are present on the road to help drivers navigate [15], are sometimes not clearly observed by the drivers. As indicated by a PRS Legislative Research report in 2015, there were around five lakh street mishaps in India, which claimed the lives of about 1.5 lakh individuals and harmed around five lakh individuals.

The work on traffic sign recognition in real-time systems [2] successfully implemented in the late 2000s. Nonetheless due to the difficulties like blurring, weather condition, faded color, unclear pictures due to vehicles motion, scene complexity, etc. it's still a challenging problem. This traffic sign recognition system is a key component of the Advanced Driver Assistance System (ADAS) and

© Springer Nature Singapore Pte Ltd. 2020
S. M. Thampi et al. (Eds.): SIRS 2019, CCIS 1209, pp. 198–207, 2020.
https://doi.org/10.1007/978-981-15-4828-4_17

intelligent vision assistance applications that provide drivers with traffic sign information.

Besides, when constructing a traffic sign recognition framework, care must be taken that the outcomes given by the context are precise because a traffic sign recognition framework must be dependable as it is utilized continuously. If not trustworthy, miss recognition may prompt mistaken tasks and lamentable outcomes. There are many degrees for miss recognition of signs, which must be dealt with. Variable size of traffic signals, obscuring of photographs, several lighting conditions and distinctive introduction of traffic signs are a couple of things which are dealt with. It is required to use a recognition system that is accurate and has a high processing speed. Traffic signs in different orientations and different lighting conditions makes the task of traffic sign recognition a challenge.

The inspiration driving ADAS and vision help applications in vehicles are the automation of the vehicle structures for security and improving the experience of driving. Even though there have been numerous endeavors to enhance the precision of traffic acknowledgment frameworks, they have not contrasted with a standard dataset until the arrival of the German Traffic Sign Recognition Benchmark (GTSRB) in 2011. We have utilized the datasets given by GTSRB to discover the exactness of our proposed model, which will enable us to contrast our model with others' models effectively.

This paper sorted out as pursues: Sect. 2 gives a concise audit of the current work done in the field of traffic sign recognition, Sect. 3 portrays the approach utilized in our work, Sect. 4 includes the results acquired from experimentation and discussions, and Sect. 5 gives the conclusion.

2   Literature Survey

Detection of traffic sign boards plays a vital role in the preparation of dataset. Recognition task will then consider the objects as individual images for classification. Work on traffic sign detection has been mainly revolving around three methods: Color-Based methods [18], Shape-based methods [6], and combination of both color and shape methods [14]. The final obtained bounding boxes are the subjects for recognition level. The traditional ways of traffic sign recognition are feature matching and machine learning. BRISK, SURF and SIFT are the three significant contributions to Feature Matching. The Binary Robust Invariant Scalable Key point (BRISK) identifies the point of interest and compares pixel-wise brightness [24]. This BRISK claims to be faster than Speeded Up Robust Features (SURF). The SURF acts as both, detector and descriptor [14]. As the detector detects interest points like blobs and corners in the image. Later, as descriptor from feature vector from these interest points. Finally, the feature vectors of targets are compared against the ones that exist in the dataset. Instead, templates are stored and are used for comparison [4].

Machine learning techniques are trained efficiently in classifying various classes of traffic sign recognition [1]. Using Artificial Neural Networks, and Multilayer perceptron neural networks worked well in conditions like abnormal light conditions and shadow environments [13,18]. SVM is the majorly used classifier combining with various feature extractors prior. Zernike moment features that are invariant to translation, scale and rotation are tried combining with SVM in [21]. Also, the HOG feature, invariant to scale used in pedestrian detection are used to train SVM [16] for traffic sign recognition. Modification of template using Genetic algorithm [11] representing scale, rotation, blur, and intensities classify without the requirement of the training phase. The LBP feature extraction added prior to SVM observed to be robust to grey scale and rotational invariance [23]. Hu moments with neural networks achieved better results with low computational costs [8]. Using the concept of transfer learning, the pre-trained model *VGG-16* has shown about 96% accuracy on GTSRB dataset [20]. False positives removed by SVM classifier, and the rest is classified based HOG features gave an accuracy of 97.75% [22]. Multi-task CNN on the detected images by there own model and has produced most exquisite precision [12]. Hinge loss trained convolutional neural networks [10] and Spatial Transformer Networks [19] have even stood as the best achievements in traffic signboard classification. As per our knowledge very few contributions regarding traffic sign recognition upon the time taken to classify the image. [22] justify the trade-off and shows that the simple model developed takes a time quantum of 3 ms achieved an accuracy of 97.7% with less time consumption (Fig. 1).

Fig. 1. Detection of sign boards as individual objects from image

3   Proposed Methodology

Convolutional neural networks have proved to be exceptional for image classification tasks. But deep convolutional neural networks need a large amount of data to train on. Because of the nonattendance of a comprehensively accessible dataset with enough picture information, execution of various methodologies couldn't be looked at for traffic sign recognition until the arrival of German Traffic Sign Recognition Benchmark (GTSRB) [17] and the German Traffic Sign Detection Benchmark (GTSDB) [9] which are two openly accessible broad datasets for the correlation of methodologies for Traffic sign recognition and detection tasks individually.

Our work considered GTSRB dataset to compare the performance of the pre-trained ResNet50 [7] model, and our proposed model. The GTSRB dataset has forty-three different classes of images and more than 50000 pictures in total. It is a sizeable lifelike database suited for single-image multi-class classification problem. According to [17], it reflects the substantial variations in the visual appearance of signs due to distance, illumination, weather conditions, partial occlusions, and rotations. Physical traffic sign instances are unique within the dataset. It includes many types of complex traffic signs such as sign tilt, uneven lighting, occlusions, and distortions. Hence this is the right choice for training models those which can be deployed for practical purposes.

A parallel dataset of grayscale images is also built segmenting the color images for further experimentations.

3.1   Image Preprocessing

The images in the GTSRB dataset is of size ranging from 15×15 to 250×250. All these images are warped to 32×32 to feed the proposed CNN model as well as the ResNet50. The color (RGB) images Dataset is used in the first experimentation, the segmented grayscale images dataset obtained after the preprocessing phase is used for training the Proposed CNN model in second experimentation. The colored images were converted to grayscale by taking the weighted average of the 3 channels RGB. Then these grayscale images were normalized so that the pixel values were between 0.1 and 0.9.

3.2   Proposed CNN Architecture

The proposed model for traffic sign classification is eleven layered Convolutional neural network architecture with nine two dimensional Convolution layers that were partitioned into four blocks followed two fully connected layers. Inside each block, a chain of layers with the same size of filters in each layer are used. As block one is with two layers, 32 filters in each. Similarly, second and third blocks sharing three layers each with 64 and 128 filters. The fourth block is only with a single convolution layer. In all the four convolution blocks, the last layer is combined with the max-pooling. In the end, the softmax was used to obtain the final output probabilities of all the image classes. The activation function

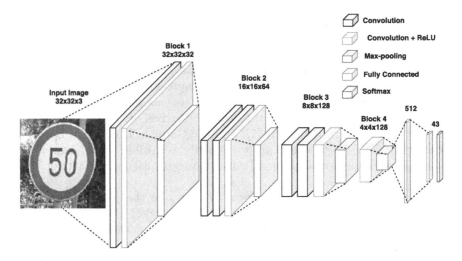

Fig. 2. Proposed convolution neural network architecture

Rectified linear Unit (ReLu) followed by batch normalization is used for decision making to get the features as the outcome of each of the block. The structure diagram is shown in Fig. 2. Images from the GTSRB dataset are considered as input to models. For each input image, the convolution layers extract the edge features acting as features extractors. ReLU operation is then performed on these output filter maps to introduce non-linearity. This information is further filtered, and only the maximum pixel values are passed onto the next stack using the max-pooling operation. After passing the convolution phase, the features obtained are the feature maps are flattened out and given as input to the fully connected layers. The classification of feature maps into target labels is being taken care of by this phase.

4   Experimentation and Results

The images from the GTSRB dataset first warped to 32×32. The dataset has 39209 colored images in the training set and 12630 colored images in the testing set. Images in the testing set are unseen images and serve the purpose of testing the model. The training set further randomly split into the training and the validation set in the ratio 80:20. The training set is used to tune the weights of the neural network, and the validation set is used to calculate the loss of the network after every iteration.

4.1   Proposed Model

Training our proposed CNN architecture is being done in two ways, one with the color (RGB) images from the GTSRB dataset and the second with the segmented

Table 1. Comparison of accuracies of ResNet50 and the proposed model

CNN model	Training acc.	Testing acc.
ResNet 50	99.4%	95%
Proposed CNN (Color)	99.9%	98.66%
Proposed CNN (Grayscale)	99.8%	97.48%

grayscale images of the same. The batch size of 128, categorical cross entropy loss function and the Adam optimizer parameters used in experimentation. In the first approach, test accuracy 98.66% obtained at 100 epochs. The intuition behind this is that the human eye classifies the signs based on the color and shape. Hence, retaining the color information in the data gets the trained neural network closer to that of the human brain. Instead, the preprocessing phase is being carried out on the color images from the GTSRB dataset, and the outcome segmented grayscale images are considered to train our proposed model in the second case. Grayscale inputs result in 97.48% accuracy, which can have the major advantageous even in adverse light conditions. As expected, colored data produced better results as compared to grayscale data.

Table 1 summarises the training and testing accuracies that are obtained for the Proposed CNN models with color and grayscale images along with ResNet50. In comparison, the results based on testing time for the proposed model and the ResNet50 model on 12630 images and the total time taken for classifying all images by images count gives the computational time for each image. We observed that the proposed model is almost twice as fast as the ResNet50 during classification.

4.2   ResNet 50 Model

ResNet is a pre-trained model (Winner of ILSVRC 2015 in image classification, detection, and localization) which is intensively used in the classification of Images. Using the concept of transfer learning, we have experimented with ResNet50 model. The ResNet50 structure is with 50 layers in deep built to classify images into 1000 object categories for the ImageNet competition. Here in our experimentation, the number of classes to be classified is 43. So, last fully-connected (FC) layers of this model pruned and fine-tuned adding one more FC layer to match the target labels. Only parameters at the fully-connected layers which are on top are considered to be trainable parameters, and the rest

Table 2. Classification responsive time comparison for the two models

CNN model	Testing accuracy	Response time per image
Proposed model	98.66%	0.296 ms
ResNet 50	95%	0.554 ms

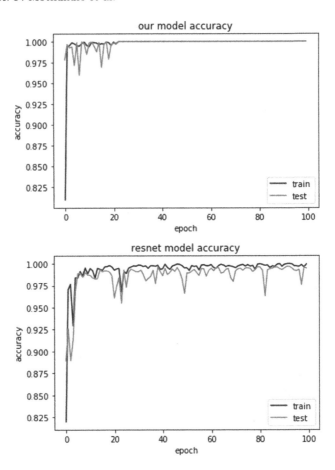

Fig. 3. Comparison of training and testing accuracies of the Proposed Convolution and ResNet50 models

are treated as Non-training parameters. The batch size and optimization functions are the same as in the proposed model. The behavior of learning observed for 100 epochs. Results show how fast the proposed model achieves the global minimum, where the pre-trained model struggles in achieving it. The accuracy obtained from ResNet50 is 95%. Table 2 summarises the computational responsive time taken for classification per image on GPUs worked with color images. We observed that the training and testing accuracy for the proposed model obtain stability earlier before 20 epochs whereas ResNet50 model has some inconsistencies in the testing loss and accuracies even after epoch 80 (Figs. 3 and 4).

The comparison with the existing state of the art models in classification responsive time is shown in Table 3 where proposed model tried with available system configuration.

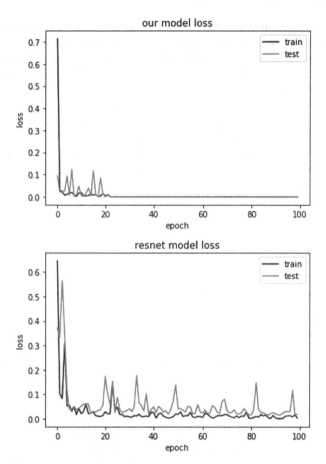

Fig. 4. Comparison of training and testing Error rate of the Proposed Convolution and ResNet50 models for 100 epochs

Table 3. Comparison of test accuracies and classification time with Existing models

	Model [22]	Model [3]	Proposed model
Accuracy (in %)	97.75%	99.461%	98.66%
Time (in milliseconds)	3 ms	11.4 ms	3.6 ms
Hardware	Quadcore	GPU	1 Core, 2 Threads, **Machine** 3.7 GHz CPU Xeon Processor 2.3 GHz

5   Conclusion

The proposed Convolution neural network model of traffic sign recognition system has shown results outperformed pre-trained ResNet50 model, both in terms of classification accuracy as well in responsive time. Hence, we suggest that CNN architectures with suitable number of layers in depth like proposed CNN model suits for fine-grained classification such as traffic signboard recognition. Because the real-world environmental systems designed for specific task. Further, fine-tuning the proposed model makes it efficient to be deployed in Advanced Driver Assistance Systems. Achieving unsupervised pre-training of early layers of CNN, which acts as feature extraction stages will be challenging for easy learning. Variety of datasets, including blur perturbations, brightness, contrast, and shear can be trained to overcome real-world distortion effects. Studies on different input resolutions of signboards help much to improve efficiency. Involvement of ensemble methods with multiple networks and considering both colored and non-colored image datasets alleviate situations day or night, and signboard may be colored or not.

References

1. Balali, V., Golparvar-Fard, M.: Evaluation of multiclass traffic sign detection and classification methods for us roadway asset inventory management. J. Comput. Civil Eng. **30**(2), 04015022 (2015)
2. Berkaya, S.K., Gunduz, H., Ozsen, O., Akinlar, C., Gunal, S.: On circular traffic sign detection and recognition. Expert Syst. Appl. **48**, 67–75 (2016)
3. CireşAn, D., Meier, U., Masci, J., Schmidhuber, J.: Multi-column deep neural network for traffic sign classification. Neural Netw. **32**, 333–338 (2012)
4. Farhat, W., Faiedh, H., Souani, C., Besbes, K.: Real-time recognition of road traffic signs in video scenes. In: 2016 2nd International Conference on Advanced Technologies for Signal and image Processing (ATSIP), pp. 125–130. IEEE (2016)
5. Fleyeh, H.: Traffic sign detection and recognition. In: Computer Vision and Imaging in Intelligent Transportation Systems, pp. 343–373. Wiley (2017)
6. Garcia-Garrido, M.A., Sotelo, M.A., Martin-Gorostiza, E.: Fast traffic sign detection and recognition under changing lighting conditions. In: 2006 IEEE Intelligent Transportation Systems Conference, pp. 811–816. IEEE (2006)
7. He, K., Zhang, X., Ren, S., Sun, J.: Deep residual learning for image recognition. In: Proceedings of the IEEE Conference on Computer Vision and Pattern Recognition, pp. 770–778 (2016)
8. Hossain, M.S., Hasan, M.M., Ali, M.A., Kabir, Md.H., Ali, A.B.M.S.: Automatic detection and recognition of traffic signs. In: 2010 IEEE Conference on Robotics, Automation and Mechatronics, pp. 286–291. IEEE (2010)
9. Houben, S., Stallkamp, J., Salmen, J., Schlipsing, M., Igel, C.: Detection of traffic signs in real-world images: the German traffic sign detection benchmark. In: The 2013 International Joint Conference on Neural Networks (IJCNN), pp. 1–8. IEEE (2013)
10. Jin, J., Kun, F., Zhang, C.: Traffic sign recognition with hinge loss trained convolutional neural networks. IEEE Trans. Intell. Transp. Syst. **15**(5), 1991–2000 (2014)
11. Kobayashi, M., Baba, M., Ohtani, K., Li, L.: A method for traffic sign detection and recognition based on genetic algorithm. In: 2015 IEEE/SICE International Symposium on System Integration (SII), pp. 455–460. IEEE (2015)

12. Luo, H., Yang, Y., Tong, B., Fuchao, W., Fan, B.: Traffic sign recognition using a multi-task convolutional neural network. IEEE Trans. Intell. Transp. Syst. **19**(4), 1100–1111 (2018)
13. Nguwi, Y.-Y., Kouzani, A.Z.: Automatic road sign recognition using neural networks. In: The 2006 IEEE International Joint Conference on Neural Network Proceedings, pp. 3955–3962. IEEE (2006)
14. Oruklu, E., Pesty, D., Neveux, J., Guebey, J.-E.: Real-time traffic sign detection and recognition for in-car driver assistance systems. In: 2012 IEEE 55th International Midwest Symposium on Circuits and Systems (MWSCAS), pp. 976–979. IEEE (2012)
15. Ouerhani, Y., Alfalou, A., Desthieux, M., Brosseau, C.: Advanced driver assistance system: road sign identification using VIAPIX system and a correlation technique. Opt. Lasers Eng. **89**, 184–194 (2017)
16. Ben Romdhane, N., Mliki, H., Hammami, M.: An improved traffic signs recognition and tracking method for driver assistance system. In: 2016 IEEE/ACIS 15th International Conference on Computer and Information Science (ICIS), pp. 1–6. IEEE (2016)
17. Stallkamp, J., Schlipsing, M., Salmen, J., Igel, C.: The German traffic sign recognition benchmark: a multi-class classification competition. In: IJCNN, vol. 6, p. 7 (2011)
18. Supreeth, H.S.G., Patil, C.M.: An approach towards efficient detection and recognition of traffic signs in videos using neural networks. In: 2016 International Conference on Wireless Communications, Signal Processing and Networking (WiSPNET), pp. 456–459. IEEE (2016)
19. Uhrig, J., Schneider, N., Schneider, L., Franke, U., Brox, T., Geiger, A.: Sparsity invariant CNNs. In: 2017 International Conference on 3D Vision (3DV), pp. 11–20. IEEE (2017)
20. Wang, C.: Research and application of traffic sign detection and recognition based on deep learning. In: 2018 International Conference on Robots & Intelligent System (ICRIS), pp. 150–152. IEEE (2018)
21. Xing, M., Chunyang, M., Yan, W., Xiaolong, W., Xuetao, C.: Traffic sign detection and recognition using color standardization and Zernike moments. In: 2016 Chinese Control and Decision Conference (CCDC), pp. 5195–5198. IEEE (2016)
22. Yang, Y., Luo, H., Xu, H., Wu, F.: Towards real-time traffic sign detection and classification. IEEE Trans. Intell. Transp. Syst. **17**(7), 2022–2031 (2016)
23. Zhang, H., Wang, B., Zheng, Z., Dai, Y.: A novel detection and recognition system for Chinese traffic signs. In: Proceedings of the 32nd Chinese Control Conference, pp. 8102–8107. IEEE (2013)
24. Zheng, Z., Zhang, H., Wang, B., Gao, Z.: Robust traffic sign recognition and tracking for advanced driver assistance systems. In: 2012 15th International IEEE Conference on Intelligent Transportation Systems, pp. 704–709. IEEE (2012)

Development of a Novel Database in Gujarati Language for Spoken Digits Classification

Nikunj Dalsaniya, Sapan H. Mankad$^{(\boxtimes)}$, Sanjay Garg, and Dhuri Shrivastava

CSE Department, Institute of Technology, Nirma University, Ahmedabad, India
nikunj5390.dalsaniya@gmail.com, {sapanmankad,16bce043}@nirmauni.ac.in,
gargsv@gmail.com

Abstract. India is the country with lot of diversity, however, lack of sufficient data in Indian languages hinders testing and deployment of various commercial applications. We introduce a novel audio database of isolated Gujarati digits in this work. This database contains recordings of digits spoken by 20 users from 5 different regions of Gujarat in practical environments. To the best of our knowledge, this is the first publicly available Gujarati spoken digits database. We implement a naive neural network classifier on the statistical features derived from the audio data, and test its performance on the newly designed Gujarati database and an existing English language database of spoken digits. Cross-corpus experiments are also conducted to investigate the generalization capabilities of our approach. Results indicate that current systems are language and database specific; and they are still in immature stage to bring generalization capability across the different databases.

Keywords: Spoken digits classification · Gujarati · speech

1 Introduction

Speech is the most natural means of communication among human beings. In any human-computer interaction system, Automatic Speech Recognition (ASR) plays very important part. It is there in our smart phone, gadgets and smart devices like smart watch, personal assistants like Amazon Echo Dot, Google Home etc. From getting live weather reports to playing songs according to mood, dialing a number or sending a message to someone, these personal assistants are handy. Human life has become cooler and becoming cooler than ever due to these advancements and features. The main reason behind such rapid development is active research in machine learning and artificial intelligence, which has improved and made speech recognition reasonably accurate in some of the applications. However, these technologies rely on sufficient amount of data, and in adverse conditions, it becomes difficult to get acceptable performance. Thus, the performance of these data-driven applications is highly dependent on the database under consideration [8].

© Springer Nature Singapore Pte Ltd. 2020
S. M. Thampi et al. (Eds.): SIRS 2019, CCIS 1209, pp. 208–219, 2020.
https://doi.org/10.1007/978-981-15-4828-4_18

1.1 Motivation

Gujarati is an Indian language with a rich cultural heritage. It is a native language for the people of state of Gujarat. Apart from various states of India, it is spoken by over 60 million people across the world. Even after a wide spread of the language, there is a lack of any standard audio database in Gujarati language which can be used for several speech based commercial and societal applications. However, through crowdsourcing, companies like Google and Microsoft[1] have started to build corpus in every under-resourced language, still there is a wide gap between the availability of benchmark resources to develop systems which can ensure reasonable performance. Due to lack of sufficient publicly available database, it becomes difficult to develop efficient systems.

Gujarati is considered among under-resourced languages in the Speech Processing community. The purpose of the database presented in this work is to present a starting point to the researchers to test their algorithms on Gujarati language so that systems or commercial applications built using this may be useful for regional real life applications. Moreover, this will help the farmers and laymen to get experienced with it in their mother tongue.

One of the applications of spoken digits classification is in developing Intelligent Voice Response Systems (IVRS) in which user can just respond to questions using voice commands instead of pressing a button. Another typical application is in remote phone banking applications where a user can provide his account number to a banking system equipped with such algorithm and it will detect the spoken digits so that appropriate account details can be fetched. Voice assisted dialing [25] can help blind and physically challenged people to dial a specific phone number through voice commands. Some other applications include airline and train reservation, Robot control and navigational command systems and interactive spoken dialogue systems. A class of applications that use a numeric language such as recognizing and understanding credit card and telephone numbers, zip codes, social security numbers is presented in [24]. Thus, these voice enabled services can offer effortless and user-friendly service to their clients.

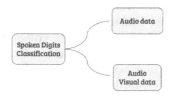

Fig. 1. Two approaches for spoken digits classification

Spoken Digit classification is carried out in two ways as depicted in Fig. 1: (i) using speech features from audio recordings (ii) using visual cues obtained

---

[1] https://azure.microsoft.com/en-in/services/cognitive-services/directory/speech/.

through lip movements from audio-visual data. In this work, we focus on the recognition of digits spoken in English and Gujarati language using audio only approach.

Following are the main contributions of this paper:

1. A novel spoken digits dataset in Gujarati language is presented.
2. An attempt has been made to explore whether feature engineering and classification approaches are language specific. Cross-corpus experiments have been conducted to test the generalization capability of the proposed approach.

The remainder of this paper is as follows. In the next section, we discuss the related work carried out in this area. The datasets used in this work are described in Sect. 3. In Sect. 4, short-term spectral features extracted from audio files are discussed. Section 5 presents the experimental scenario and discusses our observations. Finally, the last section concludes the paper.

2   Related Work

In this section, we briefly review the existing databases for spoken numerals, and studies conducted on such databases for classification task.

The first comprehensive audio visual database of spoken digits, known as Extended XM2VTS [19] consists of recordings from 295 participants of different age and gender, carried out over the duration of 5 months. The audio recordings comprise of two continuous digits strings from each speaker. The BANCA database [4] consists of utterances in four different languages, namely, English, Spanish, Italian and French. The purpose of this database was to test the person verification performance in controlled, degraded and adverse scenarios. Although less number of speakers, the CUAVE database [22] has a variety of utterances from multiple sessions. Some notable spoken digits databases are summarized in Table 1.

Arabic Spoken digits classification was attempted in [15] using Gaussian copula approach. Another audio visual digit recognition task was performed by Jain et al. [16] using Discrete Cosine Transform and Local Binary Patterns. The task of sequence recognition using Italian spoken digits was performed in [7] with radial basis functions. An RNN based spoken Arabic digit recognizer was proposed in [2]. Another noted work is by [6] on a set of 100 spoken numerals of English language obtained from a single speaker. They implemented the task using neural networks. A decision level fusion of two different hidden Markov model (HMM) based systems was applied for audio-visual digit recognition in [20] on a single-speaker system on audio stream corrupted by additive multi-speaker babble noise. A VLSI chip for speech recognition was developed by Kim et al. [17] using multi-layer perceptron. Another recent work for Arab Digit recognition has been proposed with the use of bi-directional recurrent end-to-end neural network classifiers in [28]. Gujarati handwritten numeral recognition was carried out in [10] with an application to optical character recognition.

Table 1. Summary of the databases for Spoken Digits classification

Paper	Database	Speakers	Utterances	Language
[19]	XM2VTS[a]	295	885	English
[13]	VALID[b]	106	1590	English
[4]	BANCA[c]	208	29952	Multiple
[22]	CUAVE[d]	36	7000	English
[18]	AVICAR[e]	86	59000	English
[11]	AusTalk	1000	24000	English
[27]	Censrec-1	42	3234	Japanese
[3]	OuluVs2	53	1590	English
[23]	AVDigits	53	795	English
[5]	Inhouse	5	250	Marathi
[9]	Inhouse	82	NA	Brazilian Portuguese
[21]	Inhouse	150	NA	Pashto

[a] http://www.ee.surrey.ac.uk/CVSSP/xm2vtsdb/
[b] http://ee.ucd.ie/validdb/
[c] http://www.ee.surrey.ac.uk/Research/VSSP/banca
[d] http://ece.clemson.edu/speech
[e] http://www.ifp.uiuc.edu/speech/AVICAR/

It can be seen from Table 1 that majority of the work is focused on English language database generation. Being a country with diverse culture and several languages, there is a need for databases in regional languages for promoting the research work. An inhouse database for Marathi spoken digits was prepared in [5] for studying lip reading movement and digit classification. Spoken Bengali numerals classification was conducted on a locally collected dataset by [26] using dynamic time warping (DTW) and minimum distance classifier. A corpus for Gujarati language may help substantially in regional applications. To the best of our knowledge, there is no any publicly available dataset for spoken Gujarati numerals. This motivated us to design this database.

3   Audio Datasets

In this work, we experimented with two different datasets of spoken digits. For digits spoken in English language, we used Free Spoken Digit Dataset (FSDD[2]). For Gujarati spoken digits, to the best of our knowledge, no standard dataset is available like English numerals database. We collected audio samples from native speakers of five major regions of the state of Gujarat, and designed our own dataset for Gujarati numerals. Details of both the datasets is given in Table 2.

Figure 2 shows the geographical location of these five regions in the map of Gujarat state. Distribution of recordings from these regions is depicted in Table 3.

---

[2] Available from: https://github.com/Jakobovski/free-spoken-digit-dataset.

Table 2. Statistics of both databases used in this work

Database	English	Gujarati
Sampling rate	8 kHz	44.1 kHz
# speakers	4	20
# utterances	2000	1940

Fig. 2. Geographical map of Gujarat state with five major regions (adapted from [1])

3.1 Data Collection

All regions have different accents and different dilect of Gujarati language. Speech Samples are collected of Gujarati 0 to 9 digits from 20 speakers of different regions, gender and age groups. Total 20 speakers (14 male and 6 female) were asked to pronounce 0 to 9 digits in a continuous stream form. This trial was repeated for each speaker for approximately ten times, thus we obtained 100 digits uttered by the same speaker. Important parameters like environment in which sound was recorded, different accents/pronunciation of different speakers from different regions and different age groups and gender etc. were considered to add diversity in the corpus.

Table 3. Distribution of voice samples belonging to each region

Zone	# speakers	# of utterances
Central Gujarat	5	430
North Gujarat	5	500
South Gujarat	4	400
Saurashtra	5	510
Kutch	1	100
Total	20	1940

Table 4. Pronunciation of English and Gujarati numerals

English Numeral	Pronunciation	Gujarati Numeral	English Pronunciation	Gujarati Translation
0	Zero	૦	Shoonya	શૂન્ય
1	One	૧	Ek	એક
2	Two	૨	Be	બે
3	Three	૩	Tran	ત્રણ
4	Four	૪	Char	ચાર
5	Five	૫	Paanch	પાંચ
6	Six	૬	Chh	છ
7	Seven	૭	Saat	સાત
8	Eight	૮	Aath	આઠ
9	Nine	૯	Nav	નવ

Each sample was collected using mobile recorder in .wav format in practical acoustic environments, and naming convention of each and every file was: <RegionID><SpeakerID><TrialID><Digit>. Thus, for a file with the name $R1S1T1D1.wav$, $R1$ indicates region 1, $S1$ is speaker 1 from that region, $T1$ is trial 1, and $D1$ suggests digit 1. An illustration of pronunciation and translation of Gujarati digits is given in Table 4. For academic and research purpose, we have kept this dataset publicly available at this link[3].

3.2   Preprocessing

Audio pre-processing mainly consists of two stages. (i) Pre-edit and processing raw audio file and smoothing of a signal, and (ii) unwanted noise removal. After this, segmentation of each signal was performed to split one big audio file (containing continuously recorded ten digits as can be seen from Fig. 3) into individual digit specific recording file. Then we used spectral noise filter to remove background noise present in the signal. We used Wavepad audio editor[4] for both these tasks.

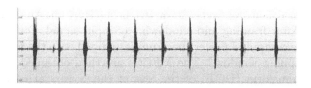

Fig. 3. Time domain representation of the audio file consisting of the utterance of one speaker speaking digits 0 through 9

An illustrative representation of spectrum and spectrogram for a single digit is shown in Fig. 4.

---

[3] https://github.com/Nikunj1729/free-spoken-gujarati-digit-dataset.
[4] https://www.nch.com.au/wavepad/index.html.

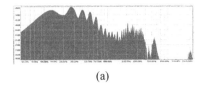
(a)
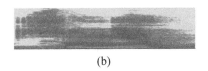
(b)

Fig. 4. An illustration of digit 0 uttered by speaker 1 from region 1 (R1S1T1D0.wav) in form of (a) spectrum and (b) spectrogram

A comparison between two dimensional PCA and t-SNE based visualization for both English and Gujarati dataset is demonstrated in Fig. 5. The features used for this visualization are mid-term statistical features. The distribution shows the arrangement of data points in feature space, indicating that it is not easy to separate the classes even after applying dimensionality reduction.

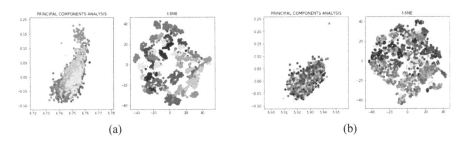
(a)   (b)

Fig. 5. Visualization in two-dimensional PCA and t-SNE space for (a) English and (b) Gujarati dataset

4   Feature Representation

Short-term spectral features were extracted using PyAudioAnalysis [14] from raw audio files. After framing and windowing of input signal, Fourier transform is performed. The computed power spectrum is mapped onto the mel scale, using triangular overlapping windows. The MFCCs are then obtained by taking the discrete cosine transform (DCT) of the logs of the powers at each of the mel frequencies. Figure 6 illustrates the process of extracting these MFCC features.

Frame level features were extracted wherein every frame was represented by short-term features of 34 dimensions which are described in Table 5. A detailed description of these features is available in [14]. First and second order statistics (in form of mean and standard deviation) were computed from each feature, referred to as mid-term features. This gave 68-dimensional feature vector of statistical parameters. Standardization and scaling of data using StandardScaler method of sklearn library was performed. We used this 68-dimensional feature vector for all the experiments.

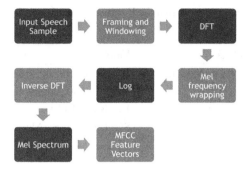

Fig. 6. Mel Frequency Cepstral Coefficients (MFCC) feature extraction process

Table 5. Short-term features extracted in our experiments

# of dimensions	Feature
1	Zero Crossing Rate (ZCR)
1	Energy
1	Entropy of energy
1	Spectral Centroid
1	Spectral Spread
1	Spectral Entropy
1	Spectral Flux
1	Spectral Rolloff
13	MFCC
12	Chroma vector
1	Chroma Deviation

5   Experiments

5.1   Architecture of the Classifier

It has been reported in literature [12] that deep learning models outperform traditional classifiers for complex tasks such as word or sentence recognition, but when it comes to simple tasks such as digit recognition, they perform similarly to traditional classifiers; hence we implemented classical artificial neural networks for classifying spoken digits.

A grid based search with 5-fold cross validation was performed to determine the optimal hyper-parameters for configuring the neural network classifier. Parameters under consideration for this search are listed in Table 6. After obtaining the best parameter values, the classifier was applied on both English and Gujarati datasets. The implementation approach is described in Fig. 7.

Table 6. Parameters used in grid search. Parameters with emphasis indicate optimal values.

Parameters	Values
# neurons in hidden layer	10, 20, 30, 40, 50, **60**, 70
Epochs	50, 100, 150, 200, 250, **300**
Activation function	**Logistic**, ReLU
Solver	SGD, **LBFGS**, ADAM

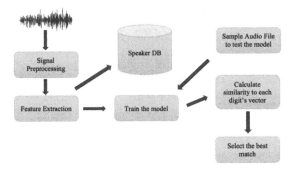

Fig. 7. The proposed architecture for spoken digit classification system

5.2   Observations

The confusion matrix depicting the number of correct predictions for each class for two different scenarios is given in Fig. 8.

From Fig. 8b, it can be observed that detecting spoken 4 seems quite easy in cross-corpus scenario.

Table 7. Performance of the system for within-corpus and cross-corpus scenario

Scenario	Accuracy (%)	Average precision	Average recall	Average F-measure
Eng/Eng	93.67	0.938	0.936	0.94
Eng/Guj	9.97	0.11	0.09	0.10
Guj/Guj	75.6	0.751	0.757	0.75
Guj/Eng	10.17	0.1	0.11	0.10

To investigate the generalization capability of the classifier, cross-corpus experiments were also conducted. Results are given in Table 7 in which scenario A/B indicates the system trained on data A, and tested on dataset B. Performance has been measured using average precision, recall and F measure values in addition to accuracy. It can be seen that training and testing using the same (language) database gives reasonable accuracy, whereas the performance

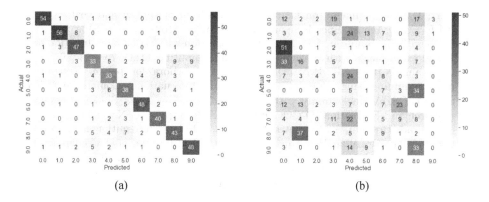

Fig. 8. Confusion matrix for (a) Guj/Guj and (b) Guj/Eng scenario

for cross-corpus scenario wherein the training and testing is on different datasets is extremely poor.

6   Conclusion

In this paper, we presented a novel publicly available audio dataset of spoken digits in Gujarati language and reported some preliminary results on it. A comparison was also carried out of the classification algorithm with English language database. The cross-corpus experiments show that there is a huge scope of improvement to explore features and techniques to bring generalization capability into existing system. It can be seen that one of the classes has considerably less number of samples than rest of the classes, hence this dataset may also be used to test the systems which can handle class-imbalanced problems.

References

1. Gujarat - region wise result, May 2019. https://specials.manoramaonline.com/News/2017/gujarat-election-results/indepth-ml.html
2. Alotaibi, Y.A.: Spoken Arabic digits recognizer using recurrent neural networks. In: Proceedings of the Fourth IEEE International Symposium on Signal Processing and Information Technology, pp. 195–199, December 2004
3. Anina, I., Zhou, Z., Zhao, G., Pietikäinen, M.: OuluVS2: a multi-view audiovisual database for non-rigid mouth motion analysis. In: 2015 11th IEEE International Conference and Workshops on Automatic Face and Gesture Recognition (FG), vol. 1, pp. 1–5, May 2015
4. Bailly-Bailliére, E., et al.: The BANCA database and evaluation protocol. In: Kittler, J., Nixon, M.S. (eds.) AVBPA 2003. LNCS, vol. 2688, pp. 625–638. Springer, Heidelberg (2003). https://doi.org/10.1007/3-540-44887-X_74
5. Brahme, A., Bhadade, U.: Marathi digit recognition using lip geometric shape features and dynamic time warping. In: TENCON 2017 – 2017 IEEE Region 10 Conference, pp. 974–979, November 2017

6. Burr, D.J.: Experiments on neural net recognition of spoken and written text. IEEE Trans. Acoust. Speech Signal Process. **36**(7), 1162–1168 (1988)
7. Ceccarelli, M., Hounsou, J.T.: Sequence recognition with radial basis function networks: experiments with spoken digits. Neurocomputing **11**(1), 75–88 (1996)
8. Chhayani, N.H., Patil, H.A.: Development of corpora for person recognition using humming, singing and speech. In: 2013 International Conference Oriental COCOSDA Held Jointly with 2013 Conference on Asian Spoken Language Research and Evaluation (O-COCOSDA/CASLRE), pp. 1–6, November 2013
9. de Andrade Bresolin, A., Neto, A.D.D., Alsina, P.J.: Digit recognition using wavelet and SVM in Brazilian Portuguese. In: 2008 IEEE International Conference on Acoustics, Speech and Signal Processing, pp. 1545–1548, March 2008
10. Desai, A.A.: Gujarati handwritten numeral optical character reorganization through neural network. Pattern Recogn. **43**(7), 2582–2589 (2010)
11. Estival, D., Cassidy, S., Cox, F., Burnham, D., et al.: AusTalk: an audio-visual corpus of Australian English (2014)
12. Fernandez-Lopez, A., Sukno, F.M.: Survey on automatic lip-reading in the era of deep learning. Image Vis. Comput. **78**, 53–72 (2018)
13. Fox, N.A., O'Mullane, B.A., Reilly, R.B.: VALID: a new practical audio-visual database, and comparative results. In: Kanade, T., Jain, A., Ratha, N.K. (eds.) AVBPA 2005. LNCS, vol. 3546, pp. 777–786. Springer, Heidelberg (2005). https://doi.org/10.1007/11527923_81
14. Giannakopoulos, T.: pyaudioanalysis: An open-source python library for audio signal analysis. PLoS One **10**(12) (2015)
15. Hammami, N., Bedda, M., Farah, N.: Probabilistic classification based on Gaussian copula for speech recognition: application to spoken Arabic digits. In: 2013 Signal Processing: Algorithms, Architectures, Arrangements, and Applications (SPA), pp. 312–317, September 2013
16. Jain, A., Rathna, G.N.: Visual speech recognition for isolated digits using discrete cosine transform and local binary pattern features. In: 2017 IEEE Global Conference on Signal and Information Processing (GlobalSIP), pp. 368–372, November 2017
17. Kim, K.-C., Han, I.-S., Lee, J.-H., Lee, H.-S., Yi, Y.N.: Spoken digit recognition using URAN (universally reconstructable artificial neural-network) VLSI chip. In: Proceedings of 27th Asilomar Conference on Signals, Systems and Computers, vol. 1, pp. 822–825, November 1993
18. Lee, B., et al.: AVICAR: audio-visual speech corpus in a car environment. In: Proceedings of the INTERSPEECH, Jeju Island, Korea, pp. 2489–2492 (2004)
19. Messer, K., Matas, J., Kittler, J., Jonsson, K.: XM2VTSDB: the extended M2VTS database. In: Second International Conference on Audio and Video-Based Biometric Person Authentication, pp. 72–77 (1999)
20. Meyer, G.F., Mulligan, J.B., Wuerger, S.M.: Continuous audio-visual digit recognition using N-best decision fusion. Inf. Fusion **5**(2), 91–101 (2004). Robust Speech Processing
21. Nisar, S., Shahzad, I., Khan, M.A., Tariq, M.: Pashto spoken digits recognition using spectral and prosodic based feature extraction. In: 2017 Ninth International Conference on Advanced Computational Intelligence (ICACI), pp. 74–78, February 2017
22. Patterson, E.K., Gurbuz, S., Tufekci, Z., Gowdy, J.N.: CUAVE: a new audio-visual database for multimodal human-computer interface research. In: 2002 IEEE International Conference on Acoustics, Speech, and Signal Processing, vol. 2, pp. II-2017–II-2020, May 2002

23. Petridis, S., Shen, J., Cetin, D., Pantic, M.: Visual-only recognition of normal, whispered and silent speech. In: 2018 IEEE International Conference on Acoustics, Speech and Signal Processing (ICASSP), pp. 6219–6223, April 2018

24. Rahim, M., Riccardi, G., Saul, L., Wright, J., Buntschuh, B., Gorin, A.: Robust numeric recognition in spoken language dialogue. Speech Commun. **34**(1), 195–212 (2001). Noise Robust ASR

25. Salmela, P., Lehtokangas, M., Saarinen, J.: Neural network based digit recognition system for voice dialing in noisy environments. Inf. Sci. **121**(3–4), 171–199 (1999)

26. Ghanty, S.K., Shaikh, S.H., Chaki, N.: On recognition of spoken Bengali numerals. In: 2010 International Conference on Computer Information Systems and Industrial Management Applications (CISIM), pp. 54–59, October 2010

27. Tamura, S., et al.: CENSREC-1-AV: an audio-visual corpus for noisy bimodal speech recognition. In: AVSP (2010)

28. Zerari, N., Abdelhamid, S., Bouzgou, H., Raymond, C.: Bi-directional recurrent end-to-end neural network classifier for spoken Arab digit recognition. In: 2018 2nd International Conference on Natural Language and Speech Processing (ICNLSP), pp. 1–6, April 2018

Modeling Vehicle Fall Detection Event Using Internet of Things

Nikhil Kumar[1]($\boxtimes$), Anurag Barthwal[1], Debopam Acharya[2], and Divya Lohani[1]

[1] Department of Computer Science and Engineering, Shiv Nadar University, Gautam Buddha Nagar, Greater Noida, Uttar Pradesh, India
{nk438,ab414,divya.lohani}@snu.edu.in
[2] School of Computing, DIT University, Dehradun, Uttarakhand, India
dr.debopamacharya@dituniversity.edu.in

Abstract. Research in the realm of accident prediction and analysis is mostly confined to the study of road accidents caused by motor vehicle collisions where the event of vehicle fall from high altitude is mostly ignored. An IoT system has been introduced in this work, which uses a budget smartphone and off-the-shelf sensors to accurately detect and notify vehicle fall event. The proposed system is low cost, reliable and can be easily retrofitted to any type of vehicle. It uses the vehicle speed, absolute linear acceleration and altitude to build an SVM classifier based model to detect the fall event. The proposed IoT system is found to be accurate with a MAPE of 1.8%.

Keywords: Vehicle fall detection · Accident detection · Context-aware · Internet of Things · Support Vector Machine · Absolute linear acceleration

1 Introduction

According to Global Status Report on Road Safety 2018 by World Health Organization (WHO), the total number of deaths due to road accidents is excessively high with more than 1.35 million people dying each year worldwide [1]. For all kind of death causes, road accident is the eighth biggest reason for the death of people and the biggest reason for the death of children and youth aged below 29 years [1]. According to the golden hour principle (the relationship between the time taken for the treatment after an accident and the death rate), timely notification reduces the time for medical treatment after the accident and significantly decreases the mortality rate [2]. Research in the area of accident prediction has mainly focused on the use of information and communication technologies to detect vehicle collisions. Such accident detection systems fail to report the accident of a vehicle due to a fall from a bridge, fly-over or a hill. A significant percentage of road accidents has been recorded when the vehicle either goes off the road or falls-off from an altitude. Such events may occur when the vehicle is at an altitude from the ground, plying on a bridge, fly-over, elevated highway or is traversing a hilly road. According to the road accident database management system (RADMS) report of the government of the State of Himachal Pradesh (HP) in India [3], from April 2018 to June

© Springer Nature Singapore Pte Ltd. 2020
S. M. Thampi et al. (Eds.): SIRS 2019, CCIS 1209, pp. 220–233, 2020.
https://doi.org/10.1007/978-981-15-4828-4_19

2018, a total of 9076 accidents were reported in HP, out of which 1850 accidents were run-off-the-road accidents. In more than 80% of reported ran-off-the-road accidents, vehicle fell into the gorge. To the best of our knowledge, no research work has been carried out to detect and report such fall events.

In this work, the authors present an end-to-end IoT system which can detect the fall of a vehicle from an altitude. The IoT system can generate an instant alert to the nearby emergency medical services (EMS), relatives of the victims and police so that immediate help can be provided. The novelty of the approach lies in (a) its capability of using low-cost hardware and its ability to be retrofitted to any vehicle, (b) use of statistical models in accurately predicting the vehicle fall event, and (c) automatic reporting of the accident to relevant agencies.

2 Related Work

In this section, the latest research related to the detection of a road accident with the help of inputs from sensors is discussed. Jair Ferreira Júnior et al. [4] have per-formed driver behavior profiling by using different smartphone sensors and classification algorithms. The results have been compared to determine the optimum meth-od to characterize driver aggressiveness profile. Aloul et al. [5] have developed a smartphone-based system that uses accelerometer data to build a predictive model based on Dynamic Time Warping (DTW) and Hidden Markov Models (HMM).

Linayage et al. [6] have proposed a Bayesian quickest change detection approach to optimize the trade-off between average detection delay and false alarm rate. To reduce the latency in reporting an accident, Dar et al. [7] have proposed a fog computing-based approach to develop a low-cost, delay-aware system for detection and reporting of an accident. A smartphone-based early recognition system has been developed by Xu et al. in [8], which can help to avoid vehicle accident by identifying inattentive driving at early stage and alerting the driver.

Park et al. [9] have used in-built sensors of the driver's smartphone to develop an event-driven solution to prevent distracted driving. Bhatti et al. [10] have used speed, pressure, gravitational force, location, and sound sensors of a commodity smartphone to develop a low cost, portable solution that detects an accident and reports it to the nearest hospital.

In all these works, sensors of the passenger's smartphone have been used to build the accident detection system. None of these works has provided analysis and solution for the scenario where the passenger vehicle falls-off from a certain height. Hence, this work presents an IoT system, which uses in-built as well as connected sensors to build an intelligent system that can accurately detect the fall of a vehicle from an elevation.

3 Context-Aware IoT System Architecture

3.1 System Architecture

To address the problem of vehicle fall detection, a novel architecture has been pro-posed as shown in Fig. 1. Modern smartphones are furnished with a variety of in-built sensors that can be used to measure different physical parameters. These devices can provide a commanding, economical and versatile research platform for data collection and edge computing. Our proposed IoT system can exploit the five standard sensors of a SAMSUNG Galaxy S8 smartphone (microphone, GPS, magnetometer, accelerometer, and gyroscope) and seven in-built (pressure, temperature, infrared, humidity, CO, illuminance, altitude) and one supplementary (CO_2) sensor of Sensordrone [11], and off-the-shelf sensor device.

In this system, the research focus will be with absolute linear acceleration, the altitude from sea level and speed of the vehicle to detect the fall of the vehicle from a

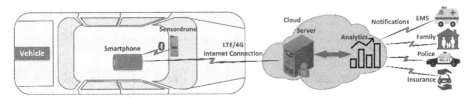

Fig. 1. Context-aware IoT system architecture

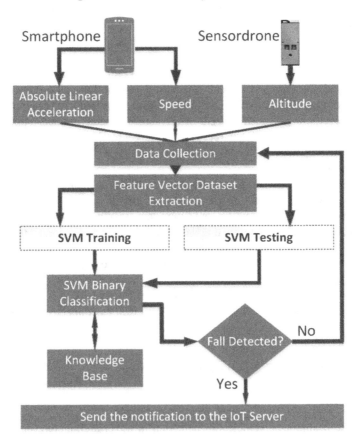

Fig. 2. Working flowchart of the system

certain height. Sensordrone's barometer communicates the atmospheric pressure to the smartphone via Bluetooth link. Atmospheric pressure is used to calculate the altitude of the vehicle with the help of a standard formula. Altitude measurement is fused with absolute linear acceleration and speed to accurately estimate the fall detection scenario.

This whole processing of sensors' streamed data is done within the smartphone itself. In the case of the vehicle fall event, the smartphone immediately sends the fall details like time, coordinates, device ID, etc. to the IoT server in the cloud for further processing. The smartphone uses a 4G/LTE connection to send this valuable information to the IoT server. After analyzing the received data, the IoT server sends the notification with the location and other relevant data to the nearest EMS, family and other emergency services such as police, insurance, etc. As shown in Fig. 2, absolute linear acceleration and speed parameters are obtained from inbuilt sensors of the smartphone while the information about altitude is derived from the Sensordrone. Features extracted from these input parameters are used for training and testing of the SVM classifier model. Occurrence of the fall event is notified to the end users. The process keeps reiterating until either the system shuts down or a vehicle fall event is detected.

3.2 Hardware Setup

Samsung Galaxy S8 Smartphone. Samsung Galaxy S8 smartphone is equipped with a 1.9 GHz Exynos 8895 octa-core processor, 4 GB of RAM and 64 GB of storage, which makes it a suitable platform for edge processing. It houses multiple standard sensors such as inertial sensors, microphone, GPS, light sensor, proximity sensor, etc. Proposed system utilizes two standard sensors of the smartphone: LM6D1 six-axis inertial sensor (with range $\pm$ 16 g) for measuring the absolute linear acceleration and BCM4774 GPS sensor for measuring the location of the fall and speed of the vehicle.

Sensordrone. Sensordrone is a small sensor hub that can be programmed according to the research needs. It can measure eleven different environmental parameters including atmospheric pressure, altitude, temperature, humidity, non-contact object temperature, CO, illuminance, proximity, capacitance, oxidizing and reducing gases. External devices can also be added using the serial port expansion. It can be connected easily to any android smartphone via Bluetooth 4.0 link. Android/Java library for Sensordrone is available at [12]. The setup of the experiment has used the barometer LPS331AP (range 26 kPa to 126 kPa) of the Sensordrone to measure the altitude of the vehicle.

RC Car. As it is not feasible to perform the experiment on a full-size, real vehicle and collect streaming data, we have used a well-configured 1:12 scaled RC car to imitate a real-life vehicle fall from an altitude. Configuration of RC car is shown in Table 1. This off-road toy monster truck is made up of high-performance ABS material and has a strong front bumper structure and very good suspension.

Table 1. RC Car configuration [13]

Vehicle Parameter	Value
Brand Name and Item No.	GPTOYS Foxx S911 RC Car
Dimension (L * W * H)	310 mm * 265 mm* 150 mm
Weight	1078 gm
Scale	1:12
Track Width	220 mm
Wheelbase	207 mm
Ground Clearance	40 mm
Max Speed	33 MPH
Transmitter	2.4 GHz 4Ch
Operating Range	80 m
Max Turning Angle	45°

3.3 Software Setup

Android application, SNUSense, has been developed for collecting and processing the received data from smartphone and Sensordrone. SNUSense has a user-friendly interface as shown in Fig. 3(a).

SNUSense collects the data and stores it in a CSV file in the memory of the smartphone. It continuously processes the streamed data and makes an inference using three attributes namely speed, absolute linear acceleration, and altitude of the vehicle. After successfully identifying the vehicle fall event, SNUSense sends the identity of the user and location of the vehicle to the SNUSense IoT server in the cloud (Google Firebase [14]). SNUSense has a broad vision and is developed by keeping in mind other kinds of vehicle accidents such as collision, rollover, fire, etc. As shown in Fig. 3(a), by exploiting sensors of smartphone and Sensordrone, SNUSense can measure time, date, location, sound, speed, linear acceleration, yaw, pitch, roll, illuminance, pressure, temperature, humidity, CO, CO_2, proximity, object temperature, and altitude.

We have developed another android application, SNUAlertApp, for receiving the alert notifications of accident as shown in Fig. 3(b). The notification contains the location of the accident, name of the driver, and type of the accident. A single tap on notification area opens the Google Map with location marker so that the user can trace the accident location easily.

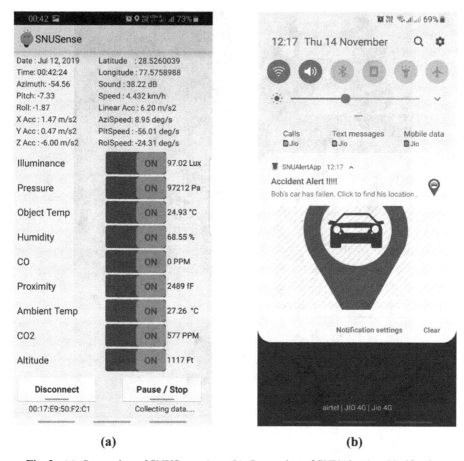

Fig. 3. (a). Screenshot of SNUSense App, (b). Screenshot of SNUAlertApp Notification

4 Testbed, Data Sensing and Pre-processing

4.1 Experiment Setup and Data Collection

A high-speed 1:12 scaled RC car has been used for the experiment to emulate real-life scenarios of vehicle fall. As shown in Fig. 4, the smartphone and the Sensordrone have been tied up with the chassis of the RC car because the optimum value of accelerometer can be obtained when it is placed near the center of gravity (CG) of the vehicle.

The experiment has been performed at 15 feet elevated running track at Indoor Sports Complex of the university (shown in Fig. 4). Multiple experiments have been performed to collect data for characterizing the vehicle fall event. RC car was operated using a 2.4 GHz 4Ch wireless transmitter and dropped on a badminton wooden court by turning it to the left on the track at a 45-degree angle. The trends of ALA, altitude, and speed of the vehicle are shown in Figs. 5, 6 and 7 respectively.

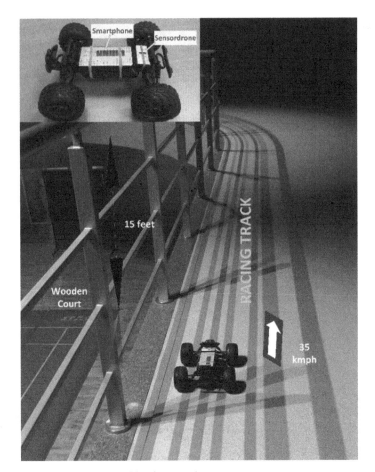

Fig. 4. Experiment Setup

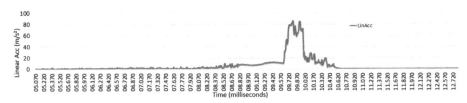

Fig. 5. Linear acceleration after fall from 15 feet elevated track

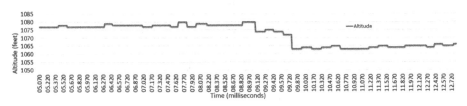

Fig. 6. Altitude measured after fall from 15 feet elevated track

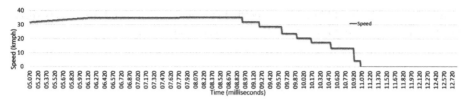

Fig. 7. Change in speed after falling from 15 feet elevated track

4.2 Parameters Used for Fall Detection

In this work, three attributes (viz. absolute linear acceleration, change in the speed and altitude of the vehicle) are used to detect the fall event of the vehicle. If we do not consider any one of them, the system may generate a false alarm.

Absolute Linear Acceleration (ALA). When a moving vehicle falls from a height, the orientation of the vehicle does not remain the same as on the road. In such a situation, it can not stay parallel to the gravitational axes and the acceleration characteristic can be distributed over two or three axes. The static acceleration value on different axes can also change with the change in the orientation of the vehicle. Therefore, it is very difficult to estimate the exact peak of the acceleration of particular axes. To deal with this problem, ALA (or Signal Magnitude Vector [15]) is calculated from all axes X, Y, and Z. ALA is independent of the fall orientation of the vehicle and shows the acceleration characteristic parallel to the gravitation. It is the resultant vector of accelerations of X, Y, and Z-axes. It is the square root of the sum of squares of accelerations at X-axes (ACC_X), Y-axes (ACC_Y) and Z-axes (ACC_Z) respectively. It is a positive quantity, which is calculated by the following equation:

$$ALA = \sqrt{(ACC_X)^2 + (ACC_Y)^2 + (ACC_Z)^2} \tag{1}$$

The measurement unit of ALA is g, where g is equal to 9.80665 m/s^2. In this work, the threshold value for ALA is considered 6 g for fall event detection because it is known from the literature that if a vehicle impact with a non-movable obstacle with more than 23 km/h (i.e. 21 feet/s) then its deceleration always crosses 5 g [16]. It can be demonstrated that when an object falls from a height of more than 10 feet, its final velocity is always more than or equal to 25 feet/second.

Speed. It is observed that when a vehicle falls from a height, it will eventually stop and its speed would eventually become zero. The system has used GPS to measure vehicle speed. Every GPS device receives NMEA (National Marine Electronics Association) sentences from the satellites, which contains data related to position, velocity and time of the device [17]. Each device category has its own NMEA sentence. Each standard NMEA sentence has a two-letter prefix (e.g., GPS receivers uses GP), all proprietary sentences start with P and are followed by three letters, which identifies the device manufacturer (e.g. Garmin sentences start with PGRM). GPRMC, The Recommended Minimum, is the NMEA sentence, which is used for obtaining the velocity by most of the

Android devices. Android library has specific functions to fetch speed from the NMEA sentence. An example of GPRMC NMEA sentence is shown below:

```
$GPRMC,122844,A,1831.336,N,07734.332,E,018.9,054.4,280919,
003.1,W*6A
```
Here, velocity is 018.9 knots, which is 35 km/h.

Altitude. When a vehicle falls from a height, altitude is most prominent attribute whose value decreases immediately. GPS could be used to measure the altitude but it is not suitable when location remains the same while the altitude is changing. No dedicated sensor is available that can measure altitude directly; it can be measured by using atmospheric pressure p and temperature. In the conducted experiment, altitude is measured using the pressure p measured by the Sensordrone. Here, we are neglecting the atmospheric temperature because it hardly changes when altitude changes less than 50 feet. Therefore, the final formula would be:

$$altitude = 44330.77 * \left(1 - \left(\frac{p}{p_0}\right)^{0.190263}\right) \qquad (2)$$

Where p_0 is the standard reference pressure measured at sea level.

4.3 Ten Millisecond Moving Maximum

Generally, the accelerometer generates signals at a very high frequency, which is more than 2000 Hz. It is very difficult to process and model the data at such a high speed because there can be extreme fluctuations in the readings and such processing becomes resource-intensive. Data pre-processing is required to deal with this situation. Iyoda et al. [2] have used a 10 ms moving average method to pre-process the data generated at 2800 Hz frequency. However, the moving average can downgrade the peak value generated by the accelerometer and cannot identify the fluctuation generated for 1 ms to 2 ms. To take the peak value into account, the system has used 10 ms moving maximum method in which the maximum value of every 10 ms interval has been recorded that is shown in Fig. 8 as ..., N-2, N-1, N, N + 1....

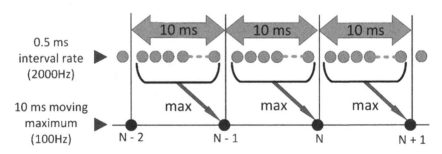

Fig. 8. Ten millisecond moving maximum

5 Statistical Modeling of Fall Detection

This work investigates support vector machines (SVM) to model the event of the fall of a vehicle from an altitude. Proposed model utilizes three inputs viz. vehicle speed, linear acceleration, and altitude to detect the occurrence of a fall event. There are only two possible outcomes of the proposed forecast model, (1) fall occurs, and (2) fall does not occur. Hence, it is deduced as a binary classification problem in which the samples are represented as points in an n-dimensional space and mapped in a way that the samples of the two categories are separated by a gap called hyperplane, which is as wide as possible.

SVM fits an optimal separating hyperplane (OSH) between the two occurrences by utilizing the training samples that are present at the edge of the class distributions - the support vectors. The hyperplane is oriented in a way that it is placed at maximum distance from the sets of support vectors. This orientation helps SVM generalize on unseen cases more accurately in comparison to the classifiers such as the neural networks that perform by minimizing the training error, such as neural networks [18]. Another advantage with SVM classification is that, in contrast with statistical classifiers such as the maximum likelihood classifiers that use all training cases to characterize the classes, only a small number of training samples present at the edge of the class distributions (support vectors) in the feature space are needed to establish the decision surface [19]. Hence, SVM classification requires a much smaller training sample size than the conventional maximum-likelihood classification models. The Naïve Bayes, KNN and random forest classifiers work faster and are more accurate only with a large number of training samples, but in our case the training samples are less. So, SVM is best suited as it requires less training samples and is the most accurate even when the data for training and testing is less [20, 21].

The hyperplane in SVM is defined as

$$W.X + B = 0 \tag{3}$$

Where X is a data-point lying on the hyperplane, W is normal to the hyperplane and B is the bias.

The linearly separable case defines a separate hyperplane for two classes: $W \cdot X_i + B \geq +1$ (for $Y_i = +1$) and $W \cdot X_i + B \leq -1$ (for $Y_i = -1$). The two equations can be combined as

$$y_i(W \cdot X_i + B) - 1 \geq 0 \tag{4}$$

The training sample points on the two hyperplanes, $W \cdot X_i + B = \pm 1$, that are parallel to the OSH, are the support vectors. The gap between the planes is 2/|W| and its maximization is achieved by the constrained optimization problem, under the inequality constraints of Eq. (4).

$$\min\left\{\frac{1}{2}\|W\|^2\right\} \tag{5}$$

To restrict the lower and upper bounds of input, slack variables $\{\xi_i\}$ are introduced, so that the Eq. (4) becomes

$$y(W.X_i + B) > 1 - \xi_i \tag{6}$$

The solutions for which ξ_i are large, are penalized by adding the penalty term $C \sum_{i=1}^{r} \xi_i$. The optimization problem from Eq. (4), under the inequality constraints of Eq. (6) thus becomes,

$$\min \left[\frac{\|W\|^2}{2} + C \sum_{i=1}^{r} \xi_i \right] \tag{7}$$

The training data X is mapped into a high-dimensional feature space H through a mapping function ϕ to allow for non-linear decision surfaces. The input data point X is represented as $\phi(X)$ in the high-dimensional space H. As the computation of $(\phi(X) \cdot \phi(X_i))$ is expensive, it is lowered in the high-dimensional space by using a kernel function such that

$$(\phi(X) \cdot \phi(X_i)) = k(X, X_i) \tag{8}$$

Solving the above equations, the decision function is obtained in the form

$$f(x) = \text{sgn} \left(\sum_{i=1}^{r} \alpha_i y_i k(X, X_i) + B \right) \tag{9}$$

Here, α_i is the Lagrange's multiplier. The output function f(x) is using the tuning function, SVR model with a cost of 2, 23 support vectors, radial basis kernel, and an epsilon value of 0.2 is used to develop our statistical model.

6 Results and Discussion

The results are obtained with the proposed fall-detection model using the database generated by SNUSense for training and testing. Using SVM tuning, we obtain the optimum SVM classifier is determined with radial basis kernel, gamma = 0.33, $\xi = 0.1$ and 25 support vectors. A total of 215 experimental observations of the fall event have been used for testing and the rest 54 have been used for testing the model. If P is the probability of occurrence, binary 1 represents the occurrence and binary 0 represents the non-occurrence of the fall event. The detection of the single fall event by our system is shown in Fig. 9.

The accuracy of the detection model has been evaluated using mean absolute percentage error (MAPE) [22]. MAPE is defined as

$$MAPE = \frac{1}{N} \sum_{i=0}^{N} \left| \frac{y_i - f(x_i)}{y_i} \right| * 100 \tag{10}$$

Where, N is the number of observations, y_i is the actual event that occurred at i^{th} observation (represented by 0 or 1), x_i is the input vector (time), and f is the forecast model.

The MAPE for our SNU database was found to be 1.80%, i.e. the model is able to accurately detect 53 out of 54 test occurrences. The detection of the fall event by the IoT system is depicted in Fig. 10.

Hence, it is concluded that our fall detection system has an accuracy of 98.2% and is capable of accurately detecting events of vehicle fall.

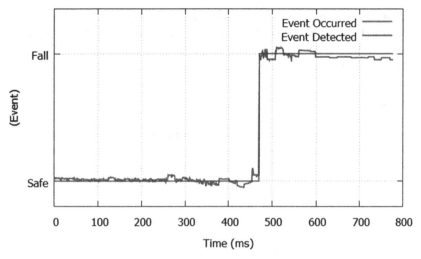

Fig. 9. Detection of a single fall event by proposed IoT system

7   Conclusion and Future Work

In this work, a reliable, affordable and precise IoT system has been presented, which can be retrofitted to any vehicle. It uses a smartphone and connected sensors to sense absolute linear acceleration, speed, and altitude of the vehicle. These parameters are used to develop an SVM based fall detection model, which has been proven to accurately detect events of vehicle fall events in this work.

As a future work, we propose to develop and evaluate an IoT accident reporting system which is capable of detecting as well as classifying the types of accident event.

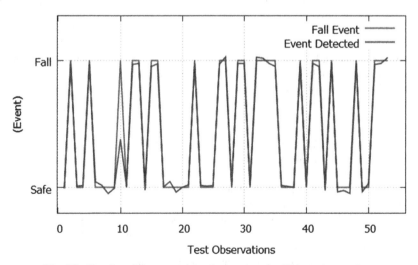

Fig. 10. Results of the event detection system for 54 test observations

The proposed system will be able to report the occurrence as well as the severity of the accident to the end users. For this purpose, we plan to use parameters such as the vehicle speed, acceleration, roll, pitch and altitude to categorize the road accident as mild, moderate or severe.

References

1. Golbal Status Report on Road Safety, World Health Organisation (WHO) (2018)
2. Iyoda, M., Trisdale, T., Sherony, R., Mikat, D., et al.: Event data recorder (EDR) developed by Toyota motor corporation. SAE Int. J. Trans. Safety **4**(1), 187–201 (2016). https://doi.org/10.4271/2016-01-1495
3. Consultancy Services for the Provision of a Road Accident Data Management System (RADMS) - Variation Order No. 4 – Operational Support and Training Completion Report - Quarter 3 from April 2018 to June 2018
4. Ferreira, J., Carvalho, E., Ferreira, B.V., de Souza, C., Suhara, Y., et al.: Driver behavior profiling: an investigation with different smartphone sensors and machine learning. PLoS One **12**(4), e0174959 (2017)
5. Aloul, F., Zualkernan, I., Abu-Salma, R., Al-Ali, H., Al-Merri, M.: iBump: smartphone application to detect car accidents. Comput. Electric. Eng. **43**, 66–75 (2015). ISSN 0045-7906
6. Liyanage, Y., Zois, D.-S., Chelmis, C.: Quickest freeway accident detection under unknown post-accident conditions. In: 6th IEEE Global Conference on Signal and Information Processing (GlobalSIP), Anaheim, CA, 26–29 November 2018 (2018)
7. Dar, B.K., Shah, M.A., Islam, S.U., Maple, C., Mussadiq, S., Khan, S.: Delay-aware accident detection and response system using fog computing. IEEE Access **7**, 70975–70985 (2019). https://doi.org/10.1109/ACCESS.2019.2910862
8. Xu, X., Yu, J., Chen, Y., Zhu, Y., Qian, S., Li, M.: Leveraging audio signals for early recognition of inattentive driving with smartphones. IEEE Trans. Mob. Comput. **17**(7), 1553–1567 (2018). https://doi.org/10.1109/tmc.2017.2772253
9. Park, H., Ahn, D., Park, T., Shin, K.G.: Automatic identification of driver's smartphone exploiting common vehicle-riding actions. IEEE Trans. Mob. Comput. **17**(2), 265–278 (2018). https://doi.org/10.1109/tmc.2017.2724033
10. Bhatti, F., et al.: A novel Internet of Things-enabled accident detection and reporting system for smart city environments. Sensors **19**(9), 2071 (2019). https://doi.org/10.3390/s19092071
11. Lohani, D., Acharya, D.: Real time in-vehicle air quality monitoring using mobile sensing. In: 2016 IEEE Annual India Conference (INDICON), Bangalore, pp. 1–6 (2016). https://doi.org/10.1109/indicon.2016.7839099
12. https://github.com/Sensorcon/Sensordrone
13. https://g-p.hk/gptoys-foxx-s911.html
14. https://firebase.google.com
15. Lee, Y., Yeh, H., Kim, K.-H., Choi, O.: A real-time fall detection system based on the acceleration sensor of smartphone. Int. J. Eng. Bus. Manage. (2018). https://doi.org/10.1177/1847979017750669
16. Kendall, J., Solomon, K.A.: Air bag deployment criteria, The Forensic Examiner (2014)
17. http://www.gpsinformation.org/dale/nmea.htm
18. Utkin, L.V., Chekh, A.I., Zhuk, Y.A.: Binary classification SVM-based algorithms with interval-valued training data using triangular and Epanechnikov kernels. Neural Netw. **80**, 53–66 (2016). ISSN 0893-6080
19. Mathur, A., Foody, G.M.: Multiclass and binary SVM classification: implications for training and classification users. IEEE Geosci. Remote Sens. Lett. **5**(2), 241–245 (2008)

20. Hmeidi, I., Hawashin, B., El-Qawasmeh, E.: Performance of KNN and SVM classifiers on full word Arabic articles. Adv. Eng. Inform. **22**(1), 106–111 (2008)
21. Madzarov, G., Gjorgjevikj, D., Chorbev, I.: A multi-class SVM classifier utilizing binary decision tree. Informatica **33**, 233–241 (2009)
22. de Myttenaere, A., Golden, B., Le Grand, B., Rossi, F.: Mean absolute percentage error for regression models. Neurocomputing **192**, 38–48 (2016). https://doi.org/10.1016/j.neucom.2015.12.114. ISSN 0925-2312

Signal and Image Processing

A Scalable Feature Selection and Opinion Miner Using Whale Optimization Algorithm

Amir Javadpour[1], Samira Rezaei[2], Kuan-Ching Li[3], and Guojun Wang[1(✉)] ⓘ

[1] School of Computer Science, Guangzhou University, Guangzhou 510006, China
a_javadpour@e.gzhu.edu.cn, csgjwang@gzhu.edu.cn
[2] Bernoulli Institute for Mathematics and Computer Science, University of Groningen, Groningen, The Netherlands
[3] Department of Computer Science and Information Engineering, Providence University, Taichung, Taiwan
kuancli@gm.pu.edu.tw

Abstract. Due to the fast-growing volume of text document and reviews in recent years, current analyzing techniques are not competent enough to meet the users' needs. Using feature selection techniques not only support to understand data better but also lead to higher speed and also accuracy. In this article, the Whale Optimization algorithm is considered and applied to the search for the optimum subset of features. As known, F-measure is a metric based on precision and recall that is very popular in comparing classifiers. For the evaluation and comparison of the experimental results, PART, random tree, random forest, and RBF network classification algorithms have been applied to the different number of features. Experimental results show that the random forest has the best accuracy on 500 features.

Keywords: Feature selection · Whale Optimization algorithm · Selecting optimal · Classification algorithm

1 Introduction

Nowadays, with the rapid growth of networks and social media, people can publish their ideas and thoughts directly on the internet. Reading and accessing to these opinions and comments on a variety of topics have been tremendously appealing [1]. On the other hand, a massive amount of generated data is not easy to handle. Users can not manually analysis these data and some summarization and classification algorithms are needed to reduce the entire data into meaningful classes. This is very important for cloud service providers as they face memory such as Cloud Computing-based Data Storage and Disaster Recovery [2–4]. Besides, as there are many contradictory ideas about a topic, one user needs to understand the positive and negative views and the reason of this contradiction in the available reviews [5–8]. This is where automatic analysis of comments and summarizing them into valuable information emerges and plays an vital role to generate scalable solutions [9–11].

© Springer Nature Singapore Pte Ltd. 2020
S. M. Thampi et al. (Eds.): SIRS 2019, CCIS 1209, pp. 237–247, 2020.
https://doi.org/10.1007/978-981-15-4828-4_20

The business world is also very dependent on this strategy as the customers prefer to read and analyze the previous customers' reviews before purchasing an item. A lot of online marketers have been created based on this need which customers want to be aware of other people's opinions about such products. Therefore, it is evident that the way to analyze and explore the ideas and beliefs of users can be useful for applications. Emotions can be defined as positive or negative feelings. Exploring the ideas is a computational technique for extracting, classifying, understanding, and obtaining the ideas expressed in various contents [11].

Karamatsis presented a system study that applies more than five dictionaries to classify comments in social into two categories [12]. Each dictionary has seven features for each message that was later used as an entry for SVM classifier. They tested their system with several datasets containing such as twits and comments on twitter. Authors in [13] used supervised statistical systems to design an emotion analyzer in the long term. The dictionary automatically generates an optimal set of features based on user's twits over a long period of time. Latent Dirichlet Allocation (LDA) is also used to cluster the document words into unsupervised learned topics using Poisson distribution in [14]. Multinomial and Dirichlet distributions are other options to use specially when the number of words is very high. For each word in the set of opinions, a specific topic is randomly selected by using the multinomial distribution.

Classifying emotions in medical domain using ant colony optimization algorithm (ACO) is done in [12]. The proposed classification framework has considered the content of medical documents which impart patients health information such as health status, obtained observations and examination results. They even might have evidence about diagnoses and interventions. Furthermore, analyzing and evaluating these data should be done wisely as it will be used to judge clinical outcomes or asses the impact of a medical condition on patient's well-being. In this study feature sets are extracted from reviews using Term Frequency-Inverse Document Frequency (TF-IDF) and then ACO has been applied. At the final step, they applied classification algorithms such as Naïve Bayes, and Support Vector Machine (SVM) to the selected features.

Besides ACO, other metaheuristic algorithms also have been widely used for feature selection in opinion mining and sentiment analysis. Authors in [13] used firefly algorithm to optimize the exploration of comments in stock market. A hybrid model of metaheuristic algorithms of Whale Optimization Algorithm (WOA) and Simulated Annealing (SA) algorithm has been used in [15]. The main idea of using SA is to improve the output of WOA by exploring the most promising regions detected by WOA algorithm. Authors evaluated their method on 18 standard benchmark datasets of UCI repository and compared with three well-known wrapper feature selection methods in the literature. Other approaches such as Genetic Algorithm and Rough Set Theory [16] has also been used to extract and identify sentiment patterns on opinions by consumers. Authors compared the quality of traditional feature selection techniques with sentiment analysis methods by applying text classification.

Researchers in [17] analyzed the spam comments which artificially either promote or devalue the quality of a specific product or service. This study is based on the fact that many of features are redundant, irrelevant and noisy. Furthermore, extracting meaningful features from in a very high dimensional feature space has a crucial role in detecting spam comments. The extracted features by a hybrid model of Cuckoo and Harmony search are used to classify comments into spam and non-spam reviews using the Naive Bayes algorithm. Another study proposed sentiment mining to analyze customers' comments to handle the unstructured format of online reviews. Researchers considered the hybrid role of typed dependency relations (TDR) and part-of-speech tagging (POST) to find existing relation between features and sentiment words. These techniques also have been used to create a list of rules in identifying relations between features [18].

Table 1. An example of preselected words in different classes

Class	A few examples of Preselected words
Positive	Easy, good, economic, Good, quality, inexpensive, Awesome, fun
Negative	eat, no, gulps, awkward, consumption, small, crazy, expensive
Neutral	sometimes, be, see Viewfinder, life, is, are, I, discover, media

The rest of this article contains the following parts. In the next section, we review the proposed method and introduce the proposed feature selection algorithm in details. Then we present the result and the discussion. We will also examine the classification algorithms, the results of their application on the optimal set of features, and results of applying them on an optimal set of features. Finally, conclusions and future suggestions are explained.

1.1 Modeling and Problem Solving

In this section, we explain the details of the proposed method. Figure 1 illustrates the main steps of our method. By importing data into our system, we should take care of any incompatibility of data and our programing and simulating software. To be able to handle texts in Matlab and apply feature selection, we need to operate data preprocessing. The software that we have used to apply data preprocessing on is Visual Studio, and the selected programming language is C#. We defined three classes to put each word in the sentences on. The defined classes are positive words, negative and neutral words. In the preprocessing step we also made some changes based on this word division. The idea behind this is to create a dataset of positive and negative words for the next steps and select most significant words among them.

Regarding the effect of feature selection in achieving higher accuracy of classification algorithms, feature selection in this research has been utilized through the method of feature selection of the Whale Optimization algorithm. Once the top features are selected, we applied different classification algorithms such as random forest, RBF network, random tree, and PART. Finally, we compare the results based on the evaluation criteria. The database used in this research is available at the Illinois University Data Center [19] and has two sections in which the users' positive and negative views are included for a mobile product. The number of positive comments in the database is 22940, and the number of negative comments is 22935, so the total number of records in this study is 45875.

1.2 Preparing of Data

Preprocessing is one of the essential steps in any machine learning algorithms. There are two text files containing negative and positive comments as our input data. We applied initial processing steps to be able to use this data for future analysis. The methodology is based on the presented method in [20], in which the words in the comments are divided into three categories of words representing a positive opinion, the words representing negative opinions and neutral words. In order to do this, two sets of data containing positive words and negative words are extracted by reviewing the opinions in the dataset, as shown in Table 1. After extracting positive and negative words, all records in the database are examined according to the positive and negative lists.

The process of achieving the optimal set of features is that according to a study [21] that explores the methods of feature selection in the users' comments. The number of repetitions per word in the entire data has been calculated, and the following criteria have been applied to emit some words from our positive, negative and neutral words.

i.   The list of words which have been repeated only once or twice in the entire dataset.
ii. In order to remove prepositions, words with repetitions more than half the number of records are deleted from the vocabulary lists.

Therefore, in the selected initial set of words, words with a frequency less than two and more than half the total number of comments are removed from the data. Also, in order to prevent any mistakes in lowercase and uppercase letters, all letters are converted into lowercase. The total number of records containing the negative and

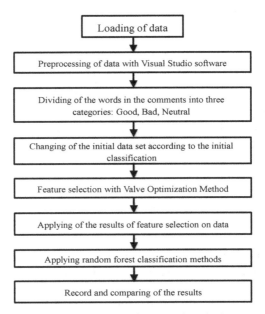

Fig. 1. Flowchart of proposed method.

positive comments are 22935 and 22940 respectively. The entire preprocessing step is implemented in Visual Studio software with C# programming language.

2   Feature Selection

The Whale optimization algorithm is a relatively new meta-heuristic algorithm that imitates humpback whales hunting behavior. The significant difference between this algorithm and other heuristic algorithms is the stimulation of random hunting behavior. In other words, this algorithm is based on the behavior of humpback whales to identify the location of prey and encircle it. At each iteration, WOA algorithm assumes that the current best candidate is the optimum solution or is close it. Since the position of the optimal design in the search space is not known a priori, the WOA algorithm assumes that the current best candidate solution is the target prey or is close to the optimum. After the best search agent is defined, the other search agents will hence try to update their positions towards the best search candidate. Other search agents update their position based on the position of the best current solution [22]. Initially, the parameters related to the Whale optimization algorithm are set according to the conditions of the problem. In the next step, it is necessary that through fitting function, we evaluate the results achieved at each stage and improve the results. Table 2 shows the parameters used in this algorithm.

Table 2. Parameters used in the Whale Optimization Algorithm.

Parameters	Description	Quantity
SearchAgents_no	The number of search factors	30
MaxIt	Maximum number of rings	100
lb	lower limit	0
ub	Upper limit	4000
dim	Number of Dimensions	100

(a)   Fit Function

In order to select the most critical features for opinion mining, several methods can be used to compute the fit function. In this paper, we used the F-Measure's criteria to evaluate the results. It is a combination of two criteria of precision and recall that are extensively considered in calculating the quality of the classifications. The concepts of FP, TP, and FN in this paper are defined according to the textual data content as follows [3]:

TN: Indicates the number of records whose actual category is negative, and the proposed system could also correctly identify this category.

TP: Represents the number of records whose real category is positive, and the proposed algorithm also correctly recognizes this category.

FP: Represents the number of records that are negative in their actual data but the proposed algorithm has mistakenly identified it as positive.

FN: Indicates the number of records whose actual data is positive, but the proposed algorithm has put this record in a negative category.

$$Precision = \frac{TP}{TP + FP} \tag{1}$$

$$Recall = \frac{TP}{TP + FN} \tag{2}$$

$$Accuracy = \frac{TP}{TP + FP + TN + FN} \tag{3}$$

$$f(measure) = \frac{2 * recall + precision}{recall + precision} \tag{4}$$

Table 3. Examining the variations of 10 words in the selected collection of words.

Repetitions	The word in the positive selection of words
The first repeat	'forever.', 'outstanding', 'numbers', 'hardly', 'efficiency', 'clam', 'full', 'reception!, 'developing', 'add'
The tenth repeat	forever.', 'outstanding', 'numbers', 'low', 'efficiency', 'clam', 'batteries', 'reception!', 'developing', 'add'
The fiftieth repeat	forever.', 'outstanding', 'camcorder', 'easy-to-use', 'efficiency', 'beautiful', 'clam' 'batteries', 'reception!', 'developing'

(b) Applying Classification Algorithms

At this stage, considering the best wordlist from the feature selection step, we will apply a random forest algorithm. The implementation of this part of the simulation has been done with the Weka open-source tool. The random forest algorithm has a huge potential to become a popular method for future classifications since its performance is comparable to other group methods, including Bagging and Boosting. This algorithm, as a voting group algorithm, produces several different decision trees as the basic categories and applies the majority vote to the original tree results. In order to evaluate the proposed method, the k-fold method has been used.

3   Simulation and Results

With the implementation of the Whale optimization algorithm in the feature selection phase, in each repetition, the algorithm tries to find the optimum solution according to the fit function. At first, a collection of words is extracted randomly from positive and negative words, and the final result is the best vocabularies that can optimize the defined fit function. The results are reported in Table 3. The number of selected features (selected words) is 100, in which 50 are positive, and 50 are related to the negative wordlist. The results of applying classification algorithms on the data, considering the set of selected features in the feature selection step have been followed up. By examining the results, it is observed that the random forest algorithm performs better than all other classifier algorithms in all of the evaluation criteria. However, the running time for this algorithm is higher compared to other methods (Fig. 2).

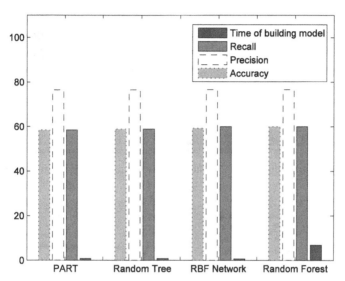

Fig. 2. Comparison of the quality indices obtained from applying classifier algorithms on 100 selected features.

(a) Change the Number of Features

In this section, we have changed the number of features selected in the Whale algorithm. The number of selected words in the Whale algorithm has been changed to 200 and 500. Due to the increasing the number of words, we expect the program's running time to increase. By comparing the obtained results, it is observed that with the increasing number of features, the accuracy of the fit function is much improved. This means that by increasing the number of features, we have been able to achieve better results from the set of features or words in our evaluation function. Table 4 and Fig. 4 shows the results of the applying classification algorithms on the selected data set with 200 and 500 words (features) (Fig. 3).

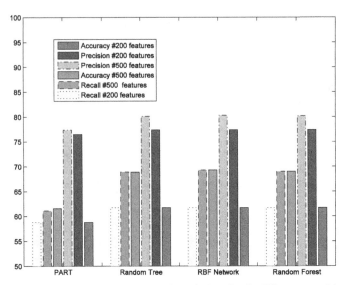

Fig. 3. Chart of changing the value of the fit function in different repetitions.

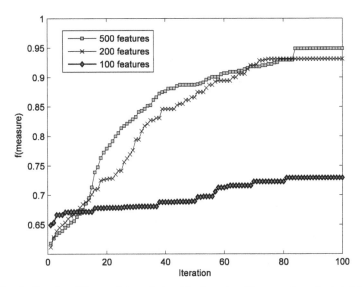

Fig. 4. Results of the applying of classification algorithms on the selected data.

4   Conclusion

Social media is the most crucial advertisement strategy for most Internet marketing companies. Although in the past advertising and public relations have played a significant role in firm brand promotion, it is social media based on virtual social networks that measure the value of organizations, companies, products, and services that have dominated in the last decades. In this research, the Whale optimization algorithm has been used to select the best set of features, aimed at reducing the dimensions of the data in the mining opinion or review classification. We tested our model over different number of features. We started with 100 features, and then we compared the classification quality by choosing 200 and 500 features. In this study, we applied different classification algorithms such as RBN network, random tree, and random forest. We recorded the accuracy of the obtained methods. We propose a hybrid model of metaheuristic algorithms for feature selection. Also, other fit functions such as information gain can be used to evaluate the obtained results (Table 5).

Table 4. Evaluation criteria derived from the applied methods for 100 features with k-fold evaluation.

Evaluation criteria	PART	Random tree	RBF network	RF
Accuracy	58.48	58.92	59.3	60
Precision	76.4	76.5	76.6	76.6
Recall	58.5	58.9	60	60
Time of building model	0.76	0.82	0.65	6.86

Table 5. Evaluation of the applied methods with k-fold evaluation.

Criteria	# Features	PART	Random tree	RBF network	RF
Accuracy	200	58.78	61.7	61.728	61.71
Accuracy	500	61.55	68.89	69.297	69.02
Precision	200	76.5	77.4	77.4	77.4
Precision	500	77.4	80.1	80.3	80.2
Recall	200	58.8	61.7	61.7	61.7
Recall	500	61.1	68.9	69.3	69

Acknowledgment. This work was supported in part by the National Natural Science Foundation of China under Grant 61632009, in part by the Guangdong Provincial Natural Science Foundation under Grant 2017A030308006, and in part by the High-Level Talents Program of Higher Education in Guangdong Province under Grant 2016ZJ01.

References

1. Fresneda, J., Gefen, D.: Automatic methods for online review classification: An empirical investigation of review usefulness—An abstract. In: Stieler, M. (ed.) Creating Marketing Magic and Innovative Future Marketing Trends. DMSPAMS, pp. 1331–1332. Springer, Cham (2017). https://doi.org/10.1007/978-3-319-45596-9_242

2. Javadpour, A., Wang, G., Rezaei, S., et al.: Detecting straggler MapReduce tasks in big data processing infrastructure by neural network. J. Supercomput. (2020). https://doi.org/10.1007/s11227-019-03136-6

3. Javadpour, A., Kazemi Abharian, S., Wang, G.: Feature selection and intrusion detection in cloud environment based on machine learning algorithms. In: 2017 IEEE International Symposium Parallel Distribution Processing with Application 2017 IEEE International Conference Ubiquitous Computing Communications, pp. 1417–1421 (2017)

4. Javadpour, A., Wang, G., Rezaei, S., Chend, S.: Power curtailment in cloud environment utilising load balancing machine allocation. In: 2018 IEEE SmartWorld, Ubiquitous Intelligence Computing, Advanced Trusted Computing, Scalable Computing Communications, Cloud Big Data Computing, Internet of People and Smart City Innovation (SmartWorld/SCALCOM/UIC/ATC/CBDCom/IOP/SCI), pp. 1364–1370 (2018)

5. Yang, W., Wang, G., Choo, K.-K.R., Chen, S.: HEPart: a balanced hypergraph partitioning algorithm for big data applications. Futur. Gener. Comput. Syst. **83**, 250–268 (2018)

6. Zhang, Q., Liu, Q., Wang, G.: PRMS: a personalized mobile search over encrypted outsourced data. IEEE Access **6**, 31541–31552 (2018)

7. Zhang, S., Wang, G., Bhuiyan, M.Z.A., Liu, Q.: A dual privacy preserving scheme in continuous location-based services. IEEE Internet Things J. **5**(5), 1 (2018)

8. Ali, F., Kwak, K.-S., Kim, Y.-G.: Opinion mining based on fuzzy domain ontology and support vector machine: a proposal to automate online review classification. Appl. Soft Comput. **47**, 235–250 (2016)

9. Javadpour, A., Memarzadeh-Tehran, H.: A wearable medical sensor for provisional healthcare. In: ISPTS 2015 – Proceedings of 2nd International Symposium Physics and Technology Sensors Dive Deep Into Sensors, pp. 293–296 (2015)

10. Javadpour, A., Memarzadeh-Tehran, H., Saghafi, F.: A temperature monitoring system incorporating an array of precision wireless thermometers. In: 2015 International Conference on Smart Sensors and Application (ICSSA), pp. 155–160 (2015)

11. Balazs, J.A., Velásquez, J.D.: Opinion mining and information fusion: a survey. Inf. Fusion **27**, 95–110 (2016)

12. Karampatsis, R.-M., Pavlopoulos, J., Malakasiotis, P.: AUEB: two stage sentiment analysis of social network messages. In: SemEval@COLING (2014)

13. Zhu, X., Kiritchenko, S., Mohammad, S.: NRC-Canada-2014: recent improvements in the sentiment analysis of tweets. In: Proceedings of the 8th International Workshop on Semantic Evaluation (SemEval 2014), pp. 443–447 (2014)

14. Santosh, D.T., Babu, K.S., Prasad, S.D.V., Vivekananda, A.: Opinion mining of online product reviews from traditional LDA topic clusters using feature ontology tree and sentiwordnet. Int. J. Educ. Manag. Eng. **6**(6), 34 (2016)

15. Mafarja, M.M., Mirjalili, S.: Hybrid whale optimization algorithm with simulated annealing for feature selection. Neurocomputing **260**, 302–312 (2017)

16. Ahmad, S.R., Bakar, A.A., Yaakub, M.R.: Metaheuristic algorithms for feature selection in sentiment analysis. In: 2015 Science and Information Conference (SAI), pp. 222–226 (2015)

17. Rajamohana, S.P., Umamaheswari, K., Keerthana, S.V.: An effective hybrid cuckoo search with harmony search for review spam detection. In: 2017 Third International Conference on Advances in Electrical, Electronics, Information, Communication and Bio-Informatics (AEEICB), pp. 524–527 (2017)

18. Ahmad, S.R., Yaakub, M.R., Bakar, A.A., et al.: Detecting relationship between features and sentiment words using hybrid of typed dependency relations layer and POS tagging (TDR Layer POS Tags) algorithm. Int. J. Adv. Sci. Eng. Inf. Technol. **6**(6), 1120–1126 (2016)

19. The University of Illinois at Chicago: Opinion Mining, Sentiment Analysis, and Opinion Spam Detection (2017). https://www.cs.uic.edu/~liub/FBS/sentiment-analysis.html

20. Zhang, K., Xu, H., Tang, J., Li, J.: Keyword extraction using support vector machine. In: Yu, J.X., Kitsuregawa, M., Leong, H.V. (eds.) WAIM 2006. LNCS, vol. 4016, pp. 85–96. Springer, Heidelberg (2006). https://doi.org/10.1007/11775300_8

21. Forman, G.: An extensive empirical study of feature selection metrics for text classification. J. Mach. Learn. Res. **3**, 1289–1305 (2003)

22. Mirjalili, S., Lewis, A.: The whale optimization algorithm. Adv. Eng. Softw. **95**, 51–67 (2016)

Reliability and Connectivity Improve the Ranking Principle

Harshit Pandey[1]([⊠]), Priya Ranjan[2], Arjun Singh[3],
and Malay Ranjan Tripathy[4]

[1] Fiserv, Noida, India
harshit.pandey@fiserv.com, harshitpandey16@gmail.com
[2] EEE, ASET, AUUP, Noida, UP, India
[3] ASET, AUUP, Noida, UP, India
[4] ECE, ASET, AUUP, Noida, UP, India

Abstract. The world of networks inherently being complex, is a primitive based on graphs. It uses adjacency matrix for abstracting the topology of underlying expanse. This structure of graphs, single-handedly decides the entries of the adjacency matrix, and thus, the overall topology of the system being described. In this research, we open new horizons at the cost of conventions of traditional adjacency formulation. We argue that fresh perspectives need to be plugged into as adjacency entries for modeling, which need not necessarily be binary in nature, rather a real number between one and zero. This adjacency entry between one and zero can be interpreted as a probability of successful transmission in a communication network for example. Also, we roll out the viable idea that number of connections together with probability of successful transmission for a link combined, rules the crux of ranking nodes as decided by the HITS algorithm. Broader perspectives have been discovered, deviating from traditional approaches of capturing networks and an innovative computational insight has been presented to illustrate the relationship between reliability of links, connectivity and leadership in the community. In particular, results presented here are in agreement with work done in the area of trust, reputation and leadership in the sense that leaders should be able to attain highest amount of activity.

Keywords: HITS (Hyperlink Induced Topic Search) · SVD (Singular Value Decomposition) · Hubs · Authorities · Single layer networks · Adjacency matrix · Reliability · Connectivity

1 Introduction

Analysis complexity soars with increased number of candidates forming a structure/system. Networks, which flourish on innumerable nodes, inherit the same fate. As data is becoming the most valuable commodity in today's life, every transmission and reception of data packets accurately and appropriately

© Springer Nature Singapore Pte Ltd. 2020
S. M. Thampi et al. (Eds.): SIRS 2019, CCIS 1209, pp. 248–262, 2020.
https://doi.org/10.1007/978-981-15-4828-4_21

demands high attention. The Internet, due to its enormous utility and its trade-offs being more in the positive, makes the exchange of resources encrypted as data packets, convenient and reliable. However, this transfer depends on the topology of the network, where nodes and links play an eminent role in deciding where congestion bottlenecks might appear and where the shortest paths might exist! Multifaceted explorations in Networks, though strenuous and intricate, yield results which are much looked out for, considering the developments in science. This kind of research has to be in sync with the state-of-the-art technical developments, and needs fresh perspectives to engender innovation. Models of networks cover a wide spectrum depending on what approach is being deployed. However, majority of scenarios consider the base framework to be intact, with variations in node numbers and links.

Network modeling and analysis, has rich literature to start with, and gets complicated pretty soon if one wants to delve into details. Congestion avoidance and control as 'packet-conservation' and 'self clocking' mechanism based on nonlinear control principles of Lyapunov [1] are discussed by Van et al. in their landmark paper which saved the internet as we know it [3]. Kelly's [2] work forms the base of mathematical modeling for the internet, with rate allocations, traffic rerouting, and stability with optimization issues, fairness criterion and effects of a delayed feedback in a local sense. Stochastically evolving networks with random interactions in game theories discussed in [4] and works of Ranjan et al. in [5,6] talk of Kelly's framework being leveraged to provide conditions of global stability with arbitrary delays and evolving a feedback system model for analysis of parametric sensitivities in transport protocol dynamical interactions of sender and receiver. Further, due to ever expanding data packets, their timely delivery with multiple optimizations, pareto-efficiency, apt throughput, evolving topologies and their priorities etc. render multilayer networks extremely intriguing. Works in [7–11], talk of multilayer networks with- optimal rate allocations, their survivability, mathematical formulations, their in-depth analysis and their architectures. As complexity of networks increases, a strong tool to represent time varying and multilayer networks together can be found in Multi Aspect graph formalism [12]. This framework justifies the complexities of multilayered real life networks by considering a number of indices that pitch in for pragmatic understanding. Work in [12] describes the MAG's in their rigorous mathematical and programmable framework. From inception of modeling of two-node networks to MAG's, there is a natural question of reliability in any network organization and operation, which has not received the attention it deserves. Reliability has been defined in various ways, where [13] and [14], both discuss their own takes on reliable throughput and interconnections.

In the real time domain analysis, apart from the strength of connectivity between nodes [14], the total number of links associated with a node at an instance also plays a rooted role in defining scores that in turn are an indicator of the rankings. Literature for rankings can be found in [15–17] where specifically, HITS search algorithm is explained in [18] and its modern version for temporal multilayer settings [19]. For a node to top in the race of rankings, the reliability factor of links as discussed in [14] also needs to be accompanied by parameters of interest that might play an essential role in determining scores.

This work addresses some of the major questions outlined below:

1. Taking this reliability perspectives explained in [14], what remains to be explored, is the type of the connections formed among the nodes, keeping the number of connections at that node in mind while mapping rankings in the community and what could be a possible relationship among reliability of links, connectivity and leadership ranks?
2. What we observe is a phenomenon of revolution, where lower rank nodes get highly ranked as their reliability increases. The question we would like to investigate is that, what are the real life implications of such a revolution, and are there real data sets available, to validate the theoretical framework developed in this work?

Specifically, this research aims to explore that computational ranking of nodes in a network based on hyperlink induced topic search algorithm and functional relationship with multiple parameters considered. Specifically, we argue that there is a fresh need to consider multiple parameters for rankings, given that this question needs wide attention in contemporary discourse. This research applies the mathematical tool, the Singular Value Decomposition (SVD) for generating the scores of hub and authority nodes in a network [20]. In our work, a phenomenon of revolution is observed, where node rankings improves drastically due to the effect of 'reliability of connection' and 'connectivity' combined. We consider reliability of connections along with number of connections put together, and computationally demonstrate that the top scores vary as a combination of both, reliability and connectivity. Perceptions we gather from this work, break conventions, where the thought that more number of node connections alone lead to better rankings, fails, and the evolution of top scorers happens due to a complex interaction of underlying parameters. The real life debate of either having stronger connections with powerful people or having more numbers of connections leverage the opportunities for a person, is given an innovative direction by this work, as we delve more into the intricate complexities and demonstrate as to what actually matters in computational ranking also matters in real life! This work is intriguing and motivating at the same time because it has a real life implications to improve one's social standing. Intuition from this could be further directed into believing that there might be cases wherein, apart from these two parameters of interest in deciding top scores, there could be many others, thus leaving the research open and challenging both for the research fraternity and for the social network ranking analysis [19]!

2 Mathematical Representation

2.1 HITS and SVD

The Hyperlink Induced Topic Search (HITS) aims to recognize web pages as good authorities and hubs. This algorithm aims to find pages on the web having higher authority and relating aptly to the query. A page with relevant information is an authority whereas a page pointing to an authority is a hub [15]. The algorithm works by hooking two numbers to a page, a hub and an authority weight being updated iteratively. Every page on the web is hooked with two weights:

1. Authority weight a_j
2. Hub weight h_j

Web pages with higher values of a_j are better authorities whereas hubs are pages with higher value of h_j. Vectors for weights of hub and authority are denoted by v and u and their upgrade at each iteration is given by:

$$u = A'v \tag{1}$$

$$v = Au \tag{2}$$

Iterating for k times gets the realization as:

$$u_k = AA'v_{k-1} \tag{3}$$

$$v_k = A'Au_{k-1} \tag{4}$$

To analyze the hub and authority scores for the graph considered, the Singular Value Decomposition (SVD) tool is applied to the adjacency matrix denoted by A. The singular values of the matrix A of size $j * k$ are realized as the square roots of the eigen values of the matrix $A'A$ which is symmetric $(k * k)$. These singular values are enumerated with their multiplicities in decreasing order as: $\sigma_1 \geq \sigma_2 \geq \sigma_3...\sigma_k$. Taking a square matrix $A \in M_{n,n}(C), \lambda_1(A),.........\lambda_n(A)$ represent the eigenvalues of A which are in fact the roots in C of the main characteristic polynomial given by:

$$det(A - ZI) \in C[Z]$$

Then, eigen values of the matrix A can be seen as: $|\lambda_1(A)| \geq,.........,\geq |\lambda_n(A)|$. Taking the singular value decomposition into account, each matrix $A \in M_{m,n}(K)$, there inherently exists some K unitary matrices, given by $U(m * m)$ and $V(n * n)$ together with a sequence of real numbers $s_1 \geq ... s_{mn} \geq 0$, such that $U*AV = diag(s1,...........s_{mn}) \in M_{m,n}(K)$. The first column of matrices U and V give us the hub and authority scores [20].

2.2   Reliability and Connectivity

The traditional method of adjacency construction is with entries as only binary digits, that is, either 0 or 1. Thus, the entries A(ij) of any adjacency A can be visualized as:

$$A_{ij} = \begin{cases} 1, & \text{if i connects to j} \\ 0, & \text{otherwise.} \end{cases} \tag{5}$$

The conventional form of employing only 0 and 1 infer that connections are rigid and exist for an entry as 1 and do not exist for entry 0. In the real physical time analysis, this does not hold true. There are situations when there exists a connection, but a weak one, as the link is only partially capable of sending the data. This would mean that the transmission is not faithful. Hence, we can plug in reliability as an aspect for formulating the adjacency for networks to faithfully replicate the real timed physical scenarios. We consider the adjacency being formulated on grounds on reliable transmissions, where entries might be a probability. In accordance to this, we define the connections as:

p, the strength of the link of connection
$1 - p$, the drop probability

where p is a real number between 0 and 1. This value of p is assigned depending upon the strength of connection.

Also, together with the entries being an aspect of transmission probabilities, we plug into the system, a parameter of connectivity, which means, different number of connections with different nodes, leads to the adjacency being varying in the number of entries it holds, depending on the number of connections.

3   Model Formulation

For the purpose of understanding networks, they need to be modeled mathematically, so that connections of the network can be observed as parameters that can be computed. The best way to model networks is using conventional graph theory, specifically the -Adjacency matrix. Depending on the type of model inherited, the adjacency can be dynamic (where connections change at every epoch), or it can be static (where network connections do not change with time). Node rankings are computed based on this adjacency, where a small change in the adjacency entries, can transform a slave node into a master node by giving higher scores and thus imitating a Revolution like phenomenon! In this work, we consider single layer networks for simplicity with nodes in the integer powers of the number 2, where simulations are begun starting with networks of 4 nodes, and without the loss of generality, scaling up these simulations to networks formed by 16 nodes. Two types of networks are considered for a particular node number, namely Symmetric and Asymmetric. Symmetric means that same number of connections exist for all nodes, however, on the contrary, Asymmetric means that the number of connections to nodes in the network are not the same. Thus, simulations are run over networks of 4, 8 and 16 nodes with two versions for each, symmetric and asymmetric.

The peculiar aspect of models considered for this work is that, in every model, the probability of connection from node 1 and 2 is a varying probabilistic quantity. This means that if, node 1 connects to node 2 with probability $p1$, then may be node 2 connects to node 1 with probability $p2$. This can be imagined as a two way process, where going from A to B has a broader path, while returning from B to A may be a narrower path and vice versa. For all other nodes, except the nodes being pointed by node 1 and node 2, the probability of connection remains unity as we wish to investigate the impact of $p1$ and $p2$ on overall node rankings.

$p1$ and $p2$ are varied from 0 to 1 with a step size of 0.1 at every instance, and corresponding node rankings are computed and plotted. What we are interested in, is observing the changes in the dominance patterns which arise for authorities and hubs, when connection probabilities $p1$ and $p2$ are varied and how the number of connections to a node plays an important role in deciding its overall rank. On each model, we operate the SVD tool on its adjacency and find rankings based on incrementing the connection probability of node 1 and 2. This leads us to realizations that help address the patterns observed as an aftermath of the SVD operation, to clearly figure out the Revolution like phenomenon observed.

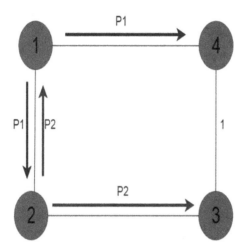

Fig. 1. Four node symmetric

The symmetric models of networks in this case, need not be of much attention, as this work aims to see the combined effect of connectivity and strength of connections together. However, in the symmetric case, the parameter of connectivity gets ruled out since all networks would have same number of connections. This type of a network analysis has already been done in our previous work [13]. For ease of understanding, a 4 node symmetric model has been incorporated as seen in Fig. 1 Fig. 2, Fig. 3 and Fig. 4 are models of asymmetric networks, built on exact same guidelines as discussed above for 4, 8 and 16 nodes respectively.

The adjacency for the models considered is seen below. The number of entries in the adjacency have the parameters of interest, namely in the form of p1 and p2 and the number of times the entries have a unity value where the former represents strength of connection and the latter represents connectivity.

$$A_{4,sym} = \begin{pmatrix} 0 & P1 & 0 & P1 \\ P2 & 0 & P2 & 0 \\ 0 & 1 & 0 & 1 \\ 1 & 0 & 1 & 0 \end{pmatrix} \tag{6}$$

$$A_{4,asym} = \begin{pmatrix} 0 & P1 & P1 & P1 \\ P2 & 0 & P2 & 0 \\ 1 & 1 & 0 & 1 \\ 1 & 0 & 1 & 0 \end{pmatrix} \tag{7}$$

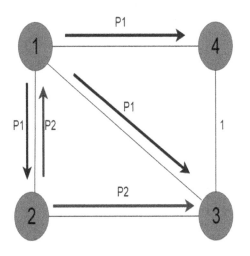

Fig. 2. Four node asymmetric

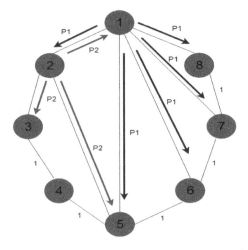

Fig. 3. Eight node asymmetric

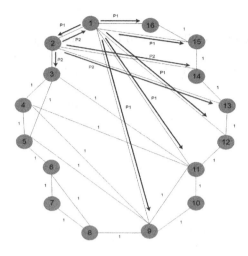

Fig. 4. Sixteen node asymmetric

$$
A_{8,asym} =
\begin{pmatrix}
0 & P1 & 0 & 0 & P1 & P1 & P1 & P1 \\
P2 & 0 & P2 & 0 & P2 & 0 & 0 & 0 \\
0 & 1 & 0 & 1 & 0 & 0 & 0 & 0 \\
0 & 0 & 1 & 0 & 1 & 0 & 0 & 0 \\
1 & 1 & 0 & 1 & 0 & 1 & 0 & 0 \\
1 & 0 & 0 & 0 & 1 & 0 & 1 & 0 \\
1 & 0 & 0 & 0 & 0 & 1 & 0 & 1 \\
1 & 0 & 0 & 0 & 0 & 0 & 1 & 0
\end{pmatrix}
\tag{8}
$$

$$
A_{16,asym} =
\begin{pmatrix}
P2 & 0 & P2 & 0 & 0 & 0 & 0 & 0 & 0 & 0 & 0 & P2 & P2 & 0 & 0 & 0\\
0 & 1 & 0 & 1 & 1 & 0 & 0 & 0 & 0 & 1 & 0 & 0 & 0 & 0 & 0 & 0\\
0 & 0 & 1 & 0 & 1 & 0 & 0 & 1 & 0 & 1 & 0 & 0 & 0 & 0 & 0 & 0\\
0 & 0 & 1 & 1 & 0 & 1 & 0 & 0 & 1 & 0 & 0 & 0 & 0 & 0 & 0 & 0\\
0 & 0 & 0 & 0 & 1 & 0 & 1 & 1 & 0 & 0 & 0 & 0 & 0 & 0 & 0 & 0\\
0 & 0 & 0 & 0 & 0 & 1 & 0 & 1 & 0 & 0 & 0 & 0 & 0 & 0 & 0 & 0\\
0 & 0 & 0 & 0 & 0 & 1 & 1 & 0 & 1 & 0 & 0 & 0 & 0 & 0 & 0 & 0\\
1 & 0 & 0 & 1 & 1 & 0 & 0 & 1 & 0 & 1 & 1 & 0 & 0 & 0 & 0 & 0\\
0 & 0 & 0 & 0 & 0 & 0 & 0 & 1 & 0 & 1 & 0 & 0 & 0 & 0 & 0 & 0\\
1 & 0 & 1 & 1 & 0 & 0 & 0 & 1 & 1 & 0 & 1 & 0 & 0 & 0 & 0 & 0\\
1 & 0 & 0 & 0 & 0 & 0 & 0 & 0 & 1 & 0 & 1 & 0 & 0 & 0 & 0 & 0\\
0 & 1 & 0 & 0 & 0 & 0 & 0 & 0 & 0 & 1 & 0 & 1 & 0 & 0 & 0 & 0\\
1 & 0 & 0 & 0 & 0 & 0 & 0 & 0 & 0 & 0 & 1 & 0 & 1 & 0 & 0 & 0\\
0 & 1 & 0 & 0 & 0 & 0 & 0 & 0 & 0 & 0 & 0 & 1 & 0 & 1 & 0 & 0\\
1 & 0 & 0 & 0 & 0 & 0 & 0 & 0 & 0 & 0 & 0 & 0 & 1 & 0 & 1 & 0\\
1 & 0 & 0 & 0 & 0 & 0 & 0 & 0 & 0 & 0 & 0 & 0 & 0 & 1 & 0 & 1
\end{pmatrix}
\tag{9}
$$

4   Results and Discussions

Two parameters of interest pitch into simulations for the models considered. Results obtained are presented in a tabular fashion for all values of p_1 and p_2. A 2-D graph, where axes denote step increments, and values shown in the cells correspond to each row and column header, are the node numbers of those nodes, that stand highest in rankings! Color changes denote the points, where, transition of leadership or switching in dominance occurs. These color changes mark the point of Revolution and manifest due to both connectivity and reliability being innately present and playing an eminent role in deciding who is ranked highest!

For the ease of understanding of results presented in this section, let us consider Fig. 1 as reference. Let us assume two cases, one where p_1 is higher and p_2 is lower and the other where p_2 is higher and p_1 is lower. We understand the connectivity weights being pointed inward, which will help us decipher authority results, as authorities are the ones, being pointed to! Let us consider the first case, where p_1 is much higher that p_2. In this case, by simple link connectivity mathematics, we see that the links have following in-degrees: Node 1: $1 + p_2$; Node 2: $1 + p_1$; Node 3: $1 + p_2$ and Node 4: $1 + p_1$. Now since p_1 is higher than p_2, then we see that Node 2 and 4 have higher weights than Node 1 and 3. That means, Node 2 and node 4 are better authorities in this case. Similarly, in the case where p_2 is higher than p_1, Node 1 and 3 will be better authorities than Node 2 and 4. This realization is for authorities as we are considering incoming weights. This understanding can be scaled for all types of nodes considering outgoing link weights, and then it would explain hub results. This realization can be absorbed without the loss of generalization for any case presented.

Figures 5 and 6 represent the node ranking changes on a 2-D view for a four node symmetric graph. Initially, the computations provide the best authority as node 1 & 3, as they are being pointed upon by the best rated hub, which is node 4 (by computational analysis). As P1 is incremented, the weightage of link strengths pointing towards nodes 2 & 4 are higher than others, thus making node 2 & 4 best authorities, while node 3 tops as a hub, as it is pointing towards the best authorities. This fashion continues until node 1 & 3 become equally good hubs and nodes 2 & 4 together top the list of authorities. Transitions can be seen by color changes. These transitions are in actual fact, Revolutions!

Figures 7 and 8 represent the node ranking changes on a 2-D view for a four node asymmetric graph. In this result, the scores are seen to be stable for most of the cases. The reason is simple: Node 3 is the best hub, as it points to node 1 and 4 with a strength of unity with total of three connections. Node 1 is top authority, as it is pointed upon by three links, two from node 3 and one from node 2. Note: As node 1 and 3 have maximum number of connections, they top the list. However, this scenario might change if strength of connection is much weaker than other links, irrespective of connections. Revolution occurs when P1 acquires a significantly bigger value, which makes node 2 & 4 as best authorities, because node 1 & 3 (which have maximum connections) point to node 2 & 4 with maximum strength!

Node 1 retains to be strongest authority, as it has total of five connections pointing to it, with no competition. Figures 9 and 10 represent the node ranking changes on a 2-D view for a eight node asymmetric graph. Since node 5 has a total of four connections, each with a unity weight, it tops the ranks for hubs till the threshold for P1 = 0.8. After this, as P1 increments, it is observed that node 1 tops as hub. This is because, now node 1 has five outwardly connections,

	p1=0.1	p1=0.2	p1=0.3	p1=0.4	p1=0.5	p1=0.6	p1=0.7	p1=0.8	p1=0.9	p1=1
p2=0.1	1&3	2&4	2&4	2&4	2&4	2&4	2&4	2&4	2&4	2&4
p2=0.2	1&3	2&4	2&4	2&4	2&4	2&4	2&4	2&4	2&4	2&4
p2=0.3	1&3	1&3	2&4	2&4	2&4	2&4	2&4	2&4	2&4	2&4
p2=0.4	1&3	1&3	1&3	2&4	2&4	2&4	2&4	2&4	2&4	2&4
p2=0.5	1&3	1&3	1&3	1&3	2&4	2&4	2&4	2&4	2&4	2&4
p2=0.6	1&3	1&3	1&3	1&3	1&3	2&4	2&4	2&4	2&4	2&4
p2=0.7	1&3	1&3	1&3	1&3	1&3	1&3	2&4	2&4	2&4	2&4
p2=0.8	1&3	1&3	1&3	1&3	1&3	1&3	1&3	2&4	2&4	2&4
p2=0.9	1&3	1&3	1&3	1&3	1&3	1&3	1&3	1&3	1&3	2&4
p2=1	1&3	1&3	1&3	1&3	1&3	1&3	1&3	1&3	1&3	2&4

Fig. 5. Four node symmetric authorities

	p1=0.1	p1=0.2	p1=0.3	p1=0.4	p1=0.5	p1=0.6	p1=0.7	p1=0.8	p1=0.9	p1=1
p2=0.1	4	3	3	3	3	3	3	3	3	1&3
p2=0.2	4	3	3	3	3	3	3	3	3	1&3
p2=0.3	4	4	3	3	3	3	3	3	3	1&3
p2=0.4	4	4	4	3	3	3	3	3	3	1&3
p2=0.5	4	4	4	4	3	3	3	3	3	1&3
p2=0.6	4	4	4	4	4	3	3	3	3	1&3
p2=0.7	4	4	4	4	4	4	3	3	3	1&3
p2=0.8	4	4	4	4	4	4	4	3	3	1&3
p2=0.9	4	4	4	4	4	4	4	4	4	1&3
p2=1	2&4	2&4	2&4	2&4	2&4	2&4	2&4	2&4	2&4	1&3

Fig. 6. Four node symmetric hubs

	p1=0.1	p1=0.2	p1=0.3	p1=0.4	p1=0.5	p1=0.6	p1=0.7	p1=0.8	p1=0.9	p1=1
p2=0.1	1	1	1	1	1	1	1	1	2&4	2&4
p2=0.2	1	1	1	1	1	1	1	1	2&4	2&4
p2=0.3	1	1	1	1	1	1	1	1	2&4	2&4
p2=0.4	1	1	1	1	1	1	1	1	2&4	2&4
p2=0.5	1	1	1	1	1	1	1	1	1	2&4
p2=0.6	1	1	1	1	1	1	1	1	1	2&4
p2=0.7	1	1	1	1	1	1	1	1	1	2&4
p2=0.8	1	1	1	1	1	1	1	1	1	2&4
p2=0.9	1	1	1	1	1	1	1	1	1	2&4
p2=1	1	1	1	1	1	1	1	1	1	2&4

Fig. 7. Four node asymmetric authorities

	p1=0.1	p1=0.2	p1=0.3	p1=0.4	p1=0.5	p1=0.6	p1=0.7	p1=0.8	p1=0.9	p1=1
p2=0.1	3	3	3	3	3	3	3	3	3	1 & 3
p2=0.2	3	3	3	3	3	3	3	3	3	1 & 3
p2=0.3	3	3	3	3	3	3	3	3	3	1 & 3
p2=0.4	3	3	3	3	3	3	3	3	3	1 & 3
p2=0.5	3	3	3	3	3	3	3	3	3	1 & 3
p2=0.6	3	3	3	3	3	3	3	3	3	1 & 3
p2=0.7	3	3	3	3	3	3	3	3	3	1 & 3
p2=0.8	3	3	3	3	3	3	3	3	3	1 & 3
p2=0.9	3	3	3	3	3	3	3	3	3	1 & 3
p2=1	3	3	3	3	3	3	3	3	3	1 & 3

Fig. 8. Four node asymmetric hubs

each with a strength weight strong enough to pull it up and make it the best hub of all. Here, in this result, the roles of both Reliability and Connectivity are seen clearly.

	p1=0.1	p1=0.2	p1=0.3	p1=0.4	p1=0.5	p1=0.6	p1=0.7	p1=0.8	p1=0.9	p1=1
p2=0.1	1	1	1	1	1	1	1	1	1	1
p2=0.2	1	1	1	1	1	1	1	1	1	1
p2=0.3	1	1	1	1	1	1	1	1	1	1
p2=0.4	1	1	1	1	1	1	1	1	1	1
p2=0.5	1	1	1	1	1	1	1	1	1	1
p2=0.6	1	1	1	1	1	1	1	1	1	1
p2=0.7	1	1	1	1	1	1	1	1	1	1
p2=0.8	1	1	1	1	1	1	1	1	1	1
p2=0.9	1	1	1	1	1	1	1	1	1	1
p2=1	1	1	1	1	1	1	1	1	1	1

Fig. 9. Eight node asymmetric authorities

Figures 11 and 12 represent the node ranking changes on a 2-D view for a sixteen node asymmetric graph. In the mathematical analysis of this model, node 11 maintains to be the strongest/top rated authority, while hub scores fluctuate and show revolutions between node 9 & 11, both of which have equal number of connections. Due to complex network dynamics as a whole, the results are seen to revolutionize the ranking of nodes that contend for Leadership!

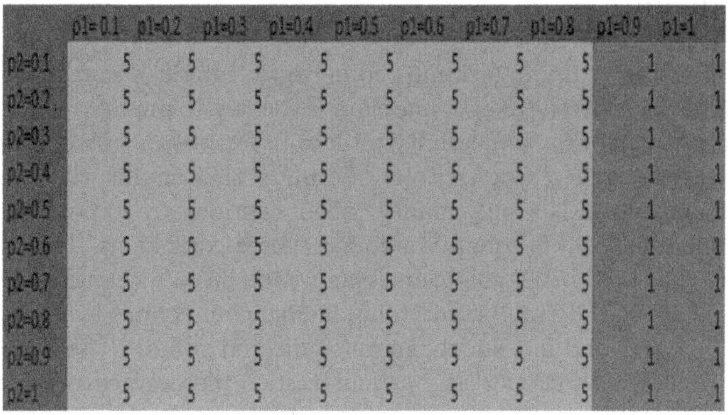

	p1=0.1	p1=0.2	p1=0.3	p1=0.4	p1=0.5	p1=0.6	p1=0.7	p1=0.8	p1=0.9	p1=1
p2=0.1	5	5	5	5	5	5	5	5	1	1
p2=0.2	5	5	5	5	5	5	5	5	1	1
p2=0.3	5	5	5	5	5	5	5	5	1	1
p2=0.4	5	5	5	5	5	5	5	5	1	1
p2=0.5	5	5	5	5	5	5	5	5	1	1
p2=0.6	5	5	5	5	5	5	5	5	1	1
p2=0.7	5	5	5	5	5	5	5	5	1	1
p2=0.8	5	5	5	5	5	5	5	5	1	1
p2=0.9	5	5	5	5	5	5	5	5	1	1
p2=1	5	5	5	5	5	5	5	5	1	1

Fig. 10. Eight node asymmetric hubs

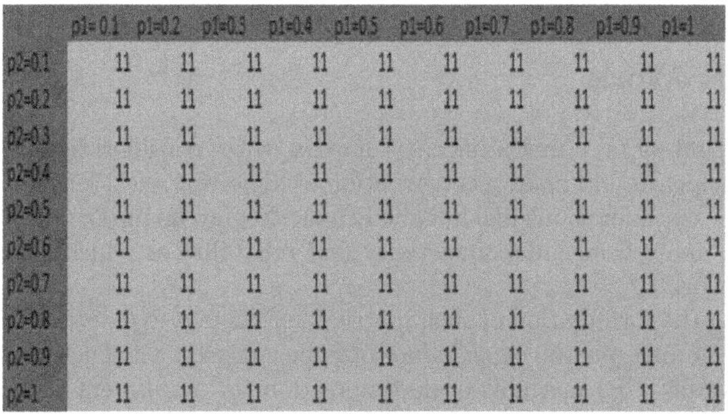

	p1=0.1	p1=0.2	p1=0.3	p1=0.4	p1=0.5	p1=0.6	p1=0.7	p1=0.8	p1=0.9	p1=1
p2=0.1	11	11	11	11	11	11	11	11	11	11
p2=0.2	11	11	11	11	11	11	11	11	11	11
p2=0.3	11	11	11	11	11	11	11	11	11	11
p2=0.4	11	11	11	11	11	11	11	11	11	11
p2=0.5	11	11	11	11	11	11	11	11	11	11
p2=0.6	11	11	11	11	11	11	11	11	11	11
p2=0.7	11	11	11	11	11	11	11	11	11	11
p2=0.8	11	11	11	11	11	11	11	11	11	11
p2=0.9	11	11	11	11	11	11	11	11	11	11
p2=1	11	11	11	11	11	11	11	11	11	11

Fig. 11. Sixteen node asymmetric authorities

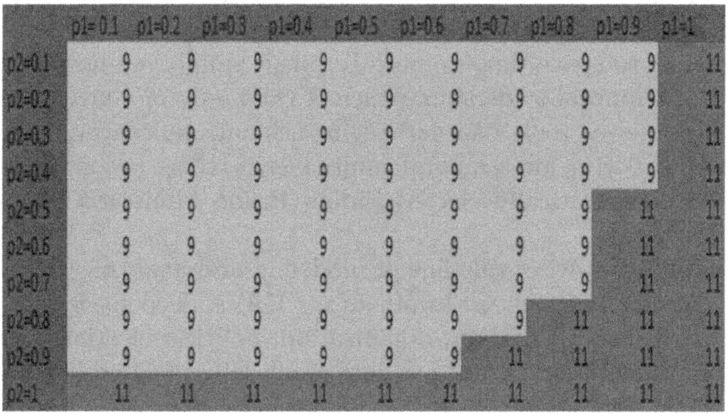

	p1=0.1	p1=0.2	p1=0.3	p1=0.4	p1=0.5	p1=0.6	p1=0.7	p1=0.8	p1=0.9	p1=1
p2=0.1	9	9	9	9	9	9	9	9	9	11
p2=0.2	9	9	9	9	9	9	9	9	9	11
p2=0.3	9	9	9	9	9	9	9	9	9	11
p2=0.4	9	9	9	9	9	9	9	9	9	11
p2=0.5	9	9	9	9	9	9	9	9	11	11
p2=0.6	9	9	9	9	9	9	9	9	11	11
p2=0.7	9	9	9	9	9	9	9	9	11	11
p2=0.8	9	9	9	9	9	9	9	11	11	11
p2=0.9	9	9	9	9	9	9	11	11	11	11
p2=1	11	11	11	11	11	11	11	11	11	11

Fig. 12. Sixteen node asymmetric hubs

5 Conclusion

In conclusion, we say that, in a connected network, rankings change with respect to system dynamics. Strength of connection is the key to quality per se, for nodes and the holistic network. Connectivity pitches in for instances when the varying probability of strength of connections fluctuate, thus giving these rankings a whole new formulation. Not only number of connections are important for better rankings, their reliability is equally important for top rankings. This gives a real life takeaway that only being connected is not enough, but strength/reliability of connections is also vitally important. Only having connections is not enough, but the number matters and indeed, the more connected you are, the higher you are likely to go functional societies! In particular, its a pleasant surprise to discover that a paper on reputation, trust and dynamics of leadership in community of practice also concludes that leadership is all about higher level of activity which more reliable members of the community are able to attain as compared to less reliable members [18].

6 Future Work

1. Formulation of a novel adjacency matrix with consideration of multiple aspects which would involve construction of hypermatrices/tensors, for example, other considerations like financial capacity may be important for electing a leader, apart from link connectivity and reliability as which are studied in current work.
2. Following the formulation of such matrices effectively, explorations need to be done for finding out the eigen values of hypermatrices via Tucker and Canonical Polyadic (CP) methods of decomposition for assessment and evaluation of their eminent characteristics.
3. Calculation of singular value decompositions for both deterministic/random hypermatrices depending on the context.
4. Explorations need to be done considering situations where the network itself is highly dynamic and thus the entries of hypermatrices are anisotropic in nature leading to space time graphs. To put it simply, we need to introduce, a concept of temporal leadership, which is valid only in a given time interval $[t_1, t_2]$ where newer tools like multi-Dimensional, multilayer, nonlinear and dynamic HITS* [19] for temporal multilayer settings become relevant. For example, a democratically elected Indian Prime Minister's tenure is valid only for five years!
5. Talking application-wise, this line of modeling and analysis would give clear results in trade offs for networks of lets say UAVs, deep space sensors, under water networks, social network organization, political elections, merit based office appointments etc., which are part of all daily life, and how we organize our functional societies.

Dedications. This work is dedicated to fond memories of late Dr. Vashishtha Narayan Singh whose life and work has always inspired us to take on difficult research challenges.

References

1. Lyapunov, A.M.: The general problem of the stability of motion. Int. J. Control **55**(3), 531–534 (1992)
2. Kelly, F.: Mathematical modelling of the Internet. In: Engquist, B., Schmid, W. (eds.) Mathematics Unlimited – 2001 and Beyond, pp. 685–702. Springer, Heidelberg (2001). https://doi.org/10.1007/978-3-642-56478-9_35
3. Jacobson, V.: Congestion avoidance and control. ACM SIGCOMM Comput. Commun. Rev. **18**(4), 314–329 (1988)
4. Skyrms, B., Pemantle, R.: A dynamic model of social network formation. In: Gross, T., Sayama, H. (eds.) Adaptive Networks. UCS, pp. 231–251. Springer, Heidelberg (2009). https://doi.org/10.1007/978-3-642-01284-6_11
5. Ranjan, P., La, R.J., Abed, E.H.: Global stability conditions for rate control with arbitrary communication delays. IEEE/ACM Trans. Netw. (TON) **14**(1), 94–107 (2006)
6. Pandey, H., Ranjan, P., Pushp, S., Tripathy, M.R.: Optimal rate allocation for multilayer networks. In: Kulkarni, A.J., Satapathy, S.C., Kang, T., Kashan, A.H. (eds.) Proceedings of the 2nd International Conference on Data Engineering and Communication Technology. AISC, vol. 828, pp. 651–659. Springer, Singapore (2019). https://doi.org/10.1007/978-981-13-1610-4_66
7. Liakh, S.: Multilayer network survivability (2009)
8. De Domenico, M., et al.: Mathematical formulation of multilayer networks. Phys. Rev. X **3**(4), 041022 (2013)
9. Kivelä, M., Arenas, A., Barthelemy, M., Gleeson, J.P., Moreno, Y., Porter, M.A.: Multilayer networks. J. Complex Netw. **2**(3), 203–271 (2014)
10. Lehman, T., et al.: Multilayer networks: an architecture framework. IEEE Commun. Mag. **49**(5), 122–130 (2011)
11. Wehmuth, K., Fleury, É., Ziviani, A.: MultiAspect Graphs: algebraic representation and algorithms. Algorithms **10**(1), 1 (2016)
12. Rodionov, A., Rodionova, O.: Random hypernets in reliability analysis of multilayer networks. In: Mastorakis, N., Bulucea, A., Tsekouras, G. (eds.) Computational Problems in Science and Engineering. LNEE, vol. 343, pp. 307–315. Springer, Cham (2015). https://doi.org/10.1007/978-3-319-15765-8_17
13. Ranjan, P., Pandey, H., Tripathy, M.R., Tan, C.-M., Pushp, S.: Reliability ranking of nodes: a case of revolution. In: 2018 Progress in Electromagnetics Research Symposium (PIERS-Toyama), pp. 1542–1549. IEEE (2018)
14. PageRank (1998). https://en.wikipedia.org/wiki/PageRank. Accessed 03 Mar 2018
15. HITS algorithm (1999). https://en.wikipedia.org/wiki/HITS_algorithm. Accessed 03 Mar 2018
16. SALSA algorithm (2001). https://en.wikipedia.org/wiki/SALSA_algorithm. Accessed 03 Mar 2018
17. Kolda, T., Bader, B.: The TOPHITS model for higher-order web link analysis. In: Workshop on Link Analysis, Counterterrorism and Security, vol. 7, pp. 26–29 (2006)

18. Muller, P.: Reputation, trust and the dynamics of leadership in communities of practice. J. Manage. Governance **10**(4), 381–400 (2006). https://doi.org/10.1007/s10997-006-9007-0
19. Arrigo, F., Tudisco, F.: Multi-dimensional, multilayer, nonlinear and dynamic hits. In: Proceedings of the 2019 SIAM International Conference on Data Mining, pp. 369–377. Society for Industrial and Applied Mathematics (2019)
20. http://web.mit.edu/18.06/www/Fall12/Pset%209/ps9_f12_MATLAB_solutions. pdf. Accessed 10 Aug 2019

Distributed Detection
of a Non-cooperative Target
with Multiplicative Fading

Domenico Ciuonzo[1](✉) and Pierluigi Salvo Rossi[2]

[1] University of Naples "Federico II", Naples, Italy
domenico.ciuonzo@unina.it
[2] Norwegian University of Science and Technology (NTNU), Trondheim, Norway
salvorossi@ieee.org

Abstract. We tackle distributed detection of a non-cooperative target with a Wireless Sensor Network (WSN) made of tiny and inexpensive sensor nodes. When the target is present, sensors observe an (unknown) deterministic signal with attenuation depending on the distance between the sensor and the (unknown) target positions, multiplicative fading (accounting for both line-of-sight and non-line-of-sight components), and additive Gaussian noise. To model energy-constrained operations usually encountered in an Internet of Things (IoT) scenario, local one-bit quantization of the raw measurement is performed at each sensor. The Fusion Center (FC) receives quantized sensor observations through error-prone binary symmetric channels and is in charge of performing a more-accurate global decision. Such model leads to a two-sided test with nuisance parameters (i.e. the target position x_T) observable solely in the case of $\mathcal{H}_1$ hypothesis. After introducing the Generalized Likelihood Ratio Test (GLRT) for the problem, the appealing Davies' framework is exploited to design a generalized form of the Rao test which obviates GLRT high complexity requirements. Equally important, a rationale for threshold-optimization (resorting to a heuristic principle) is proposed and confirmed via simulations. Finally, the aforementioned rules are compared in terms of detection rate in practical scenarios.

Keywords: Data fusion · Decentralized detection · GLRT · IoT · Rao test · Threshold optimization · Wireless sensor networks

1 Introduction

The Internet of Things (IoT) paradigm envisages billions of tiny devices with sensing, computation, and communicating capabilities to be used in numerous areas of everyday life [1]. These include Industry 4.0, smart cities and homes, precision-agriculture, healthcare, surveillance and security [2], just to name a few. In all these "verticals", such devices are required to (a) measure the environment, (b) allow interaction with the physical world and (c) use Internet infrastructure

© Springer Nature Singapore Pte Ltd. 2020
S. M. Thampi et al. (Eds.): SIRS 2019, CCIS 1209, pp. 263–275, 2020.
https://doi.org/10.1007/978-981-15-4828-4_22

to provide services for data analytics, information transfer, and applications usage [3]. Wireless Sensor Networks (WSNs) constitute the sensing & actuation arm of the IoT and have attracted significant interest thanks to their flexibility and reduced costs [4,5]. Decentralized detection is a key collective inference task for a WSN, which has been heavily investigated in the last decades [6].

Unfortunately, stringent bandwidth and energy constraints in WSNs hamper full-precision reporting by sensors. As a consequence, each node usually reports one bit to the Fusion Center (FC) regarding the inferred hypothesis. In such a case, the optimal sensor-individual decision procedure (from both Bayesian and Neyman-Pearson standpoints) corresponds to the local Likelihood-Ratio Test (LRT) being quantized into one-bit [7,8]. Still, the design complexity of quantizer thresholds grows exponentially [9,10] and, equally important, the evaluation of sensor LRT is precluded by ignorance of target parameters [10]. Hence, the bit reported is either the outcome of a naive quantization of the raw measurement [11,12] or exemplifies the inferred binary-valued event (obtained via a sub-optimal detection statistic [13]). In both situations, FC gathers sensors bits and fuses them via a wisely-designed rule to improve (single-)sensor detection capability.

The optimum strategy to fuse the sensors' bits at the FC, under conditional independence assumption, is a weighted sum, with weights depending on unknown target parameters [6], except for some peculiar cases [14]. Then, simple fusion strategies, based on *simple* decisions' count rule or *simplifying* sensing model assumptions (at design stage), have been initially put forward to circumvent such unavailability [15–18]. Still, when the model is parametrically-specified (with some parameters unknown), the FC is in charge to tackle a composite test of hypotheses and the Generalized LRT (GLRT) is usually taken as the natural design solution [19]. Indeed, GLRT-based fusion of quantized data has been extensively studied in WSN literature [12,20,21], especially for decentralized detection of: (*i*) a cooperative target with unknown location, (*ii*) an uncooperative target modelled by observation coefficients assumed *known*, and (*iii*) an unknown source at unknown position (*uncooperative target*). Although case (*iii*) represents the most interesting and challenging (due to the least knowledge requirements), only a few works have recently dealt with it [11,21–23]. In [21], a GLRT was derived for revealing a target with unknown position and emitted power and compared to the so-called counting rule, the optimum rule and a GLRT aware of of target emitted power, showing a marginal loss of the proposed rule with respect to the latter GLRT form. Unfortunately, the designed GLRT requires a grid search on both the target location and emitted power (or signal) domains. Therefore, as a computationally simpler solution, generalized forms of score tests have been proposed for non-cooperative detection of either *deterministic* [11] or *stochastic* target emissions [23].

More recently, [24] and [25] have recently addressed the challenging multiplicative fading scenario in a decentralized estimation and detection problems. The latter setup indeed can be seen a generalization of both deterministic and stochastic emission models, and is able to model complicated propagation mechanisms, such as Rician models or imperfectly-estimated small-scale fading. However, for the multiplicative fading scenario, *only the simpler case (ii) has been addressed* (namely, detecting an unknown source with known observation coefficients).

To fill this gap, we focus on decentralized detection of a non-cooperative target with a spatially-dependent emission (signature), with emitted signal modelled as unknown and deterministic (as opposed to [22,23]). More specifically, the received signal at each individual sensor experiences *multiplicative fading*, additive Gaussian noise, and a deterministic Amplitude Attenuation Function (AAF) depending on the sensor-target distance. Each sensor observes a local measurement on the absence/presence of the target and transmits a single bit version to a FC, over noisy reporting channels (modelled as Binary Symmetric Channels, BSCs, to emulate low-energy communications), having the task of a global (more accurate) decision output. The problem considered is a two-sided parameter test with nuisance parameters present only under the alternative hypothesis [26], which thus precludes the application of usual score-based tests, such as the Rao test [19]. In order to reduce the computational complexity required by the GLRT, the FC is designed to adopt a (simpler) sub-optimal fusion rule based on a generalization of the Rao test [11], and a *novel* quantizer threshold design is proposed herein, based on a *heuristic rationale* developed resorting to the performance of Position-Clairvoyant (PC) Rao in asymptotic form. The resulting design is *sensor-individual*, considers the channel status between each sensor and the FC, and depends upon neither the target strength nor its position, thus allowing *offline* computation. More important, we show zero-threshold choice is optimal according to the latter criterion. Finally, simulation results are provided to compare these rules in terms of performance and complexity in practical scenarios.

Paper Organization: Section 2 describes the system model; Sect. 3 develops GLR and G-Rao tests for the problem introduced; then, Sect. 4 focuses on optimization of the quantizer; numerical results are reported and discussed in Sect. 5; finally, concluding remarks (briefly highlighting further directions of research) are provided in Sect. 6.

List of Employed Math Notations: Bold letters in lower-case indicate vectors, with a_n representing the nth component of a; $\mathbb{E}\{\cdot\}$ and $(\cdot)^T$ are the expectation and transpose operators, respectively; the unit (Heaviside) step function is denoted with; $p(\cdot)$ and $P(\cdot)$ differentiate probability density functions (pdf) and probability mass functions (pmf), respectively; we denote a Gaussian pdf having mean μ and variance σ^2 with $\mathcal{N}(\mu, \sigma^2)$ is used to; $\mathcal{Q}(\cdot)$ (resp. $p_{\mathcal{N}}(\cdot)$) denotes the complement of the cumulative distribution function (resp. the pdf) of a normal random variable in its standard form, i.e. $\mathcal{N}(0,1)$; finally, the symbol $\sim$ (resp. $\overset{a}{\sim}$) corresponds to "distributed as" (resp. to "asymptotically distributed as").

2 System Model

We focus on a binary test of hypotheses in which a set of nodes $k \in \mathcal{K} \triangleq \{1, \ldots, K\}$ is displaced to monitor a given area to decide the absence ($\mathcal{H}_0$) or presence ($\mathcal{H}_1$) of a non-cooperative source with incompletely-specified spatial signature, and signal attenuation depending on the sensor-source distance and multiplicative fading, namely:

$$\begin{cases} \mathcal{H}_0 : & z_k = w_k \\ \mathcal{H}_1 : & z_k = g(\boldsymbol{x}_T, \boldsymbol{x}_k) \, h_k \, \theta + w_k \end{cases}, \tag{1}$$

In other terms, when the target is present (i.e. $\mathcal{H}_1$), we assume that its radiated signal θ, modelled as *unknown deterministic*, is isotropic and experiences (distance-dependent) path-loss, multiplicative fading and additive noise, before reaching individual sensors. In the test of hypotheses in Eq. (1), $z_k \in \mathbb{R}$ denotes the observation of kth sensor, whereas $w_k \sim \mathcal{N}(0, \sigma_{w,k}^2)$ and $h_k \sim \mathcal{N}(1, \sigma_{h,k}^2)$ the measurement noise and the multiplicative fading term, respectively. Additionally, $\boldsymbol{x}_T \in \mathbb{R}^d$ denotes the *unknown* position of the target, while $\boldsymbol{x}_k \in \mathbb{R}^d$ denotes the *known* kth sensor position (obtained via standard self-localization procedures). Both $\boldsymbol{x}_T$ and $\boldsymbol{x}_k$ *uniquely* determine the value of $g(\boldsymbol{x}_T, \boldsymbol{x}_k)$, generically denoting the AAF. We underline that we do not place any specific restriction regarding the AAF modelling the spatial signature of the target to be detected. In view of the spatial separation of the sensors, we hypothesize that the contributions due to noise and fading terms w_ks and h_ks are statistically independent.

Accordingly, the measured signal z_k is *conditionally* distributed as:

$$\begin{cases} z_k | \mathcal{H}_0 \sim & \mathcal{N}(0, \sigma_{w,k}^2) \\ z_k | \mathcal{H}_1 \sim & \mathcal{N}\left(g(\boldsymbol{x}_T, \boldsymbol{x}_k)\,\theta, \; g^2(\boldsymbol{x}_T, \boldsymbol{x}_k)\,\sigma_{h,k}^2\,\theta^2 + \sigma_{w,k}^2\right) \end{cases}, \tag{2}$$

Then, to cope with stringent energy and bandwidth budgets in realistic IoT scenarios, the kth sensor quantizes z_k within one bit of information, i.e. $d_k \triangleq u\,(z_k - \tau_k)$, $k \in \mathcal{K}$, where τ_k represents the quantizer threshold (to be designed). For simplicity, we confine the focus of this paper to deterministic quantizers, while leaving the investigation of probabilistic quantizers to future studies. Additionally, with the aim of modeling a reporting phase with constrained energy, we assume that kth sensor bit d_k is transmitted over a BSC to the FC. Hence, due to non-ideal transmission, a possibly-erroneous $\hat{d}_k \neq d_k$ may be observed. In detail, we assume that bit flipping $\hat{d}_k = (1 - d_k)$ may occur with probability $P_{e,k}$, standing for the known bit-error probability of kth link. For notational convenience, we collect the WSN decisions received by the FC compactly as $\hat{\boldsymbol{d}} \triangleq \begin{bmatrix} \hat{d}_1 \cdots \hat{d}_K \end{bmatrix}^T$.

In view of the aforementioned assumptions, the bit probability under $\mathcal{H}_1$ is given by

$$\alpha_k(\theta, \boldsymbol{x}_T) \triangleq (1 - P_{e,k})\beta_k(\theta, \boldsymbol{x}_T) + P_{e,k}(1 - \beta_k(\theta, \boldsymbol{x}_T)), \tag{3}$$

where $\beta_k(\theta, \boldsymbol{x}_T) \triangleq \mathcal{Q}([\tau_k - g(\boldsymbol{x}_T, \boldsymbol{x}_k)\theta]/\sqrt{g^2(\boldsymbol{x}_T, \boldsymbol{x}_k)\sigma_{h,k}^2\theta^2 + \sigma_{w,k}^2})$. On the other hand, the bit probability when $\mathcal{H}_0$ holds is obtained as $\alpha_{k,0} \triangleq \alpha_k(\theta = 0, \boldsymbol{x}_T)$ (see Eq. (3)), simplifying into:

$$\alpha_{k,0} = (1 - P_{e,k})\beta_{k,0} + P_{e,k}(1 - \beta_{k,0}), \tag{4}$$

where $\beta_{k,0} \triangleq \beta_k(\theta = 0, \boldsymbol{x}_T) = \mathcal{Q}(\tau_k/\sqrt{\sigma_{w,k}^2})$.

We highlight that the unknown target position $\boldsymbol{x}_T$ can be observed at the FC *only* when the *signal is present*, i.e. $\theta \neq \theta_0$ ($\theta_0 = 0$). Thus, we cast the problem as a two-sided parameter where parameters of nuisance ($\boldsymbol{x}_T$) are identifiable only under $\mathcal{H}_1$ [26]. In this paper, the pair $\{\mathcal{H}_0, \mathcal{H}_1\}$ corresponds to $\{\theta = \theta_0, \theta \neq \theta_0\}$. The objective of our study is tantamount to a simple test derivation (from a computational viewpoint) deciding for $\mathcal{H}_0$ (resp. $\mathcal{H}_1$) when the statistic $\Lambda(\hat{\boldsymbol{d}})$ is below (resp. above) the threshold γ_{fc}, and the design of the quantizer (i.e. an optimized τ_k, $k \in \mathcal{K}$) for each sensor.

Accordingly, we will evaluate FC system performance through the well-known detection ($P_D \triangleq \Pr\{\Lambda > \gamma_{\text{fc}}|\mathcal{H}_1\}$) and false alarm ($P_F \triangleq \Pr\{\Lambda > \gamma_{\text{fc}}|\mathcal{H}_0\}$) probabilities, respectively. In the previous definitions Λ denotes the generic decision statistic implemented at the FC.

3 Fusion Rules

The GLR represents a widespread technique for composite hypothesis testing [21], with its implicit form given by

$$\Lambda_{\text{GLR}}(\hat{\boldsymbol{d}}) \triangleq 2\ln\left[\frac{P(\hat{\boldsymbol{d}}; \hat{\theta}_1, \widehat{\boldsymbol{x}}_T)}{P(\hat{\boldsymbol{d}}; \theta_0)}\right], \tag{5}$$

where $P(\widehat{\boldsymbol{d}}; \theta, \boldsymbol{x}_T)$ represents the decision vector likelihood as a function of $(\theta, \boldsymbol{x}_T)$. On the other hand, $(\hat{\theta}_1, \widehat{\boldsymbol{x}}_T)$ are the Maximum Likelihood (ML) estimates under $\mathcal{H}_1$, i.e.

$$(\hat{\theta}_1, \widehat{\boldsymbol{x}}_T) \triangleq \arg\max_{(\theta, \boldsymbol{x}_T)} P(\hat{\boldsymbol{d}}; \theta, \boldsymbol{x}_T), \tag{6}$$

with $\ln P(\hat{\boldsymbol{d}}; \theta, \boldsymbol{x}_T)$ being the logarithm of the likelihood function vs. $(\theta, \boldsymbol{x}_T)$, whose explicit form is [21,22]

$$\ln P(\hat{\boldsymbol{d}}; \theta, \boldsymbol{x}_T) = \sum_{k=1}^{K}\left\{\widehat{d}_k \ln[\alpha_k(\theta, \boldsymbol{x}_T))] + (1 - \widehat{d}_k)\ln[1 - \alpha_k(\theta, \boldsymbol{x}_T)]\right\}, \tag{7}$$

and an analogous functional holds for $\ln P(\hat{\boldsymbol{d}}; \theta_0)$ by substituting $\alpha_k(\theta, \boldsymbol{x}_T)) \rightarrow \alpha_{k,0}$. It is clear from Eq. (5) that Λ_{GLR} requires a maximization problem to be tackled. Sadly, an explicit expression for the pair $(\hat{\theta}_1, \widehat{\boldsymbol{x}}_T)$ is not available.

This increases GLR complexity, since grid discretization of $(\theta, \boldsymbol{x}_T)$ is usually leveraged, see e.g. [21].

On the other hand, Davies' work represents an alternative approach for capitalizing the two-sided nature of the considered hypothesis test [26], allowing to generalize Rao test to the more challenging scenario of nuisance parameters observed only under $\mathcal{H}_1$. In fact, Rao test is based on ML estimates of nuisances under $\mathcal{H}_0$ [19], that sadly cannot be obtained, because they are not observable in our case. In detail, if $\boldsymbol{x}_T$ were available, Rao fusion rule would represent an effective, yet simple, fusion statistic for the corresponding problem testing a two-sided hypothesis [19]. Unfortunately, since $\boldsymbol{x}_T$ is not known in the present setup, we rather obtain a Rao statistics family by varying such parameter. Such technical difficulty is overcome by Davies through the use of the supremum of the family as the decision statistic, that is:

$$
\Lambda_{\mathrm{GRao}} \triangleq \max_{\boldsymbol{x}_T} \frac{\left(\frac{\partial \ln P(\hat{\boldsymbol{d}}; \theta, \boldsymbol{x}_T)}{\partial \theta} \right)^2 \Big|_{\theta = \theta_0}}{I(\theta_0, \boldsymbol{x}_T)}, \tag{8}
$$

where $I(\theta, \boldsymbol{x}_T) \triangleq \mathbb{E}\left\{ \left(\partial \ln \left[P(\hat{\boldsymbol{d}}; \theta, \boldsymbol{x}_T) \right] / \partial \theta \right)^2 \right\}$ represents the Fisher Information (FI) assuming $\boldsymbol{x}_T$ known. Henceforth, the above decision test will be referred to as Generalized Rao (G-Rao), to highlight the usage of Rao as the basic statistic within the umbrella proposed by Davies [22]. The closed form of Λ_{GRao} is drawn resorting to the explicit forms of the score function and the FI, as stated via the following lemmas, whose proof is omitted for brevity.

Lemma 1. *The score function* $\partial \ln \left[P(\hat{\boldsymbol{d}}; \theta, \boldsymbol{x}_T) \right] / \partial \theta$ *for decentralized detection a non-cooperative target model with multiplicative fading obeys the following expression:*

$$
\frac{\partial \ln \left[P\left(\hat{\boldsymbol{d}}; \theta, \boldsymbol{x}_T \right) \right]}{\partial \theta} = \sum_{k=1}^{K} \left\{ \frac{\hat{d}_k - \alpha_k(\theta, \boldsymbol{x}_T)}{\alpha_k(\theta, \boldsymbol{x}_T)[1 - \alpha_k(\theta, \boldsymbol{x}_T)]} (1 - 2P_{e,k}) \times \right.
$$
$$
\left. p_{\mathcal{N}} \left(\frac{\tau_k - g(\boldsymbol{x}_T, \boldsymbol{x}_k)\,\theta}{\sqrt{g^2(\boldsymbol{x}_T, \boldsymbol{x}_k)\,\sigma_{h,k}^2\,\theta^2 + \sigma_{w,k}^2}} \right) \frac{g(\boldsymbol{x}_T, \boldsymbol{x}_k)\,\sigma_{w,k}^2 + g^2(\boldsymbol{x}_T, \boldsymbol{x}_k)\sigma_{h,k}^2\,\theta\,\tau_k}{(\sigma_{w,k}^2 + g^2(\boldsymbol{x}_T, \boldsymbol{x}_k)\,\sigma_{h,k}^2\,\theta^2)^{3/2}} \right\} \tag{9}
$$

Proof. The proof can be obtained analogously as [11] by exploiting in the derivative calculation the separable form expressed by Eq. (7).

Lemma 2. *The FI* $I(\theta, \boldsymbol{x}_T)$ *for decentralized detection a non-cooperative target model with multiplicative fading has the following closed form:*

$$
I(\theta, \boldsymbol{x}_T) = \sum_{k=1}^{K} \psi_k(\theta, \boldsymbol{x}_T)\, g^2(\boldsymbol{x}_T, \boldsymbol{x}_k), \tag{10}
$$

where the following auxiliary definition has been employed

$$\psi_k(\theta, \boldsymbol{x}_T) \triangleq \frac{(1 - 2P_{e,k})^2}{\alpha_k(\theta, \boldsymbol{x}_T)\left[1 - \alpha_k(\theta, \boldsymbol{x}_T)\right]} \tag{11}$$

$$\times \frac{\left\{\sigma_{w,k}^2 + g(\boldsymbol{x}_T, \boldsymbol{x}_k)\sigma_{h,k}^2\,\theta\,\tau_k\right\}^2}{(\sigma_{w,k}^2 + g^2(\boldsymbol{x}_T, \boldsymbol{x}_k)\sigma_{h,k}^2\,\theta^2)^3}\, p_{\mathcal{N}}^2\left(\frac{\tau_k - g(\boldsymbol{x}_T, \boldsymbol{x}_k)\,\theta}{\sqrt{g^2(\boldsymbol{x}_T, \boldsymbol{x}_k)\sigma_{h,k}^2\,\theta^2 + \sigma_{w,k}^2}}\right).$$

Proof. The proof can be obtained analogously as [11], exploiting conditional independence of decisions (which implies an additive FI form) and similar derivation results as $\frac{\partial \ln[P(\widehat{\boldsymbol{d}};\theta,\boldsymbol{x}_T)]}{\partial \theta}$.

Then, combining the results in (9) and (10), the G-Rao statistic is obtained in the final form as

$$\Lambda_{\mathrm{GRao}}(\widehat{\boldsymbol{d}}) \triangleq \max_{\boldsymbol{x}_T} \Lambda_{\mathrm{Rao}}(\widehat{\boldsymbol{d}}, \boldsymbol{x}_T), \tag{12}$$

where

$$\Lambda_{\mathrm{Rao}}\left(\widehat{\boldsymbol{d}}, \boldsymbol{x}_T\right) = \frac{\sum_{k=1}^K \widehat{\nu}_k(\widehat{d}_k)\, g(\boldsymbol{x}_T, \boldsymbol{x}_k)}{\sqrt{\sum_{k=1}^K \psi_{k,0}\, g^2(\boldsymbol{x}_T, \boldsymbol{x}_k)}}, \tag{13}$$

is the Rao statistic when $\boldsymbol{x}_T$ is assumed known, and we have employed $\widehat{\nu}_k(\widehat{d}_k) \triangleq (\widehat{d}_k - \alpha_{k,0})\,\Xi_k$, $\psi_{k,0} \triangleq \alpha_{k,0}\,(1 - \alpha_{k,0})\,\Xi_k^2$ and

$$\Xi_k \triangleq \frac{(1 - 2P_{e,k})}{\alpha_{k,0}\,(1 - \alpha_{k,0})} \frac{1}{\sigma_{w,k}}\, p_{\mathcal{N}}\left(\frac{\tau_k}{\sqrt{\sigma_{w,k}^2}}\right), \tag{14}$$

as compact auxiliary definitions. We motivate the attractiveness of G-Rao with a lower (resp. a simpler) complexity (resp. implementation), as we do not require $\hat{\theta}_1$, and only a grid search with respect to $\boldsymbol{x}_T$ is imposed, that is

$$\Lambda_{\mathrm{GRao}}(\widehat{\boldsymbol{d}}) \approx \max_{i=1,\dots N_{\boldsymbol{x}_T}} \Lambda_{\mathrm{Rao}}(\widehat{\boldsymbol{d}}, \boldsymbol{x}_T[i]). \tag{15}$$

Hence, the complexity of its implementation scales as $\mathcal{O}(K\,N_{\boldsymbol{x}_T})$, which implies a significant reduction of complexity with respect to the GLR (corresponding to $\mathcal{O}(K\,N_{\boldsymbol{x}_T}\,N_\theta)$).

It is evident that Λ_{GRao} (the same applies to Λ_{GLR}, see Eqs. (5) and (13)) depends on τ_k's, via the terms $\widehat{\nu}_k(\widehat{d}_k)$ and $\psi_{k,0}$, $k \in \mathcal{K}$. Hence the threshold set, gathered within $\boldsymbol{\tau} \triangleq [\tau_1 \cdots \tau_K]^T$, can be designed to optimize performance. Next section is devoted to accomplish this purpose.

4   Design of Quantizers

We point out that the rationale in [12,27] cannot be applied to design (asymptotically) optimal deterministic quantizers, since no closed-form performance expressions exist for tests built upon Davies approach [26]. In view of this reason, we use a modified rationale with respect to [12,27] (that is resorting to a heuristic, yet intuitive, basis) and demonstrate its effectiveness in Sect. 5 through simulations, as done for similar uncooperative target detection problems in [11,23]. In detail, it is well known that the (position $\boldsymbol{x}_T$) clairvoyant Rao statistic Λ_{Rao} is distributed (under an asymptotic, weak-signal, assumption[1]) as [19]

$$\Lambda_{\mathrm{Rao}}(\boldsymbol{x}_T, \boldsymbol{\tau}) \overset{a}{\sim} \begin{cases} \chi_1^2 & \text{under} \quad \mathcal{H}_0 \\ \chi_1'^2\left(\lambda_Q(\boldsymbol{x}_T)\right) & \text{under} \quad \mathcal{H}_1 \end{cases}, \tag{16}$$

where the non-centrality parameter $\lambda_Q(\boldsymbol{x}_T) \triangleq (\theta_1 - \theta_0)^2 \, \mathrm{I}(\theta_0, \boldsymbol{x}_T)$ (underlining dependence on $\boldsymbol{x}_T$) is given as:

$$\lambda_Q(\boldsymbol{x}_T) = \theta_1^2 \sum_{k=1}^K \psi_{k,0} \, g^2(\boldsymbol{x}_T, \boldsymbol{x}_k), \tag{17}$$

with θ_1 being the true value under $\mathcal{H}_1$. Clearly the larger $\lambda_Q(\boldsymbol{x}_T)$, the better the $\boldsymbol{x}_T$−clairvoyant GLR and Rao tests will perform when the target to be detected is located at $\boldsymbol{x}_T$. Also, it is apparent that $\lambda_Q(\boldsymbol{x}_T)$ is a function of τ_k, $k \in \mathcal{K}$ (because of the $\psi_{k,0}$'s). For this reason, with a slight abuse of notation we will use $\lambda_Q(\boldsymbol{x}_T, \boldsymbol{\tau})$ and choose the threshold set $\boldsymbol{\tau}$ to maximize $\lambda_Q(\boldsymbol{x}_T, \boldsymbol{\tau})$, that is

$$\boldsymbol{\tau}^\star \triangleq \arg\max_{\boldsymbol{\tau}} \, \lambda_Q(\boldsymbol{x}_T, \boldsymbol{\tau}) \tag{18}$$

In general, such optimization would lead to a $\boldsymbol{\tau}^\star$ that will be dependent on $\boldsymbol{x}_T$ (and thus not practical).

Still, for this particular problem, the optimization requires only the solution of K *decoupled* threshold designs (hence the optimization complexity presents a linear scale with the number of sensors K), being also independent of $\boldsymbol{x}_T$ (cf. Eq. (17)), that is:

$$\tau_k^\star = \arg\max_{\tau_k} \left\{ \psi_{k,0}(\tau_k) \propto \frac{p_{\mathcal{N}}^2\left(\tau_k / \sqrt{\sigma_{w,k}^2}\right)}{\Delta_k + \mathcal{Q}(\tau_k/\sqrt{\sigma_{w,k}^2})\left[1 - \mathcal{Q}(\tau_k/\sqrt{\sigma_{w,k}^2})\right]} \right\} \tag{19}$$

where $\Delta_k \triangleq [P_{e,k}\,(1 - P_{e,k})]/(1 - 2P_{e,k})^2$. It is known from the quantized estimation literature [28,29] that for Gaussian pdf it holds $\tau_k^\star \triangleq \arg\max_{\tau_k} \psi_{k,0}(\tau_k) = 0$. Also, it has been shown in [27] that $\tau_k^\star = 0$ is still the optimal choice for any value of Δ_k, which corresponds to different possibilities of noisy ($P_{e,k} \neq 0$) reporting channels. Therefore, we employ $\tau_k^\star = 0$, $k \in \mathcal{K}$, in Eq. (13), leading to the following further simplified expression for threshold-optimized G-Rao test (denoted with $\Lambda_{\mathrm{GRao}}^\star$):

$$\Lambda_{\mathrm{GRao}}^\star \triangleq \max_{\boldsymbol{x}_T} \frac{4\left[\sum_{k=1}^K (1 - 2\,P_{e,k})\,(1/\sigma_{w,k})\,g(\boldsymbol{x}_T, \boldsymbol{x}_k)\,(\hat{b}_k - \frac{1}{2})\right]^2}{\sum_{k=1}^K (1 - 2\,P_{e,k})^2\,(1/\sigma_{w,k}^2)\,g^2(\boldsymbol{x}_T, \boldsymbol{x}_k)} \tag{20}$$

---

[1] That is $|\theta_1 - \theta_0| = c/\sqrt{K}$ for a certain value $c > 0$ [19].

which is considerably simpler than the GLRT, as it obviates solution of a joint optimization problem w.r.t. $(\boldsymbol{x}_T, \theta)$. Furthermore, the corresponding optimized non-centrality parameter, denoted with $\lambda_Q^\star(\boldsymbol{x}_T)$, is given by:

$$\lambda_Q^\star(\boldsymbol{x}_T) \triangleq 4\theta_1^2 \sum_{k=1}^{K} \left[(1 - 2 P_{e,k})^2 \left(1 / (2\pi\sigma_{w,k}^2) \right) g(\boldsymbol{x}_T, \boldsymbol{x}_k)^2 \right]. \tag{21}$$

5   Simulation Results

Accordingly, in this section we delve into performance comparison of G-Rao and GLR tests. Additionally, we will provide an assessment of the threshold-optimization developed in Sect. 4. With this aim, a 2-D area ($\boldsymbol{x}_T \in \mathbb{R}^2$) is considered, in which the presence of a non-cooperative target in the surveillance region $\mathcal{A} \triangleq [0, 1]^2$ (i.e. a square) is monitored by a WSN composed of $K = 49$ sensor nodes. For simplicity the sensors are arranged according to a regular square grid which covers the whole $\mathcal{A}$. With reference to the sensing model, we assume $w_k \sim \mathcal{N}(0, \sigma_w^2)$, $k \in \mathcal{K}$ (also w.l.o.g. we set $\sigma_w^2 = 1$). Also, the AAF chosen is $g(\boldsymbol{x}_T, \boldsymbol{x}_k) \triangleq 1 / \sqrt{1 + (\|\boldsymbol{x}_T - \boldsymbol{x}_k\| / \eta)^\alpha}$ (i.e. a power-law), where we have set $\eta = 0.2$ (viz. approximate target extent) and $\alpha = 4$ (viz. decay exponent). Finally, we define the target Signal-To-Noise Ratio (SNR), including multiplicative fading effects, as SNR $\triangleq 10 \log_{10}(\theta^2 (1 + \sigma_h^2) / \sigma_w^2)$ and the LoS/NLoS ratio as $\kappa \triangleq 10 \log_{10}(1 / \sigma_h^2)$. Initially, we assume ideal BSCs, i.e. $P_{e,k} = 0$, $k \in \mathcal{K}$.

According to Sect. 3, the implementation of Λ_{GLR} and Λ_{GRao} leverages grid search for θ and $\boldsymbol{x}_T$. Specifically, the search space of the target signal θ is assumed to be $S_\theta \triangleq [-\bar{\theta}, \bar{\theta}]$, where $\bar{\theta} > 0$ is such that the SNR = 20 dB. The vector collecting the points on the grid is then defined as $\begin{bmatrix} -\boldsymbol{g}_\theta^T & 0 & \boldsymbol{g}_\theta^T \end{bmatrix}^T$, where $\boldsymbol{g}_\theta$ collects target strengths corresponding to the SNR dB values $-10 : 2.5 : 20$ (thus $N_\theta = 25$). Secondly, the search support of $\boldsymbol{x}_T$ is naturally assumed to be coincident with the monitored area, i.e. $S_{\boldsymbol{x}_T} = \mathcal{A}$. Accordingly, the 2-D grid is the result of sampling $\mathcal{A}$ uniformly with $N_{\boldsymbol{x}_T} = N_c^2$ points, where $N_c = 100$. The 2-D grid points are then obtained by regularly sampling $\mathcal{A}$ with $N_{\boldsymbol{x}_T} = N_c^2$ points, where $N_c = 100$. In this setup, the evaluation of G-Rao requires $N_c^2 = 10^4$ grid points, as opposed to $N_c^2 N_\theta = 2.5 \times 10^5$ points for GLR, i.e. a complexity decrease of *more than twenty times*.

First, in Fig. 1 we show P_D (under $P_F = 0.01$) versus a common threshold choice for all the sensors $\tau_k = \tau$, $k \in \mathcal{K}$, for a target whose location is randomly drawn according to a uniform distribution within $\mathcal{A}$. We consider two LoS-NLos conditions, namely $\kappa = 0$ dB (moderate LoS component) and $\kappa = 10$ dB (strong LoS component), whereas we fix the sensing SNR = 10 dB. It is apparent that in *both* LoS-NLoS conditions the choice $\tau_k^\star = 0$ represents a nearly-optimal solution, since the optimal value of τ found numerically depends on the polarity of θ, which is *unknown*. This both applies to GLR and G-Rao as well. We then proceed with the choice $\tau_k^\star = 0$ in following results.

Secondly, in Fig. 2 we provide a P_D comparison (for $P_F = 0.01$) of considered rules (for a target whose position is randomly generated within $\mathcal{A}$ at each run,

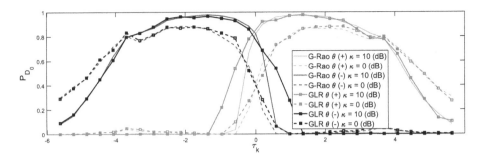

Fig. 1. P_D vs $\tau_k = \tau$, $P_F = 0.01$; WSN with $K = 49$ sensors, $P_{e,k} = 0$, SNR $= 10\,\text{dB}$, $\kappa \in \{0, 10\}$ (amplitude signal θ with positive/negative polarity).

similarly as Fig. 1) versus SNR (dB), in order to assess their detection sensitivity as a function of the received signal strength of the non-cooperative target. We consider two relevant scenarios of LoS-NLos ratio, namely $\kappa \in \{0, 10\}$ dB, and different quality of the BSC ($P_{e,k} = P_e \in \{0, 0.1\}$). From inspection of the figure, we conclude that both rules perform very similarly for a, different LoS-NLos conditions, varying reporting channel status, and over the whole SNR range.

Thirdly, in Fig. 3, we report P_D (under $P_F = 0.01$) profile vs. target location $\boldsymbol{x}_T$ (for SNR $= 5\,\text{dB}$ and $\kappa \in 0\,\text{dB}$, $P_{e,k} = 0$), to draw a detailed overview of detection capabilities over the whole monitored area $\mathcal{A}$ and underline possibly *blind spots*. Remarkably, G-Rao test performs only *negligibly worse* than the GLRT, and moderately worse in comparison to a test based on θ-clairvoyant GLR. Unfortunately, the latter assumes *the unrealistic* knowledge of θ and thus merely constitutes an *upper-limiting bound* on performance achievable. Finally, we notice that both GLR and G-Rao have a similar $P_D(\boldsymbol{x}_T)$ profile, and its "shape" highlights a lower detection rate at the edge of the monitored area. Such result can be ascribed to the *regularity* of the WSN arrangement in the monitored area $\mathcal{A}$.

6   Concluding Remarks and Further Directions

In this paper, a distributed scheme using a WSN for detection of a non-cooperative target was developed. More specifically, we considered a target emitting an unknown deterministic signal (θ) at unknown location ($\boldsymbol{x}_T$), and designed a generalized version of the Rao test (G-Rao) as an attractive (low-complexity) alternative to GLRT (the latter requiring a grid search on the whole space ($\theta, \boldsymbol{x}_T$)) for a general model with (*i*) multiplicative fading, (*ii*) quantized measurements, (*iii*) non-ideal and non-identical BSCs. Since $\boldsymbol{x}_T$ is a nuisance parameter present only under $\mathcal{H}_1$ (i.e. when $\theta \neq 0$), the G-Rao statistic arises from maximization (w.r.t. $\boldsymbol{x}_T$) of a family of Rao decision statistics, obtained by assuming $\boldsymbol{x}_T$ known, based on the framework proposed by [26].

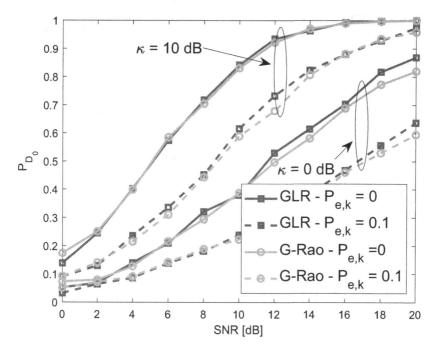

Fig. 2. P_{D_0} vs. sensing SNR (dB), when the FC false-alarm probability is set to $P_{F_0} \in$ 0.01. A WSN with $K = 49$ sensors is considered, with sensor thresholds set as $\tau_k^\star = 0$, whose decisions are sent over BSCs with $P_{e,k} = P_e \in \{0, 0.1\}$ and with $\kappa \in \{0, 10\}$ (dB).

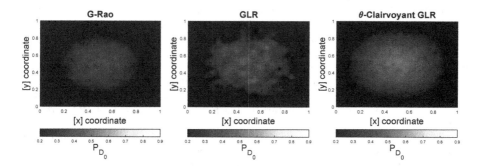

Fig. 3. G-Rao (left), GLR (center) and θ-clairvoyant GLR (right) P_D heatmaps vs. source position $\boldsymbol{x}_T$, when the FC false-alarm probability is set to $P_F = 0.01$. A WSN with $K = 49$ sensors, having sensing SNR = 5 dB and $\kappa = 0$ (dB) is considered. Corresponding decisions are sent over BSCs with $P_{e,k} = 0.1$. The sensor thresholds are set as $\tau_k^\star = 0$.

Furthermore, we developed an effective criterion (originating from performance expressions having a semi-theoretical background) to design sensor thresholds of G-Rao in an optimized fashion, resulting in a *zero-threshold* choice. This result was leveraged to optimize the performance of G-Rao and GLR tests. Numerical results underlined the close performance of G-Rao test to the GLRT in the scenarios investigated, and a small (yet reasonable) loss of G-Rao compared to a test based on θ-clairvoyant GLR. Also, it was shown through simulations that the G-Rao test, achieves practically the same performance as the GLRT in the cases considered.

Our future work will consist of investigating design of fusion rules in more challenging target scenarios, such as multiple moving sources, multi-bit quantizers and burstiness (time-correlation) of reporting channels.

References

1. ITU-T Rec. Y.2060 (06/2012): Overview of the Internet of Things (IoT), June 2012
2. Palmieri, F.A.N., Ciuonzo, D.: Objective priors from maximum entropy in data classification. Inf. Fusion **14**(2), 186–198 (2013)
3. Jin, J., Gubbi, J., Marusic, S., Palaniswami, M.: An information framework for creating a smart city through Internet of Things. IEEE Internet Things J. **1**(2), 112–121 (2014)
4. Chong, C.Y., Kumar, S.P.: Sensor networks: evolution, opportunities, and challenges. Proc. IEEE **91**(8), 1247–1256 (2003)
5. Ciuonzo, D., Salvo Rossi, P. (eds.): Data Fusion in Wireless Sensor Networks: A Statistical Signal Processing Perspective. Control, Robotics & Sensors, Institution of Engineering and Technology (IET) (2019)
6. Varshney, P.K.: Distributed Detection and Data Fusion, 1st edn. Springer-Verlag, New York (1996). https://doi.org/10.1007/978-1-4612-1904-0
7. Hoballah, I.Y., Varshney, P.K.: Distributed Bayesian signal detection. IEEE Trans. Inf. Theory **35**(5), 995–1000 (1989)
8. Reibman, A.R., Nolte, L.W.: Optimal detection and performance of distributed sensor systems. IEEE Trans. Aerosp. Electron. Syst. **1**, 24–30 (1987)
9. Tsitsiklis, J.N.: Decentralized detection. Adv. Stat. Signal Process. **2**(2), 297–344 (1993)
10. Viswanathan, R., Varshney, P.K.: Distributed detection with multiple sensors - Part I: fundamentals. Proc. IEEE **85**(1), 54–63 (1997)
11. Ciuonzo, D., Salvo Rossi, P., Willett, P.: Generalized Rao test for decentralized detection of an uncooperative target. IEEE Signal Process. Lett. **24**, 678–682 (2017)
12. Fang, J., Liu, Y., Li, H., Li, S.: One-bit quantizer design for multisensor GLRT fusion. IEEE Signal Process. Lett. **20**(3), 257–260 (2013)
13. Ciuonzo, D., Salvo Rossi, P.: Decision fusion with unknown sensor detection probability. IEEE Signal Process. Lett. **21**(2), 208–212 (2014)
14. Ciuonzo, D., Romano, G., Salvo Rossi, P.: Optimality of received energy in decision fusion over Rayleigh fading diversity MAC with non-identical sensors. IEEE Trans. Signal Process. **61**(1), 22–27 (2013)
15. Aalo, V.A., Viswanathan, R.: Multilevel quantisation and fusion scheme for the decentralised detection of an unknown signal. Proc. IEE Radar Sonar Navig. **141**(1), 37–44 (1994)

16. Chen, B., Jiang, R., Kasetkasem, T., Varshney, P.K.: Channel aware decision fusion in wireless sensor networks. IEEE Trans. Signal Process. **52**(12), 3454–3458 (2004)
17. Ciuonzo, D., Romano, G., Salvo Rossi, P.: Channel-aware decision fusion in distributed MIMO wireless sensor networks: decode-and-fuse vs. decode-then-fuse. IEEE Trans. Wireless Commun. **11**(8), 2976–2985 (2012)
18. Niu, R., Varshney, P.K.: Performance analysis of distributed detection in a random sensor field. IEEE Trans. Signal Process. **56**(1), 339–349 (2008)
19. Kay, S.M.: Fundamentals of Statistical Signal Processing, Volume 2: Detection Theory. Prentice Hall PTR, Upper Saddle River (1998)
20. Niu, R., Varshney, P.K.: Joint detection and localization in sensor networks based on local decisions. In: 40th Asilomar Conference on Signals, Systems and Computers, pp. 525–529 (2006)
21. Shoari, A., Seyedi, A.: Detection of a non-cooperative transmitter in Rayleigh fading with binary observations. In: IEEE Military Communications Conference (MILCOM), pp. 1–5 (2012)
22. Ciuonzo, D., Salvo Rossi, P.: Distributed detection of a non-cooperative target via generalized locally-optimum approaches. Inf. Fusion **36**, 261–274 (2017)
23. Ciuonzo, D., Salvo Rossi, P.: Quantizer design for generalized locally optimum detectors in wireless sensor networks. IEEE Wireless Commun. Lett. **7**(2), 162–165 (2018)
24. Zhu, J., Lin, X., Blum, R.S., Gu, Y.: Parameter estimation from quantized observations in multiplicative noise environments. IEEE Trans. Signal Process. **63**(15), 4037–4050 (2015)
25. Wang, X., Li, G., Varshney, P.K.: Distributed detection of weak signals from one-bit measurements under observation model uncertainties. IEEE Signal Process. Lett. **26**(3), 415–419 (2019)
26. Davies, R.D.: Hypothesis testing when a nuisance parameter is present only under the alternative. Biometrika **74**(1), 33–43 (1987)
27. Ciuonzo, D., Papa, G., Romano, G., Salvo Rossi, P., Willett, P.: One-bit decentralized detection with a Rao test for multisensor fusion. IEEE Signal Process. Lett. **20**(9), 861–864 (2013)
28. Papadopoulos, H.C., Wornell, G.W., Oppenheim, A.V.: Sequential signal encoding from noisy measurements using quantizers with dynamic bias control. IEEE Trans. Inf. Theory **47**(3), 978–1002 (2001)
29. Rousseau, D., Anand, G.V., Chapeau-Blondeau, F.: Nonlinear estimation from quantized signals: quantizer optimization and stochastic resonance. In: Proceedings of the 3rd International Symposium on Physics in Signal and Image Processing, pp. 89–92 (2003)

Data Aggregation Using Distributed Compressive Sensing in WSNs

Deepa Puneeth$^{(\boxtimes)}$ and Muralidhar Kulkarni

National Institute of Technology Karnataka, Surathkal, Mangalore 575025, India
deepapuneethk@gmail.com, mkul@nitk.edu.in

Abstract. The theory of Distributed Compressive Sensing (DCS) is best suited for Wireless Sensor Network (WSN) applications, as sensors are randomly distributed in the area of interest. The inter-signal and intra-signal correlations are explored in DCS through the concept of joint sparse models. In this paper, we have analysed joint sparse models and reconstruction of the signal has been done using joint recovery techniques. The reconstruction performance is achieved using joint recovery (S-OMP) and separate recovery (OMP) mechanisms utilising synthetic signals that possess the inherent qualities of natural signals. Further, we employ DCS on real data to evaluate the data reduction. DCS proves to be a better data aggregation technique depending upon the amount of intra and inter correlation between data signals. Simulation results shows that, even in less sparse environments, DCS performs better than separate recovery, which is well suited for real signals. Simulations also prove that with DCS, we can further reduce the number of measurements (compressed vector length), required for data reconstruction as compared to separate recovery. It has been shown that nearly 50% reduction in the data required for reconstruction in case of synthetic signals and 27% in case of real signals.

Keywords: Distributed Compressed Sensing (DCS) · Temporal-spatial correlation · Joint sparsity · Joint recovery · Joint Sparse Models (JSM) · Simultaneous OMP (S-OMP) · Separate recovery

1 Introduction

WSN consists of a numerous nodes, randomly deployed in a geographical area that sense the environmental condition (temperature, humidity etc.) and collectively work to manage, and route the sensed signal to the sink. Data aggregation is a fundamental action in Wireless Sensor Network (WSN's), where sensors are accountable for accumulating all the sensed values and finally delivering them to the sink. Aging of WSN can be minimized by, compressing the data of individual sensor node, which cumulatively serves the purpose. We use Compressive Sensing (CS) method to compress the data, further we emphasize with the method of Distributed Compressed Sensing (DCS), we can explore inter/intra signal correlation with the concept of joint sparsity.

© Springer Nature Singapore Pte Ltd. 2020
S. M. Thampi et al. (Eds.): SIRS 2019, CCIS 1209, pp. 276–290, 2020.
https://doi.org/10.1007/978-981-15-4828-4_23

Compressed Sensing outperformed conventional compression methods, and justifies a better trade-off between the quality of reconstruction and minimum power consumption. The high-dimensional signal which is sparse either in the acquired domain or transfer domain say $x \in R^N$, can be compressed by projecting the signal into a low dimensional signal $y \in R^M$. In order to carry this out the signal vector x, has to be multiplied with a projection matrix $A \in R^{M \times N}$. At the receiver the estimation of $\hat{x}$ has to be done from an under determined system since, matrix has fewer equations than unknowns.

Compressed Sensing theory affirms that, the probability of having an unique solution, depends on the sparsity of the signal vector x. The chances of having an unique solution increases, if the signal is sufficiently sparse. In a WSN, based on CS scheme the sensed signals are compressed using linear projections and then transmitted. At the base station estimation of the signal is done, through CS based recovery algorithms. In general, if we want to explore inter-correlation among the signals, then we need to collect the samples in a single location and perform the compression. With the concept of Distributed Compressive Sensing (DCS), there is no need to gather the samples in one location, rather than that it needs a joint recovery at the decoding point. The transmission procedure is same as done in compressed sensing i.e. transmit the individual sensor data after dimensionality reduction using a measurement matrix. The transmitted data of individual sensors will be same as done in compressed sensing. But the difference lies in the recovery procedure. At the receiving end, reconstruction takes place by exploring inter and intra signal correlation using one among the Joint Sparse Models (JSM) [2,3]. The compressed data is successfully reconstructed at the receiver, using the concept of joint sparsity.

The rest of the sections are as follows: Sect. 2 gives a brief idea of the previous works, based on joint sparse models and joint recovery. The calculation of correlations (intra and inter) are described in Sect. 3. Section 4 gives a brief idea of Distributed Compressive sensing and further details about the available models of joint sparsity. It also gives an insight of the models used in order to jointly reconstruct the signals. Section 5 summarizes the results based on synthetic and real signals. Finally the concluding remarks given in Sect. 6.

2   Related Work

A rich literature is available for collection of data ensemble in WSNs, most of the methods employed exploits the correlation (intra/inter) among the collected data. The basic idea behind DCS theory can be found in [3], in which primary focus is on, compressing the vector length, which further effects on communication cost of the signal to be transmitted. It also explains about the different joint sparsity models, relating the models with practical scenario and modeling the framework through suitable joint sparsity model. With graphical model and proof of theorems it also analyses the theoretical bound on measurement rates which is essential to guarantee the perfect recovery of the signal through the compressed sparse signal. Paper [11] proposes common-innovation subspace pursuit

(CISP), to estimate the common and innovation support sets separately, in order to minimize the reconstruction error and computing time. Joint sparsity using joint OMP and joint Subspace Pursuit (SP) has been explored in [13], but not experimented by considering real data. The work described in paper [7] explains DCS for WSN which can be widely applicable in sensor network environment. By considering different sensor network datasets, implemented joint sparsity model to recover the sensor signals. Even though the signals are not perfectly sparse the Joint Sparse Model (JSM) provides a better approximation to explore the intra/inter correlations which exists in the collected data from sensors. Paper [1] describes an aggregation method for WSN, making use of temporal and spatial data dependency which will be present among the sensor nodes. In this paper these dependencies are used in order to reconstruct the missing data, from the sensors. Authors in [17] proposes a DCS method which is based on the covariance information of the uncompressed samples but does not experiment with real data.

Implementation of a multi-channel EEG monitoring, based on CS is explained in paper [6]. Recovery of the multi-channel signals through greedy based system. Paper [4] explains two framework DCS and Kronecker Compressive Sensing (KCS) to reduce the amount of data, and to improve the network lifetime. In these above mentioned papers basically, they mention the JSM models which helps to reduce the data exploring the correlation which exist among the nodes. We have made an attempt to analyze the models which is best suited for WSN. The effect of two sparse components say, innovation sparsity and common sparsity has been studied.

We demonstrate how joint sparsity and joint recovery helps to reduce energy drain by, reducing the transmission cost. We apply the concept to synthetic and real data (indoor and outdoor) also by considering EEG data.

3   Intra and Inter Correlation Effects

The sensors are deployed randomly in the region of interest and the data values sensed from these sensors have either spatial or temporal correlation. In WSN, aggregation and compression are the fundamental means to reduce communication cost and extends network lifetime.

The inter node correlation $\Phi_{inter}(m)$ and the intra node correlation $\Phi_{intra}(m)$ which is calculated as follows [4]. Spatial correlation $\Phi_{inter}(m)$ is the amount of correlation which exists between node pairs x i.e. $m_x(k)$ and y i.e. $m_y(k)$.

$$\Phi_{inter}(m^{(\cdot)}) = \sum_{i=1}^{i=N} \sum_{j>i} \frac{(m_i^{(k)} - E[m_i])(m_j^{(k)} - E[m_j])}{\sigma_{m_i}\sigma_{m_j}} \tag{1}$$

Temporal correlation $\Phi_{intra}(m)$ is the amount of correlation which exists between node x i.e. $m_x(k)$ with itself but with a shift of t time samples i.e. $m_y(k+t)$.

$$\Phi_{intra}(m^{(\cdot)}) = \sum_{i=1}^{i=N} \frac{(m_i^{(k)} - E[x_i])(m_i^{(k+t)} - E[x_i])}{\sigma_{x_i^2}} \tag{2}$$

4 Distributed Compressive Sensing

The sensor nodes are randomly deployed in the area of interest and certain physical phenomenon are picked up by more than one sensor node. Hence, the sensed data will have certain degree of correlation either temporal or spatial. CS can be employed to the sensed data which has temporal correlation. But there might exist a spatial correlation among the sensed data, which further decreases the amount of data and improve the performance of WSN. It can exploit both correlation types if the decoding process is based on joint reconstruction methods. In a CS based system, if we want to employ inter correlation compression among the nodes, then we need an extra mechanism at the transmitter. The sensors compress the data based on intra correlation then in turn these sensors collaborate at a point to compress the data based on inter correlation. But if we want to achieve intra and inter compression without the nodes being communicated to each other or having a collaboration before transmission, then the solution can be found using Distributed compressed Sensing (DCS) technique [2,12,16].

4.1 Models Based on Joint Sparsity

The underlying theory for DCS is 'joint sparsity' of the data signals. Let us go through the different joint sparsity models with the help of which we can recover the signals based on joint recovery at the receiver. These joint sparsity models show that even without the sensors collaboration at the transmitter, it is possible to recover the data signals with reduced number of measurements. With the help of joint recovery, considerable amount of data reduction can be achieved as compared to separate recovery of data. There exist three separate models, for jointly sparse signals, and each model fits in separate class of ensembles. Most of the time signals in transformed domain is sparse thus, signals can be encoded using CS which is termed as separate recovery, which does not explore inter signal correlation.

We employ the following notation for the signal ensemble and the encoding/decoding model. The signal ensembles are denoted by x_l, with $l \in \{1,2,3,4...L\}$. There exists a sparse basis $\Psi \in R^N$ for each signal in which the signal ensemble x_l is represented sparsely. By considering suitable measurement matrix $\Phi \in R^{M_l \times N}$, the signals are compressed $y_l = \Phi x_l$ which possess the $M_l <$ N measurements of x_l. The joint sparse models (JSM) are introduced in [2,7,12,16].

The joint sparsity model can be depicted as below,

$$X = \begin{bmatrix} x_1 \\ x_2 \\ x_3 \\ \vdots \\ x_l \end{bmatrix} \qquad Y = \begin{bmatrix} y_1 \\ y_2 \\ y_3 \\ \vdots \\ y_l \end{bmatrix} \qquad \Phi = \begin{bmatrix} \Phi_1, 0 \cdots, 0 \\ 0, \Phi_2, \cdots, 0 \\ \vdots \\ 0, 0, \cdots, \Phi_L \end{bmatrix} \tag{3}$$

Signal modeling using DCS, consists of representing the signal using two components: common sparse components (present in all the signals under consideration) and sparse innovation component (pertaining to individual signal).

$$x_l = Z_C + Z_l, l \in \{1, 2, ...L\} \tag{4}$$

with $Z_C = \Psi\theta_C$, $\|\theta_C\|_0 = K$ and $Z_l = \Psi\theta_l$, $\|\theta_l\|_0 = K_l$.

The signal which is split into two components, 'Z_C' components is common to all the signal and Z_l is a special components of each signal as in Eq. 4.

Depending on the type of correlation, which exists among the sensors following joint sparsity models are configured namely Joint sparsity model (JSM)-1, JSM-2, JSM-3. In JSM-1 all the signals have similar set of sparsity (non zero components) which we can call it as common sparsity and an innovation component pertain to the individual signals. In JSM-1 the signals will share the same common non zero coefficients (the location of these elements will be same) but the amplitude and phase might be different. We can term this as common support set. These signals further possess different non-zero components (location of these elements need not be same). We can term these elements as innovation sparsity. But where as in JSM-2, All the signals coefficients are different (common sparse+innovation) but the location of these components are exactly the same. Where as in JSM-3 models, it consists of a non-sparse common component and a sparse innovation component [8].

$$x_1 = Z_C + Z_1$$
$$x_2 = Z_C + Z_2$$
$$\vdots$$
$$x_l = Z_C + Z_l, l \in \{1, 2, ...L\}$$

Where K and K_l are the corresponding values of Z_C and Z_l respectively, and $y_l = A \times x_l$.

The encoded signal further transmitted to the Base Station (BS). At the BS joint recovery of the signals are done, because of intra and inter correlation effects the original signals can be recovered with slightly reduced number of measurements than separate recovery. In order to recover the jointly sparse signals we conduct simulations, considering the synthetic signals as well as considering the real data. In both cases joint recovery proves better than separate recovery as joint decoding exploits intra and inter correlations.

4.2   Recovery of Jointly Sparse Signals

In this section we discuss the recovery algorithm for Jointly Sparse Model (JSM). Large scale WSN signals can be modeled using the joint sparsity models, where global variations (i.e. sun, temperature, humidity, wind) affects the sensors collectively but local variations such as water flow, shade, animal/human presence affects the smaller group of sensors.

Recovery of CS signals can be either based on greedy or gradient based algorithms. For our initial simulation, we consider the synthetic signals which represent the real WSN scenario. Each sensor sensed data is transmitted to the collection point, and by joint recovery procedure the intra and inter correlations are explored. Simulations are carried out using YALL1 package [18] in Matlab, which uses Basis Pursuit to recover the signals. In order to show the difference between joint recovery and separate recovery, we have used Orthogonal Matching Pursuit (OMP) [14] and Simultaneous Orthogonal Matching Pursuit (SOMP) [15] which is summarized below.

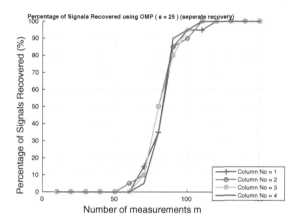

Fig. 1. Separate recovery using OMP, convergence L = 4

Separate Recovery Using OMP. In order to validate DCS, we retrieve the data at the collection point separately, i.e. decoding each sensor data but result decision is on the basis of collective recovery. We decode each branch of the sensor data in a group of say 'L' sensors but we declare success after all the sensor data are separately recovered using OMP, which is depicted in Fig. 1 for L = 4. The 's' sparse signal $x_l \in R^n$, from l sensors, where $l = \{1, 2, 3....L\}$ are encoded using the measurement matrix Φ where m < n, generating the measurements $y_l = \Phi \times x \in R^m$. Let Φ_j be the jth column, $j \in n$, $n = \{1, 2, ...N\}$. The measurement vector y_l is generated by the linear combination of 's' columns of Φ. Thus, estimation of 'x_l' is identifying those columns of Φ. In OMP this problem is solved in a greedy fashion, at each iteration of OMP algorithm selects the column which is mostly correlated with the residual of y_l. Further it eliminates the significance of this column to compute updated residual.

Recovery Using Simultaneous-OMP. In order to recover jointly sparse signals Simultaneous greedy approximation has been proposed [15]. The algorithm is much similar to other greedy methods with minor changes. This in general is known as DCS-SOMP. The algorithm is as follows. The procedure of S-OMP is similar to OMP except the fact that here there are 'L' compressed samples y_l where $l = \{1, 2, 3...L\}$. And SOMP reduces to OMP when $L = 1$.

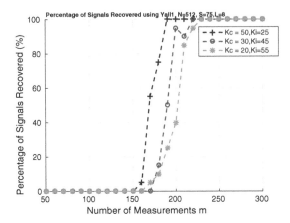

Fig. 2. Joint recovery using YALL1 L = 8, considering different values of, K_c and K_l

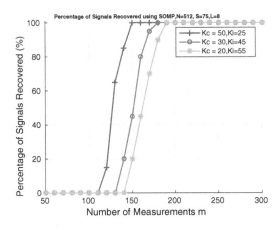

Fig. 3. Joint recovery using SOMP L = 8, considering different values of, K_c and K_l

5   Results and Analysis

To carry out the simulations, we consider the synthetic data, which is similar to the real-world data with N = 512 and L = 8. The signal under consideration comprises a common component K_c (sparse in DCT basis), which symbolizes the common temperature as well as innovation component K_l which symbolizes the abnormalities in the temperature values. Plotting the values of signal recovery versus the measurements M required to compress the signal. The success is declared depending upon the recovered signal, i.e. $\epsilon = \|\hat{x} - x\|_2 / \|x\|_2 \leq 10^{-2}$.

Figures 2 and 3 depicts the performance of YALL1 and SOMP for various values of K_c and K_l with N = 512, sparsity of the signal = 75, L = 8, across the number of measurements M.

We know that in compressive sensing as the sparsity increases the probability of exact reconstruction decreases. Thus, we try to figure out the relation of sparsity and exact reconstruction for jointly sparse signals and then compare the same with separate reconstruction. Recovery of the jointly sparse signals, using YALL1 and SOMP as in Figs. 4, 5 and 6 by letting N = 512, M = 256 and varying 'L'. As the number of sensors(L) increases even though for larger values of 's' recovery of the data is assured, as we can see in Figs. 4 and 5 using Yall1 and SOMP. But if you consider the separate recovery using OMP as in Fig. 6, there is no considerable changes, and as L increases the result is opposite as in the case of jointly sparse recovery. In both cases we try to solve $y_l = A_l X_l$, but in joint recovery the support set is shared, thus even though 's' increases recovery can be done with reduced values of measurement. But same is not true with separate recovery.

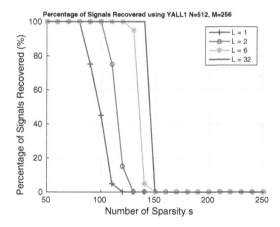

Fig. 4. Joint recovery using YALL1, for varying values of L and sparsity

Figures 7 and 8 shows the result of joint recovery using YALL1 and SOMP by varying the number of measurements and numbers of sensors(L). With the increase in the number of sensors, there is a decrement in the number of measurement required for reconstruction.

A sparse signal with sparsity s = 75 has been considered, with L varying from L = 1 4 8 32 64 as in Figs. 7 and 8. There is a drastic decrement when we consider L = 1 and 4 further when we consider L = 32 and 64, the measurement required is

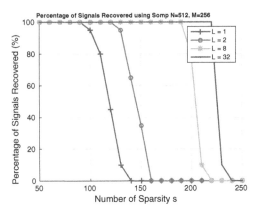

Fig. 5. Joint recovery using SOMP, for varying values of L and sparsity

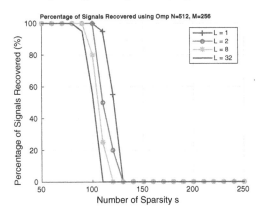

Fig. 6. Joint recovery using OMP, for varying values of L and sparsity

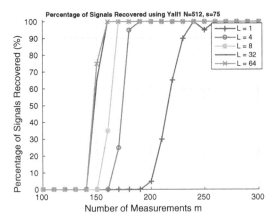

Fig. 7. Joint recovery using YALL1, for L = 1, 4, 8, 32, 64

almost same. More details about the lower bound for the measurement required for jointly sparse signals can be found in [2].

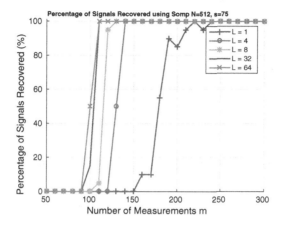

Fig. 8. Joint recovery using SOMP, for L = 1, 4, 8, 32, 64

Figure 9 shows recovery using YALL1 and OMP (separate recovery). Figure 10 using SOMP and OMP (separate recovery) by varying number sensors(L) and the number measurement (M) by letting sparsity of the signal s = 75. Thus, if the signals are correlated then joint recovery promises reduction in the measurements required for reconstruction. Thus, by joint recovery we can reduce the number of transmissions required by the sensor without communicating to each other. If the sensor signals are correlated then joint recovery has an advantage which in turn reduces the burden of sensor nodes thus helps to improve overall network lifetime. Figure 10 shows joint recovery using SOMP and

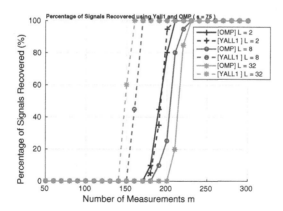

Fig. 9. Joint recovery using YALL1 and separate recovery using OMP, for L = 2, 8, 32

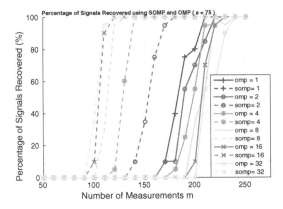

Fig. 10. Joint recovery using SOMP and separate recovery using OMP, for L = 1, 2, 4, 8, 16, 32

separate recovery using OMP for L = 1 2 4 8 16 32. As from the above results we can conclude if the recovery is separate then the number of measurements increases with number of sensors. In separate recovery only the intra correlation is explored, but in joint recovery intra as well as inter correlations are explored.

In this section we considered the data set from Intel lab, [10] which is recorded in an office environment which exhibits regular variation during day and night time. Figure 11 shows the result of joint recovery, by considering the real data set-I and Fig. 12 for dataset-II. Both dataset-I and II are indoor datasets. The signal is not exactly sparse. The signals have smooth variation in space and time. If we consider the DCT, the signal can be represented as sparse, and can be modeled using JSM-2 model.

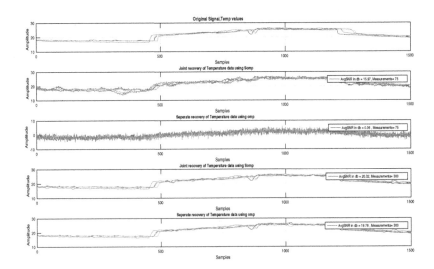

Fig. 11. Joint and separate recovery using SOMP and OMP by considering real data-I, M = 75, 300

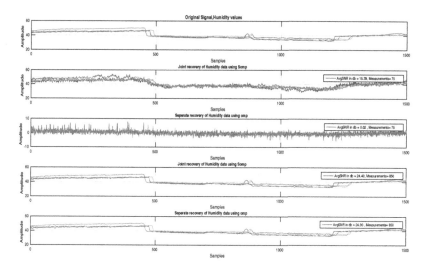

Fig. 12. Joint and separate recovery using SOMP and OMP by considering real data-II, M = 75, 850

Figure 11 depicts recovery based on joint recovery and separate recovery techniques for dataset-I, with N = 1500 and M = 75 and 300 respectively. With the SOMP as reconstruction methods we are able to recover the signal with SNR of 15.25 db but if we consider separate recovery, the data is unrecoverable. Thus, we increase the number of measurements to 300, we could successfully recover using separate reconstruction. In this particular data set an increase of 27% is required in order to reconstruct the data, when it is separate recovery. When we considered data set-II, as shown in Fig. 12 with N = 1500 and M = 75, using SOMP we are able to recover even with reduced number of measurements, SNR = 14.29 db and using separate recovery it's not possible to recover. But in this case, M = 950. This depends on the sparsity of the data in the transformed domain. In this particular data set sparsity is low thus it requires a greater number of iterations to compute the coefficients while reconstruction using OMP.

In the next section we have considered, outdoor data set taken from senorscope [9] which contains the temperature as well as humidity data sets. The plot depicts variation of SNR for different values of compressed vector for different values of L. As we can see from Figs. 13 and 14, using joint sparsity models it is able to recover the data with lesser values of y. There is an improvement in the quality of the signal when we increase the number of sensors. When we consider separate recovery, it is not possible to recover the original signal with lesser value of the compressed vector y.

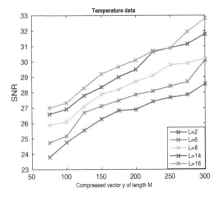

Fig. 13. Joint recovery using SOMP by considering Temperature data (outdoor)

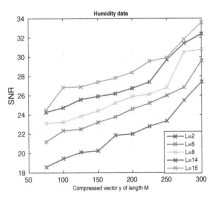

Fig. 14. Joint recovery using SOMP by considering Humidity data (outdoor)

DCS for Multi-channel EEG

The electrical activities related to the brain are measured using EEG signals. Various types of neurological disorders are detected using EEG which includes epilepsy, sleep disorders, stroke, dementia etc. The activities of brain are recorded, through the electrodes which are attached to the scalp of the person. The multi-channel EEG generates huge amount of data, which has to be either stored or transmitted. The aggregation method prior to storage/transmission motivates to employ CS to EEG signals. The performance evaluation of those CS-oriented system is dominated by two main metrics, employed recovery and the domain of sparsification. In this case for EEG signals, we consider wavelet transform as the sparsifying basis. By using CS frame work, we compress and reconstruct the multichannel EEG signals. For the simulations in MatLab, we have taken the EEG signals from EEGLab [5]. Figure 15 shows the 8 channel EEG signals with N = 256, using wavelet transform as sparsifying basis.

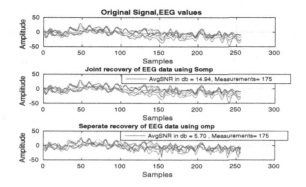

Fig. 15. Joint recovery using SOMP by considering real data-III, M = 175

6   Conclusion

In WSNs, power consumption is an important factor to reduce the energy drain, and also to increase the network lifetime. In this paper, we have shown that there is a significant reduction in the transmission power, by reducing the number of measurements after dimensionality reduction. Since we have used joint sparsity techniques, we could explore the intra as well as inter correlation among the data signals. By comparing the jointly sparse signals (using S-OMP) with separate recovery (using OMP), we have demonstrated how efficiently data can be reduced using joint sparsity techniques. Simulations were carried out by considering the synthetic signals as well as real data (including indoor and outdoor environment as well as EEG signals) and have shown that by using joint sparse model, we can further reduce the number of measurements needed to recover the data.

References

1. Alippi, C., Ntalampiras, S., Roveri, M.: Model ensemble for an effective on-line reconstruction of missing data in sensor networks. In: The 2013 International Joint Conference on Neural Networks (IJCNN), pp. 1–6. IEEE (2013)
2. Baron, D., Duarte, M.F., Sarvotham, S., Wakin, M.B., Baraniuk, R.G.: An information-theoretic approach to distributed compressed sensing. In: Proceedings 45th Conference on Communication, Control, and Computing (2005)
3. Baron, D., Duarte, M.F., Wakin, M.B., Sarvotham, S., Baraniuk, R.G.: Distributed compressive sensing. arXiv preprint arXiv:0901.3403 (2009)
4. Caione, C., Brunelli, D., Benini, L.: Compressive sensing optimization for signal ensembles in WSNs. IEEE Trans. Industr. Inf. **10**(1), 382–392 (2013)
5. Delorme, A., Makeig, S.: EEGLAB: an open source toolbox for analysis of single-trial EEG dynamics including independent component analysis. J. Neurosci. Methods **134**(1), 9–21 (2004)
6. Djelouat, H., Baali, H., Amira, A., Bensaali, F.: An adaptive joint sparsity recovery for compressive sensing based EEG system. Wirel. Commun. Mob. Comput. **2017**, 10 (2017)

7. Duarte, M.F., Wakin, M.B., Baron, D., Baraniuk, R.G.: Universal distributed sensing via random projections. In: Proceedings of the 5th International Conference on Information Processing in Sensor Networks, pp. 177–185. ACM (2006)

8. Hormati, A., Vetterli, M.: Distributed compressed sensing: sparsity models and reconstruction algorithms using annihilating filter. In: 2008 IEEE International Conference on Acoustics, Speech and Signal Processing, pp. 5141–5144. IEEE (2008)

9. Ingelrest, F., Barrenetxea, G., Schaefer, G., Vetterli, M., Couach, O., Parlange, M.: SensorScope: application-specific sensor network for environmental monitoring. ACM Trans. Sens. Netw. (TOSN) **6**(2), 17 (2010)

10. Koushanfar, F., Taft, N., Potkonjak, M.: Sleeping coordination for comprehensive sensing using isotonic regression and domatic partitions (2006)

11. Liu, J., Huang, K., Yao, X.: Common-innovation subspace pursuit for distributed compressed sensing in wireless sensor networks. IEEE Sens. J. **19**(3), 1091–1103 (2018)

12. Sarvotham, S., Baron, D., Wakin, M., Duarte, M.F., Baraniuk, R.G.: Distributed compressed sensing of jointly sparse signals. In: Asilomar Conference on Signals, Systems, and Computers, pp. 1537–1541 (2005)

13. Sundman, D., Chatterjee, S., Skoglund, M.: Greedy pursuits for compressed sensing of jointly sparse signals. In: 2011 19th European Signal Processing Conference, pp. 368–372. IEEE (2011)

14. Tropp, J.A., Gilbert, A.C.: Signal recovery from random measurements via orthogonal matching pursuit. IEEE Trans. Inf. Theory **53**(12), 4655–4666 (2007)

15. Tropp, J.A., Gilbert, A.C., Strauss, M.J.: Simultaneous sparse approximation via greedy pursuit. In: Proceedings, IEEE International Conference on Acoustics, Speech, and Signal Processing, ICASSP 2005, vol. 5, pp. 5–721. IEEE (2005)

16. Wakin, M.B., Duarte, M.F., Sarvotham, S., Baron, D., Baraniuk, R.G.: Recovery of jointly sparse signals from few random projections. In: Advances in Neural Information Processing Systems, pp. 1433–1440 (2006)

17. Wimalajeewa, T., Varshney, P.K.: Robust detection of random events with spatially correlated data in wireless sensor networks via distributed compressive sensing. In: 2017 IEEE 7th International Workshop on Computational Advances in Multi-Sensor Adaptive Processing (CAMSAP), pp. 1–5. IEEE (2017)

18. Zhang, Y.: User's guide for YALL1: your algorithms for L1 optimization. Technical report (2009)

A Modified Approach for the Removal of Impulse Noise from Mammogram Images

S. Sreedevi[1] and Terry Jacob Mathew[2,3]([✉])

[1] Department of Computer Science, Sree Ayyappa College, Eramallikkara,
Chengannur, Kerala, India
sathyansree123@gmail.com
[2] School of Computer Sciences, Mahatma Gandhi University,
Kottayam, Kerala, India
terryjacobin@gmail.com
[3] MACFAST, Tiruvalla, India

Abstract. This paper proposes an integrated approach for removing impulse noise from digital mammograms, which employs a detection method followed by a filtering mechanism. This detection is done using Modified Robust Outlyingness Ratio (MROR) mechanism and a subsequent filtering by an extended Non Local means (NL-means) framework. The pixels in mammograms are grouped into four different clusters based on MROR value, following which different decision rules are applied, to detect the impulse noise in each cluster. The NL-means filter is extended by introducing a reference image obtained from the two stage process. The performance of the proposed filter was evaluated quantitatively and qualitatively by experimental analysis and the results were compared with several existing filters. The results show that the proposed method outperforms standard procedures for impulse noise removal, even for high noise levels.

1 Introduction

Breast cancer is a health condition of prime importance, confronted by today's women and is the second leading cause of cancer death among women over 40 years of age [9]. Complete prevention of breast cancer is impossible because its root cause is still unknown. But early detection and removal can bring down the death rate. Mammography is the most effective and reliable screening method presently available for the early detection of impalpable breast tumors [25]. At times, radiologists fail to differentiate between malignant and non-malignant tumors leading to misinterpretations. In such circumstances, Computer-Aided-Detection (CAD) systems assist radiologists to make correct diagnosis with great accuracy and stability [26].

Nevertheless, mammogram images are distorted by impulse noise that may occur during image acquisition or transmission. These noises corrupt images by adding undesirable content to the original image. To perform an automatic detection of a suspicious lesion from a digitized mammogram, several preprocessing steps are to be done to improve the quality of the image and to prepare

© Springer Nature Singapore Pte Ltd. 2020
S. M. Thampi et al. (Eds.): SIRS 2019, CCIS 1209, pp. 291–305, 2020.
https://doi.org/10.1007/978-981-15-4828-4_24

it for further processing. Removal of noise is one of the important pre-processing steps in a breast cancer CAD system.

Impulse noise affects only a portion of the pixels in the original image [21] and can be removed using nonlinear filters. Among the nonlinear filters, median based and modified median based filters, such as weighted median filter [5], multistate median filter [20], center weighted median filter [13] etc. deliver better performance. But one of the main drawbacks of these filters is that, they are uniformly implemented across the image without checking the status of the pixel as corrupted or not. As a result, desirable details are lost from the original image causing quality degradation. This makes further processing difficult especially when the noise density is higher. To solve this problem, switching median filters were introduced [24]. This method incorporated a noise detection mechanism in the median filter framework to identify the corrupted pixels, so that the uncorrupted pixels are unaltered.

Traditional classification models are generally rigid and hence, fail to capture the innate uncertainty present in mammograms. However, soft expert systems are intelligent data driven applications, which can be effectively used for early diagnosis. In this context, soft computing techniques and their hybrids can be used to handle human uncertainty in diagnosis. The use of soft techniques, including fuzzy logic, soft sets, neural networks etc. have been used in combination with other machine learning techniques, quite successfully [18]. Soft expert systems are effectively used in [3, 22] for the treatment of diseases.

To improve upon the existing noise detection techniques, this paper proposes a novel detector based on the statistical measure known as Modified Robust Outlyingness Ratio (MROR), which measures the level of noise in each pixel. All pixels in the mammogram image are divided into four different clusters according to the value of MROR, and different decision rules are applied on it to detect the impulse noise. After detecting of noisy pixels from each cluster, the NL-means is further extended to clean the noisy pixels.

The rest of this paper is organized as follows. Section 2 gives the related work in noise removal for mammograms. Section 3 describes the background of the proposed work along with the details on basic definitions. Section 4 provides the two tier proposed method. In Sect. 5, the results are discussed and compared with existing filters and procedures. Finally, we conclude in Sect. 6.

2 Related Work

Researchers have developed different techniques for removing noise from mammogram images. Brownrigg [5] introduced a method which initially identified the possible noisy pixels and then replaced them by the median value or its variants. This method lacks satisfactory reproduction of finer details and edges, particularly when the noise levels are high. Mayo et al. [19] compared various existing mammographic denoising methods for low values of noise. Aroquiaraj et al. [4] proposed two new filters for removing impulse noise from mammographic images, but this method gives only a marginal increase in result than the standard median filter. Jayaraj et al. [12] developed a robust estimation technique which helped noise removal at low densities. Bo Xiong et al. [27] proposed a filter which is a combination of robust outlyingness ratio with NL means filter, where the noisy pixels are separated into various levels and then replaced by the median filter. Lai et al. [14] introduced a modified median filtering approach for removing noise and enhancing the contrast of mammogram images. Akshat et al. [11] developed a technique which produced a combined effect of suppression of high density impulse noise followed by contrast enhancement of mammographic breast lesions. Devakumari and Punithavathi [8] compared different noise removal filters and concluded that median filter is the preferable filter for preprocessing mammogram images with any type of noise.

The recent improvements in soft computing and decision making theories have generated proven models with fuzzy soft sets in medical diagnosis [2,17]. Also, see [16].

3   Background of the Proposed Work

3.1   Noise Models

The effect of impulse noise, prevalent in mammograms, is restricted to few areas within the image. Let $x_{i,j}$ and $y_{i,j}$ be the pixel values at locations (i, j) in the original and noisy image respectively and $[\eta_{min}, \eta_{max}]$ be the dynamic range of allowed pixel values. Then, for an impulse model with noise probability or noise ratio p, we have

$$y_{i,j} = \begin{cases} x_{i,j}, & \text{with probability 1-p} \\ \eta_{i,j}, & \text{with probability p} \end{cases} \tag{1}$$

where $\eta_{i,j}$ is the noise value. The two common types of impulse noise present in mammogram images are the fixed-valued impulse noise and the random-valued impulse noise [10]. Fixed valued impulse noise is also known as salt-and-pepper impulse noise where the values of the corrupted pixels are either $\eta_{min}(salt)$ or $\eta_{max}(pepper)$ with equal probability. But in random-valued impulse noise, the corrupted pixels are uniformly distributed between η_{min} and η_{max} [7].

3.2   Definition of MROR

Most of the impulse detectors are based on two-state methods to characterize each pixel as either corrupted or uncorrupted. The main objective of these two-state methods is to locate pixels that are significant outliers in comparison with their neighboring pixels. The highlight of these methods is the integrated noise detection and filtering mechanism. By this mechanism, those pixels that are identified as "noisy" would be submitted to the filtering process, while those identified as "noise-free" would remain intact.

One of the simplest method among the two-state methods is to compare the pixel's intensity with the median intensity of its neighborhood. Researchers have modified the above concept for better results. However, the main drawback of these modifications is that each pixel has to be judged under the same decision rule, without considering the level of noise in each pixel.

This work defines a new detection mechanism based on MROR, to measure the noise level in each pixel. The traditional statistical measure of the outlyingness of an observation with respect to a sample depends on the sample mean and sample standard deviation. But, this method is ineffective for very small or very large samples as it is very difficult to set the threshold value for finding the noisy pixels. These drawbacks caused by the sample mean and sample standard deviation can be avoided with the introduction of a more robust measure namely, adaptive median (AMed) and the normalized Adaptive Median Absolute Deviation (NAMAD) [15]. The new statistics can be defined as:

$$NAMAD(Y) = AMAD(Y)/0.6457 \tag{2}$$

$$AMAD(Y) = AMed\{Y - AMed(Y)\} \tag{3}$$

$$AMed(Y) = AMED(y_1, y_2, \ldots y_n) \tag{4}$$

where, $AMed$ is the adaptive median value in the window. $AMAD$ represents the Adaptive Median Absolute Deviation, $NAMAD$, the adaptive median absolute deviation of a standard normal random variable, which is equal to 0.6745 approximately and y is the vector representation of the data. The new statistic $MROR$ is defined as in Eq. 5.

$$MROR(y_{i,j}) = (y_{i,j} - AMed(Y)/NAMAD(Y) \tag{5}$$

The value of $MROR$ represents the level of noise in each pixel. As per the value of MROR, the pixels are divided in to four noisy clusters. Cluster 1 denotes $(MROR > 3)$ while Cluster 2 represents $(2 < MROR \leq 3)$. The third and fourth cluster are represented by $(1 < MROR \leq 2)$ and $(0 \leq MROR \leq 1)$ respectively. The low $MROR$ value suggests that the noise level in the pixel is very less or it is noise free. The difference between the absolute difference and the adaptive median of its neighbouring pixel is computed in the process of impulse noise detection.

Table 1 shows the distribution of pixels in the mammogram image $mdb058$ based on the $MROR$ with different noise ratios. It shows that even in a noise free image, there are still some pixels in the first noisy cluster. The number of pixels in each cluster changes with the increase in noise ratio.

3.3   Adaptive Median Filter

The standard median filter performance degrades when the spatial noise variance increases. Unlike the standard median calculation, where the size of the neighborhood is fixed, in the adaptive median calculation, the size of the neighborhood changes during operation. The general algorithm is given as Algorithm 1. In this work, the outlyingness of an observation is measured based on the adaptive median of a sample which is a significant factor in the detection of impulse noise removal.

Table 1. Representation of number of pixels in four levels of a mammogram image (mdb058) of size 512×512 with different noise ratios.

Noise ratio	Cluster 1	Cluster 2	Cluster 3	Cluster 4
Noise-free	1432	11184	57532	191996
10%	28715	8642	38066	186721
20%	42516	11227	36512	171889
30%	78313	15417	47883	120531

4   Proposed Method

The proposed method utilizes an iterative detection method followed by a filtering mechanism, where the detection system consists of two stages, coarse stage and fine stage. In each stage, different decision rules are applied to detect impulse noise from each cluster. In the coarse stage, comparatively larger thresholds are adopted to detect the noise and the adaptive median based restored image is used for the next iteration. Therefore, in the coarse stage, after a few iterations, the image becomes more and more close to the original image. This makes it easier to detect noisy pixels. In contrast with the coarse stage, comparatively smaller thresholds are used in the fine stage. Then the NL-means is extended to remove impulse noise detected in the image as given in the Subsect. 4.2.

4.1   Proposed Algorithm

The parameters are selected based on extensive experiments for getting better result. In this work the window size is selected as 2, so the actual size of the window is $(2N + 1) \times (2N + 1)$ i.e. 5×5.

For obtaining better performance, the coarse stage thresholds are selected based on the intensity of the image. The values $3r/4, r/2$ and $r/4$ for ct_1, ct_2, ct_3 are selected, where r is the range of intensity of the image. In order to select a relatively smaller threshold in the fine stage, the intensity values of the adaptive median based restored image obtained from the coarse stage are used. Thus the four threshold values $3r/4, r/2, r/4$ and $r/8$ for ft_1, ft_2, ft_3 and ft_4 are found out. The number of iterations in the coarse stage is fixed as 4, according to Tables 2 and 3. In the fine stage the number of iterations are fixed as two according to Tables 4 and 5.

The stages of the proposed impulse noise filter are shown in Fig. 1. The new detection mechanism produces a result map and a reference image for further processing. Based on the reference image and detection result, the NL-means filter is extended to the pixels marked as noisy. This method is hierarchical, progressive, iterative and anisotropic in nature. In hierarchical mode, all the pixels in the image are divided into four clusters and impulse noise is detected independently in each cluster based on the value of MROR. This method is progressive

Algorithm 1. Z_{min}, Z_{max}, Z_{med} are the minimum, maximum and median gray level value in Sxy. S_{max} is the maximum allowed size of Sxy [10] and Zxy is the gray level value at coordinate (x, y).

```
 1: procedure LEVEL A
 2:     if Z_min < Z_med < Z_max then
 3:         Go to Level B
 4:     else
 5:         Increase window_size
 6:         if window_size < S_max then
 7:             Go to level A
 8:         else
 9:             Output Z_xy
10:         end if
11:     end if
12: end procedure

13: procedure LEVEL B
14:     if Z_min < Z_xy < Z_max then
15:         Output Z_xy
16:     else
17:         Output Z_med
18:     end if
19: end procedure
```

Table 2. The number of coarse stage iterations and Peak signal-to-noise ratio (PSNR) values for a mammogram image (mdb058) with different intensities of impulse noise.

Iterations	PSNR value at different noise level			
	10%	30%	50%	70%
1	30.0015	28.4132	24.1010	17.5442
2	**31.4314**	28.9694	24.8312	17.9211
3	31.4314	**29.1318**	**25.1521**	18.4766
4	31.4314	29.1318	25.1521	**18.4841**
5	31.4118	29.1318	25.1501	18.4841

and iterative as indicated by the adoption of a coarse-to-fine and iterative framework strategy. The anisotropic nature means that different decision rules with different thresholds are used in different clusters.

4.2 Extension of Non-Local Means

The NL-means algorithm was proposed by Baud et al. [6]. It is based on the non-local averaging of all pixels in the image and was developed to denoise an image corrupted by Gaussian noise with zero mean and variance. Let x be the original image, η the gaussian noise and Y the noisy image. Then the denoised

Algorithm 2. The coarse stage detection mechanism
Input: Noisy mammogram image
Output: Adaptive median based restored image ($OutImg1$)
$OrigImage, Imgwidth, Imgheight$ are original image, image width and image height respectively.

1: **procedure** COARSE($OrigImage, Imgwidth, Imgheight$)
2:     $ct_1 = r/4, ct_2 = r/2$ and $ct_3 = 3r/4$ (for three clusters) ▷ window size is N and iteration limit set as $m = 4$
3:     flagmap $= 0$ ▷ The zeros and ones in the flagmap matrix represent good and noisy pixels respectively
4:     **for** k = 1 to m **do**
5:         **for** i = 1 to Imgwidth **do**
6:             **for** j = 1 to Imgheight **do**
7:                 Calculate the $MROR(y_{i,j})$ value using equation 5
8:                 **if** $MROR(y_{i,j}) > 3$ **then**
9:                     $d = y_{i,j} -$ adaptive median of its local window   ▷ d - absolute deviation
10:                     **if** $d > ct_1$ **then**
11:                         flagmap[i, j] = 1
12:                         AM(i, j) = adaptive median of its local window
13:                     **end if**
14:                 **else if** $2 < MROR(y_{i,j}) <= 3$ **then**
15:                     $d = y_{i,j} -$ adaptive median of its local window
16:                     **if** $d > ct_2$ **then**
17:                         flagmap[i, j] = 1
18:                         AM(i, j) = adaptive median of its local window
19:                     **end if**
20:                 **else if** $1 < MROR(y_{i,j}) <= 2$ **then**
21:                     $d = y_{i,j} -$ adaptive median of its local window
22:                     **if** $d > ct_3$ **then**
23:                         flagmap[i, j] = 1
24:                         AM(i, j) = adaptive median of its local window
25:                     **end if**
26:                 **else**
27:                     flagmap[i, j] = 0
28:                     AM(i, j) = 0
29:                 **end if**
30:                 **if** flagmap [i,j] = 1 **then**
31:                     $OutImg1[i, j] = AM(i, j)$
32:                 **else**
33:                     $OutImg1[i, j] = OrigImage(i, j)$
34:                 **end if**
35:             **end for**
36:         **end for**
37:     **end for**
38: **end procedure**

Algorithm 3. The fine stage detection mechanism

Input: Adaptive median based restored image $(OutImg1)$ obtained from the coarse stage

Output: Reference image and a flag matrix

$Imgwidth, Imgheight$ are image width and image height respectively.

1: **procedure** FINE($OutImg1$)
2:    $ft_1 = r/4, ft_2 = r/2, ft_3 = 3r/4$ and $ft_4 = r/8$, (for four clusters)   ▷ where
   $ft_1 \leq ft_2 \leq ft_3 \leq ft_4$, window size is N and iteration limit set as $m = 2$
3:    flagmap $= 0$ ▷ The zeros and ones in the flagmap matrix represent good and
   noisy pixels respectively
4:    **for** k $= 1$ to m **do**
5:       **for** i $= 1$ to Imgwidth **do**
6:          **for** j $= 1$ to Imgheight **do**
7:             Calculate the $MROR(y_{i,j})$ value using equation 5
8:             **if** $MROR(y_{i,j}) > 3$ **then**
9:                $d = y_{i,j} -$ adaptive median of its local window   ▷ d - absolute
   deviation
10:                **if** $d > ft_1$ **then**
11:                   flagmap[i, j] $= 1$
12:                   AM(i, j) = adaptive median of its local window
13:                **end if**
14:             **else if** $2 < MROR(y_{i,j}) <= 3$ **then**
15:                $d = y_{i,j} -$ adaptive median of its local window
16:                **if** $d > ft_2$ **then**
17:                   flagmap[i, j] $= 1$
18:                   AM(i, j) = adaptive median of its local window
19:                **end if**
20:             **else if** $1 < MROR(y_{i,j}) <= 2$ **then**
21:                $d = y_{i,j} -$ adaptive median of its local window
22:                **if** $d > ft_3$ **then**
23:                   flagmap[i, j] $= 1$
24:                   AM(i, j) = adaptive median of its local window
25:                **end if**
26:             **else**
27:                flagmap[i, j] $= 0$
28:                AM(i, j) $= 0$
29:             **end if**
30:             **if** flagmap [i,j] $= 1$ **then**
31:                $OutImg2[i, j] = AM(i, j)$
32:             **else**
33:                $OutImg2[i, j] = OutImg1$
34:             **end if**
35:          **end for**
36:       **end for**
37:    **end for**
38: **end procedure**

Table 3. The number of coarse stage iterations and PSNR values for different intensities of impulse noise (calculated for different categories of mammograms).

Image	Impulse noise intensity	No. of iterations in coarse stage	PSNR (DB)
mdb058 [MISC, D, M]	10%	2	32.9132
	30%	2	31.3283
	50%	3	27.1081
	70%	4	23.8859
mdb080 [CIRC, F, B]	10%	2	31.1412
	30%	3	30.3405
	50%	3	26.3258
	70%	4	20.1211
mdb141 [CIRC, F, M]	10%	2	30.9434
	30%	3	28.7383
	50%	4	25.2239
	70%	4	19.7123
mdb135 [NORM, F]	10%	2	31.0010
	30%	2	27.9559
	50%	3	24.0100
	70%	4	21.8771

Table 4. The number of fine stage iterations and PSNR values for a mammogram image (mdb058) with different intensities of impulse noise.

Iterations	PSNR value at different noise level			
	10%	30%	50%	70%
1	**32.9132**	**31.3283**	27.1021	23.8823
2	32.9132	31.3283	**27.1081**	**23.8861**
3	32.9085	31.5234	27.1083	23.8861
4	32.9052	30.2231	27.1065	23.8859

pixel $x^\wedge(i)$ at pixel i can be derived as the weighted average of all pixel values in the image.

$$x^\wedge(i) = \sum_{j \in I} w(i,j) Y(j) / \sum_{j \in I} w(i,j) \tag{6}$$

where the weights $w(i,j)$ depend on the similarity between the pixel i and j and satisfy the condition $0 \leq w(i,j) \leq 1$ and $\sum_j w(i,j) = 1$. The weights $w(i,j)$ is given as Eq. 7, where h is the decay parameter of the weights.

Table 5. The no. of fine stage iterations and PSNR values for different intensities of impulse noise.

Age	Impulse noise intensity	No. of iterations in fine stage	PSNR (DB)
mdb058 [MISC, D, M]	10%	1	32.9132
	30%	1	31.3283
	50%	2	27.1081
	70%	2	23.8861
mdb080 [CIRC,F,B]	10%	1	32.7428
	30%	1	31.9243
	50%	1	28.0431
	70%	2	24.6719
mdb141 [CIRC,F,M]	10%	1	32.5156
	30%	1	31.0842
	50%	2	29.5369
	70%	2	25.9845
mdb135 [NORM,F]	10%	1	32.9455
	30%	1	29.8743
	50%	1	27.5314
	70%	2	24.5481

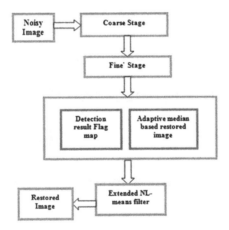

Fig. 1. Block diagram of the de-noising framework.

$$w(i,j) = exp\left(- \|Y(N_i) - Y(N_j)\|^2_{2,a}/h^2 \right) \tag{7}$$

The similarity between two pixels i and j depends on the similarity of the intensity grey level vectors $y(N_i)$ and $y(N_j)$. N_k denotes square neighborhood of fixed size centered at a pixel k, where k represents i and j in Eq. 7. The vector norm used in Eq. 7 is the euclidean distance and a is the standard deviation of the gaussian kernel. Hence, the similarity is estimated as a trailing function of the weighted euclidean distance by a gaussian kernel of zero mean and variance a.

Since the impulse noisy pixels are very different to their neighbors, implementation of original NL-means to remove impulse noise will give wrong output. In the proposed method, NL-means is extended with a reference image and a detection result map, to remove the impulse noise. Figure 1 represents the implementation process of the NL-means filter. The new detection mechanism gives two results, a final detection result and a reference image. Then NL-means is applied to the pixels marked as noisy, according to the detection result and the weights are calculated based on the reference image obtained by the detection mechanism. Through extensive experiments, the value of patch size, generally represented in the literature as k is taken as 2, smoothing level $h = 5$, $W = 7$ and the actual size as $(2k + 1) \times (2k + 1))$. Hence, the actual size of the window is $(2W + 1) \times (2W + 1)$.

5   Results and Discussion

Simulations were carried out on the MIAS database [23], containing circumscribed, speculated and ill-defined masses, and the results were compared with several existing filters such as standard median filter, adaptive median filter, ROR-local means and ROR-NL means in terms of PSNR and MSE.

Table 6. Comparison of the restoration results of mdb058 by PSNR for different methods.

Method	Noise level			
	10%	30%	50%	70%
Standard Median Filter	29.13	24.98	14.51	12.67
Adaptive Median Filter	35.76	28.44	18.92	15.86
ROR-Local Means	41.27	38.12	33.74	26.33
ROR-NL Means	42.56	38.97	34.02	26.92
Proposed method - MROR	45.27	44.37	39.93	35.09

Table 7. Mean square error (MSE) of mammogram image (mdb058) for different methods

Noise level	Standard median Filter	Adaptive median	ROR-local means	ROR-NL means	Proposed MROR
10%	14.98	12.82	8.71	8.10	6.52
30%	27.49	25.32	16.44	14.58	10.56
50%	42.11	36.22	30.03	28.99	23.17
70%	48.90	45.27	37.42	35.67	30.25

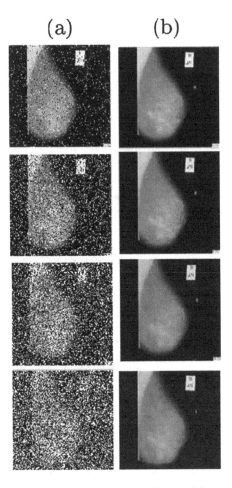

Fig. 2. The result of the proposed method on mdb058, (a) noisy image (b) denoised image.

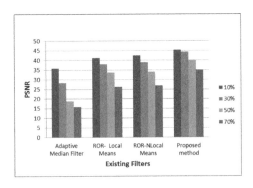

Fig. 3. Performance of different filters at different noise level based on PSNR.

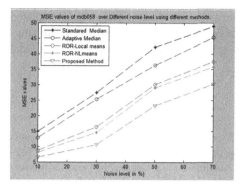

Fig. 4. Comparison of MSE values for different methods.

The proposed method was applied to various noise levels and the obtained results were superior to most of the existing methods, both in terms of image quality and quantitative measures. Impulse noise ranging from 10% to 70% was simulated on the mammogram images taken from the MIAS database followed by the application of the proposed denoising method. Figure 2(a) shows the example of a typical dense-glandular ill-defined mass (mdb 058) corrupted with impulse noise from 10% to 70%. Figure 2(b) shows the corresponding denoised image. The performance of denoising obtained in terms of PSNR is shown in Table 6 and its graphical representation is given in Fig. 3. For the image mdb058 (from MIAS database), mean square error of proposed method and some existing denoising methods at different noise levels are shown in Table 7 and its graphical representation is given in Fig. 4. A smaller mean square error value indicates that the processed output is closer to the original image.

6   Conclusion

This paper introduces a new detection mechanism by combining MROR with Extended NL-means (MROR-ENLM) filter to remove noise in mammogram images. This is an improved version of switching median filter, where the filtering is applied only to corrupted pixels in the image; while the uncorrupted pixels remain unchanged. The advantage of this method is the high noise detection capability as well as the ability to eliminate corrupted pixels in the filtering process. Through experimental analysis it has been shown that MROR-ENLM filter produced excellent results compared to conventional methods such as Standard Median Filter (SMF) and some advanced techniques, such as Adaptive Median Filters (AMF), Robust Outlyingness Ratio-Local Means (ROR-LM) and Robust Outlyingness Ratio-Non Local Means (ROR-NLM). The efficiency of the proposed algorithm is validated using images from MIAS database and that MRO-ENLM produced better results in terms of quantitative measures (PSNR, MSE).

Newer developments in soft computing can be utilized to develop intelligent decision making systems [1,17]. Proven soft computing approaches, such as hybrid fuzzy structures can be explored further, for enhancements in mammogram analysis.

References

1. Alcantud, J.C.R.: A novel algorithm for fuzzy soft set based decision making from multiobserver input parameter data set. Inf. Fusion **29**, 142–148 (2016)
2. Alcantud, J.C.R., Mathew, T.J.: Separable fuzzy soft sets and decision making with positive and negative attributes. Appl. Soft Comput. **59**, 586–595 (2017)
3. Alcantud, J.C.R., Santos-García, G., Hernández-Galilea, E.: Glaucoma diagnosis: a soft set based decision making procedure. In: Puerta, J.M., et al. (eds.) CAEPIA 2015. LNCS (LNAI), vol. 9422, pp. 49–60. Springer, Cham (2015). https://doi.org/10.1007/978-3-319-24598-0_5
4. Aroquiaraj, I.L.: Impulse noise removal from mammogram images using combiner approach. J. Glob. Res. Comput. Sci. **1**(2), 23–27 (2010)
5. Brownrigg, D.R.: The weighted median filter. Commun. ACM **27**(8), 807–818 (1984)
6. Buades, A., Coll, B., Morel, J.M.: A review of image denoising algorithms, with a new one. Multiscale Model. Simul. **4**(2), 490–530 (2005)
7. Chen, T., Wu, H.R.: Space variant median filters for the restoration of impulse noise corrupted images. IEEE Trans. Circuits Syst. II Analog. Digit. Signal Process. **48**(8), 784–789 (2001)
8. Devakumar, D., Punithavathi, V.: Comparison of noise removal filters for breast cancer detection in mammogram images. Int. J. Pure Appl. Math. **119**(18), 3863–3874 (2018)
9. Ferrini, R., Mannino, E., Ramsdell, E., Hill, L.: Screening mammography for breast cancer: American college of preventive medicine practice policy statement. Am. J. Prev. Med. **12**(5), 340–341 (1996)
10. Hwang, H., Haddad, R.A.: Adaptive median filters: new algorithms and results. IEEE Trans. Image Process. **4**(4), 499–502 (1995)
11. Jain, A., Singh, S., Bhateja, V.: A robust approach for denoising and enhancement of mammographic images contaminated with high density impulse noise. Int. J. Converg. Comput. **1**(1), 38–49 (2013)
12. Jayaraj, V., Ebenezer, D., Aiswarya, K.: High density salt and pepper noise removal in images using improved adaptive statistics estimation filter. Int. J. Comput. Sci. Netw. Secur. **9**(11), 170–176 (2009)
13. Ko, S.J., Lee, Y.H.: Center weighted median filters and their applications to image enhancement. IEEE Trans. Circuits Syst. **38**(9), 984–993 (1991)
14. Lai, S., Li, X., Biscof, W.: On techniques for detecting circumscribed masses in mammograms. IEEE Trans. Med. Imaging **8**(4), 377–386 (1989)
15. Maronna, R.A., Martin, R.D., Yohai, V.J., Salibián-Barrera, M.: Robust Statistics: Theory and Methods (with R). Wiley, Hoboken (2019)
16. Mathew, T.J., Alcantud, J.C.R.: Corrigendum to a novel algorithm for fuzzy soft set based decision making from multiobserver input parameter data set information fusion 29 (2016) 142–148. Inf. Fusion **33**(C), 113–114 (2017)

17. Mathew, T.J., Sherly, E., Alcantud, J.C.R.: An adaptive soft set based diagnostic risk prediction system. In: Thampi, S.M., Mitra, S., Mukhopadhyay, J., Li, K.-C., James, A.P., Berretti, S. (eds.) ISTA 2017. AISC, vol. 683, pp. 149–162. Springer, Cham (2018). https://doi.org/10.1007/978-3-319-68385-0_13

18. Mathew, T.J., Sherly, E., Alcantud, J.C.R.: A multimodal adaptive approach on soft set based diagnostic risk prediction system. J. Intell. Fuzzy Syst. **34**(3), 1609–1618 (2018)

19. Mayo, P., Rodenas, F., Verdu, G.: Comparing methods to denoise mammographic images. In: The 26th Annual International Conference of the IEEE Engineering in Medicine and Biology Society, vol. 1, pp. 247–250. IEEE (2004)

20. Nieminen, A., Heinonen, P., Neuvo, Y.: A new class of detail-preserving filters for image processing. IEEE Trans. Pattern Anal. Mach. Intell. **1**, 74–90 (1987)

21. Pitas, I., Venetsanopoulos, A.N.: Nonlinear Digital Filters: Principles and Applications, vol. 84. Springer, Berlin (2013)

22. Sreedevi, S., Mathew, T.J., Sherly, E.: Computerized classification of malignant and normal microcalcifications on mammograms: using soft set theory. In: 2016 International Conference on Information Science (ICIS), pp. 131–137. IEEE (2016)

23. Suckling, J., et al.: The mammographic image analysis siciety digital mammogram database. In: Proceedings of International Congress Series on Exerpta Media Digital Mammography, UK, vol. 1069, pp. 375–378 (1994)

24. Sun, T., Neuvo, Y.: Detail-preserving median based filters in image processing. Pattern Recognit. Lett. **15**(4), 341–347 (1994)

25. Takiar, R., Nadayil, D., Nandakumar, A.: Projections of number of cancer cases in India (2010–2020) by cancer groups. Asian Pac. J. Cancer Prev. **11**(4), 1045–1049 (2010)

26. Wirth, M.A.: A nonrigid approach to medical image registration: matching images of the breast. Royal Melbourne Institute of Technology (1999)

27. Xiong, B., Yin, Z.: A universal denoising framework with a new impulse detector and nonlocal means. IEEE Trans. Image Process. **21**(4), 1663–1675 (2011)

Comparison of Noninvasive Blood Glucose Estimation Using Various Regression Models

Shraddha K. Habbu[1,2,3(✉)], Manisha P. Dale[4], Rajesh B. Ghongade[5], and Shrikant S. Joshi[2]

[1] Department of E&TC Engineering, All India Shri Shivaji Memorial Society's Institute of Information Technology, Pune, India
shraddha.habbu@viit.ac.in
[2] Department of E&TC Engineering, Vishwakarma Institute of Information Technology, Pune, India
shrikant.joshi@viit.ac.in
[3] Department of E&TC Engineering, Savitribai Phule Pune University, Pune, India
[4] Department of E&TC Engineering, Modern Education Society's College of Engineering, Pune, India
mpdale@mescoepune.org
[5] Department of Electronics Engineering, Bharati Vidyapeeth's College of Engineering, Pune, India
rbghongade@gmail.com

Abstract. In this work we aim to evaluate the performance of three types of machine learning models implemented for blood glucose level (BGL) estimation. Pulse photoplethysmography signal is acquired from 611 human subjects and used for the analysis. Time and frequency domain features are extracted using (1) Frame based and (2) Single Pulse Analysis technique. These two features are used as input to train neural network, Support vector machines and Random forest models. These trained models are used for the estimation of BGL values. The BGL estimation performance of these models, (i) neural network, (ii) SVM, (iii) RF and (iv) K-fold RF are compared based on two feature sets i.e. Frame based time and frequency domain features and Single Pulse Analysis based time and frequency domain features. The performance of each system model is evaluated on the basis of, (i) Coefficient of determination i.e. R^2, (ii) Spearman's coefficient of correlation, (iii) Pearson's coefficient of correlation and (iv) Clarke error grid analysis. We observed that Single Pulse Analysis technique shows better performance as compared to Frame based technique. The highest R^2 value (0.95) for Single Pulse analysis is obtained for K-fold RF network. For Single Pulse Analysis, all other models also show comparable BGL estimation accuracy with R^2 values ranging from 0.91 to 0.94. According to Clarke error grid analysis the values that lie in class A and class B are clinically accepted. We obtained highest prediction accuracy for Single Pulse analysis with K-fold random forest with 93.2% (class A) and 6.8% (class B).

Keywords: Photoplethysmograph (PPG) · Neural network · Noninvasive blood glucose measurement NIBGM · Blood glucose level (BGL) · Support vector machines (SVM) · Random forest (RF) · Single Pulse Analysis (SPA)

© Springer Nature Singapore Pte Ltd. 2020
S. M. Thampi et al. (Eds.): SIRS 2019, CCIS 1209, pp. 306–318, 2020.
https://doi.org/10.1007/978-981-15-4828-4_25

1 Introduction

The need of self-monitoring of blood glucose levels is very crucial in diabetics. Currently glucometers that are available in market are either invasive or minimally invasive. The diabetic patients are in need of frequent monitoring of their blood glucose levels to detect the hyper or hypoglycemia events [1]. The glucometers available in market require blood sample which is obtained through puncture of the skin by a lancet. Frequent piercing causes discomfort to the patients. According to 2016 WHO survey report 422 million people were estimated infected by diabetes by 2014 which is 8.5% rise compared to 2012 survey [2]. This shows that mortality rate is increasing rapidly. With the increasing number of infected people by diabetes there is a need to develop a noninvasive method of blood glucose level measurement. However, the last couple of decades, work towards the development of a non-invasive glucose monitor has increased significantly among research groups with promising results. Many noninvasive techniques of blood glucose estimation have been studied and implemented, stating their particular benefits and drawbacks [3]. Noninvasive glucose determinations can be classified into optical and non-optical techniques. The optical properties of glucose are rather specific and these methods have shown better results and a better correlation with blood glucose content. With the advancements in technology, optical based techniques have gained lot of importance in development of biomedical instruments. Photoplethysmography (PPG) is an optical based technique which is used in the development of advanced health care. Various parameters like blood pressure (BP), respiratory rate, stroke volume, pulse transit time (PTT), heart rate variability (HRV), arterial stiffness, and blood glucose levels can be analyzed using PPG technique [1, 3–6].

Moreno et al. [6] presented a system for a simultaneous non-invasive estimate of the blood glucose level (BGL) using a Photoplethysmograph (PPG) and machine learning techniques. A comparative results of blood glucose estimation using various regression technique such as Linear Regression, Neural network, Support vector machines and Random forest are presented by the author. The coefficient of determination (R^2) obtained using linear regression (0.52), neural network is 0.54, Support Vector Machines (0.64) and 0.88 using Random forest regression technique.

Moreno et al. [6] uses 410 subjects PPG data and computed frequency domain features of fixed duration windowed PPG signal for the estimation of BGL values. The importance of the use of Single Pulse Analysis technique is motivated by the fact that the PPG signal looks similar to arterial pressure pulse but the wave contour does not remain the same [7]. In this technique we separate pulses from the entire signal and then extract time domain features of each single pulse. It captures the time varying nature of the pulse. In the current work, we used Single Pulse based time domain features.

In this paper, we present photoplethysmography based BGL estimation using two signal analysis techniques as (i) Frame based and (ii) Single Pulse Analysis. Time and Frequency domain features are computed and are used as input to train neural network, Support Vector Machine and Random Forest models and comparison of BGL estimation is performed using these models. A part of this work is funded by Board of Colleges and University Development (BCUD). Savitribai Phule Pune University, Pune. Development of data acquisition system, data collection work and preliminary experimentation work is carried out under this funding.

2 Methodology

2.1 System Implementation

Figure 1 shows the block schematic of implementation of BGL estimation system based on Frame based and Single Pulse Analysis technique. The PPG acquisition system is built using an SPO2 sensor which is operated in near infrared region [8]. This sensor consists of an optical transmitter (LED) and a receiver (pin photodiode) and is placed along the finger tip. The reflected signal is resultant of volumetric change observed in blood during systole. This change is detected by the detector and is amplified by signal conditioning circuitry. The amplified signal is PPG signal which is then interfaced to a processor to generate digitized pulsed data for recording and storing the data. The details of pulse data acquisition is explained in detail in our earlier research work reference paper [9]. The PPG data of 611individuals is recorded over a 3 min duration each. Using this system the PPG signal of the individuals are recorded along with their BGL values on the Accu-Check® [10] machine simultaneously. The range of BGL values varied from 70 to 450 mg/dl. The database details can be referred in reference [9]. From the collected data we created training dataset and testing dataset which could be used for experimentation. We extracted multiple instances of one minute window out of three minute recorded PPG signal with suitable time overlap. With this process we obtained around 1900 window segments of 1 min duration. By randomly observing the window segments we selected 1000 window segments which are noise, power line interference free and useful for training purpose and 500 samples for testing purpose. The details of training and testing dataset are provided in Table 1. DB3 is a combined training set formed by adding DB1 and DB2 consisting of 1000 samples and DBtest is the testing set consisting of 500 different samples. Preliminary experimentation were carried out using the three datasets and performance evaluation of baseline testing were carried out.

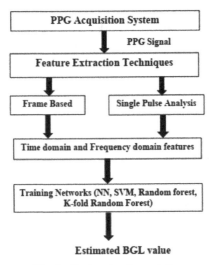

Fig. 1. System implementation

The best results were obtained using DB3 [9]. Hence DB3 dataset is used as a training dataset for further experimentation and DBtest is the testing dataset.

Table 1. Training and testing dataset

Training dataset	Testing dataset	
DB1 (500)	DB2 (500)	DBtest
DB2 (500)	DB1 (500)	DBtest
DB3 (1000)	DBtest	

2.2 Feature Extraction Techniques

The extracted PPG signal is a digitized signal of 1 min duration with a sampling rate of 100 samples per second giving a total of 6000 samples. To remove the noise components generated through power line interference, motion artifacts, muscle artifacts, high frequency artifacts and low amplitude the preprocessing of the signal is carried out. The base line interference is removed using moving average filter and the PPG signal is smoothed using Savitzky Golay filter [11]. The feature vector consists of features computed at window level and local features computed at frame level. The feature set in [6] consists of 33 dimensional features. We excluded age, weight, body mass index and oxygen saturation features to eliminate the person specific dependency. This reduced feature set with 29 features is called baseline feature set.

Frame Based Feature Set: We extracted features based on fixed duration frame with 500 samples per frame. Computed 29 baseline features as mentioned above for each frame and additional features such as pulse transit time, pulse interval and pulse amplitude of the PPG signal along with their statistical measures like mean, variance standard deviation and interquartile range are added to it. This results into 35 dimensional feature set. The details about the computation of Autoregressive coefficients, Kaiser Teager Energy, Heart rate, Spectral Entropy, Log energy profile, Pulse interval, Pulse transit time, and Pulse amplitude can be obtained in our previous research work presented in [9]. We represent this feature vector as FV and is given by

$$FV = \{FVAR_{PPG}, FVKTE, FVHr, FVHs, FVLogE, VPTT, FVPint, FVPamp\}$$

Single Pulse Analysis Feature Set: As mentioned earlier, the PPG signal looks similar to arterial pressure but wave contour of pulse does not remain same hence to capture the variations in pulse wave shape of the entire signal we implemented single pulse technique. We split the pulses by finding the local minima and extracted features of each pulse based on energy, pulse interval, pulse transit time, pulse start time, pulse end time, peak amplitude along with its statistical measures like mean, variance, standard deviation and interquartile range. This feature set consists of 28 features and is given as

$$FV_{new} = \{PFV_n^{\mu}, PFV_n^{\sigma}, PFV_n^{skew}, PFV_n^{iqr}\}$$

The details about the computation of the above feature set is explained in our research work presented in [9].

2.3   BGL Estimation

We implemented BGL estimation using time domain feature set with 35 features and Single pulse feature set with 28 features using various regression techniques. As explained in Sect. 2.1 we created training dataset consisting of 1000 samples and testing dataset of 500 samples useful for training purpose and for testing purpose. The features of training dataset along with their BGL values are then given as input to various training models like neural network, support vector machine, random forest and k-fold random forest generating the training model as explained in further sections. A nonlinear regression model is trained by each network. These networks are then used for testing dataset to estimate their BGL values. Also the performance of each network is further analyzed based on Clarke error grid analysis and coefficient of determination.

Neural Network: For neural network we used Multilayer perceptron trained by Levenberg Marquart algorithm. The Multilayer perceptron is able to approximate nonlinear relationships. We used 3 hidden layers neural network topology with 30, 20, 10 neurons in layer 1, 2 and 3 respectively to train the network using nonlinear regression. Matlab 2017b neural network toolbox is used for neural network training and testing purpose.

Support Vector Machines: SVM regression is implemented using Matlab 2017 SVM tool box. We used Gaussian Kernel based on the earlier results reported in [6] which had shown best results. The specifications of kernel (kernel type, soft margin, and kernel parameters) were set based on the quality criterion i.e. R^2 value on the testing database.

Random Forest: Random forests are a ML technique based on a Set (forest) of classification and regression trees [12], each trained in a different way. The output of the system is an aggregation of the outputs of the trees of the forest. The methodology is justified by the fact that the mathematical expression of the error of a classifier depends on the bias and variance of each of the trees of the forest. In order to take advantage of the fact that averages reduce the bias and variance, the algorithm introduces a systematic controlled variability and bias [12]. The advantages of this technique is (i) low global error rate as the mean output reduces the variance and bias of the estimate, (ii) the execution time is low as the base classifier is a tree and (iii) computations are based on comparisons at each node. Also each node compares only one feature, making the system robust with respect to the scale of the inputs and the correlations between them [6]. The best configuration (number of trees in the forest, number of features tested at each node and maximum number of samples at a given node) was determined based on the quality criterion i.e. R^2 value on the testing database. The algorithm was programmed in Matlab using *'fit-r-ensemble'* function. We also implemented cross validation on testing dataset using K-fold random forest technique with 10 folds and maximum decision tree number 50. The same was also implemented in Matlab.

3 Experimentation and Results

In this work we performed BGL estimation using two feature extraction techniques as (1) Frame based and (2) Single pulse analysis. The performance of BGL estimation is evaluated using coefficient of determination (i.e. R^2), computed via four different regression models namely neural network, SVM, Random forest and K-fold Random Forest. We also analyze the performance based on Clarke Error Grid Analysis. Clarke Error Grid Analysis is clinically accepted for validation of BGL estimation [13]. The grid is divided into 5 regions. Clarke grid scatter is plotted using Matlab function [14]. Region 'A' represents prediction within 20% of the actual BGL value. Region 'B' represents prediction more than 20% away from actual BGL but do not give false predictions. Region 'C' represents false positives of either cases of hypoglycemia or hyperglycemia. Region 'D' represents predictions that fail to detect cases of hypoglycemia or hyperglycemia. Region 'E' represents prediction errors which could wrongly classify cases of hypo or hyperglycemia. Following section gives the details about the results of BGL estimation using four different regression models for 2 features sets i.e. (1) Framed based time and frequency features and (2) Single Pulse Analysis based time and frequency features. Table 2 gives the comparative performance measures of all the four regression models for baseline feature set with the reference [6]. Table 3 gives the comparative performance measures of all the 4 regression models for time and frequency domain and Single Pulse Analysis feature set. Figures 2, 3, 4 and 5 show the computation of R^2 and Clarke Error grid Analysis using 4 different regression models for Single Pulse analysis technique.

Table 2. Comparative BGL estimation performance of 3 different regression models NN, SVM and RF of baseline testing

	NN	SVM	RF	K-fold RF
Monte Moreno [6] (33 features) R^2	0.54	0.64	0.88	–
Baseline system (29 features) R^2	0.71	0.78	0.83	0.93

Table 3. Comparative BGL estimation performance of 4 different regression models NN, SVM RF and K-fold RF for Frame based and Single Pulse Analysis.

	NN frame based	NN Single Pulse Analysis	SVM Frame Based	SVM Single Pulse Based
R^2	0.83	0.91	0.74	0.94
A	403 (80.6)	415 (83)	340 (68)	434 (86.8)
B	87 (17.4)	85 (17)	150 (30)	66 (12.2)
C	5 (1)	0	10 (2)	0
D	5 (1)	0	0	0
E	0	0	0	0
Spearman coefficient	0.886	0.948	0.909	0.970
Pearson coefficient	0.8794	0.954	0.905	0.977
	RF Frame based	RF Single Pulse Based	K-fold Frame based	K-fold Single Pulse Based
R^2	0.84	0.91	0.91	0.95
A	395(79.4)	424 (84.8)	447 (89.4)	466 (93.2)
B	96 (19.2)	75 (15)	53 (10.6)	34 (6.8)
C	3 (0.6)	0	0	0
D	6 (1.2)	1 (0.2)	0	0
E	0	0	0	0
Spearman coefficient	0.895	0.928	0.944	0.957
Pearson coefficient	0.893	0.937	0.947	0.965

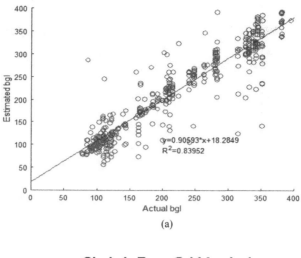

(a)

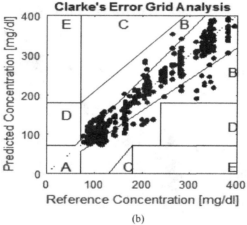

(b)

Fig. 2. Performance metric of Single Pulse analysis for nonlinear regression technique using NN (a) Coefficient of determination (i.e. R^2) and (b) Clarke error grid.

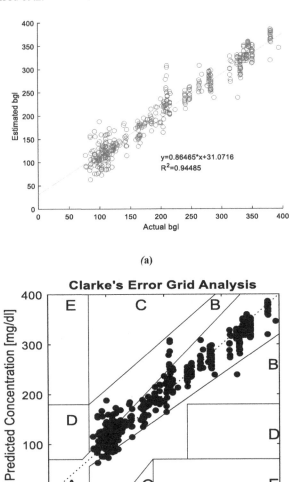

(a)

(b)

Fig. 3. Performance metric of Single Pulse analysis for nonlinear regression technique using SVM (a) Coefficient of determination (i.e. R^2) and (b) Clarke error grid.

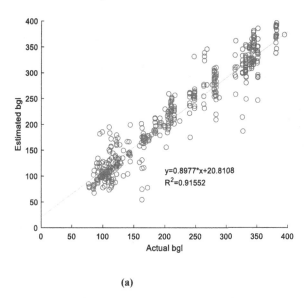

(a)

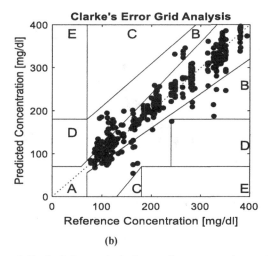

(b)

Fig. 4. Performance of Single Pulse analysis for nonlinear regression technique using RF (a) Coefficient of determination (i.e. R^2) and (b) Clarke error grid

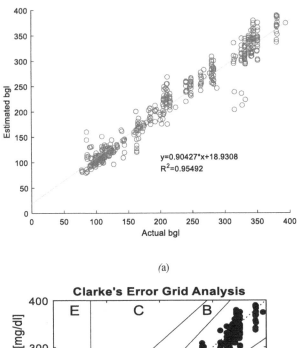

(a)

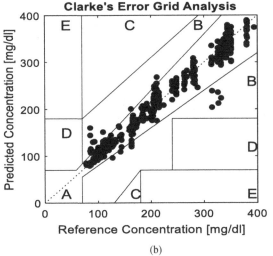

(b)

Fig. 5. Performance metric of Single Pulse analysis for nonlinear regression technique using K-fold (a) Coefficient of determination (i.e. R^2) and (b) Clarke error grid.

4   Observations and Discussion

From Table 2, it is observed that we obtained comparable baseline results with the results presented in the literature [6]. By eliminating the person specific dependent features and reducing the feature size there is improvement in R^2.

Table 3 represents comparative performance evaluation of machine learning models based on two PPG signal analysis techniques i.e. (i) Frame based and (ii) Single Pulse Analysis. Single Pulse Analysis technique outperforms Frame based technique for all performance evaluation metrics and this is also true for all machine learning models i.e. NN, SVM, RF and K-fold RF. The performance of Single Pulse Analysis shows highest accuracy via R^2, Spearman and Pearson correlation coefficient performance metrics. High value of Spearman's and Pearson's correlation coefficient shows that the estimated and actual BGL values show monotonic and linear relationship. The results obtained for Single Pulse Analysis with neural network shows improvement in BGL value estimation ($R^2 = 0.84$ for Frame based and 0.91 for Single Pulse technique) as compared with reference result ($R^2 = 0.54$) obtained by Monte et al. [6].

The performance of machine learning models (NN, SVM, RF, K- fold RF) for Single Pulse analysis via R^2 are comparable and lie in range 0.91 to 0.95. According to Clarke error grid analysis we obtained the highest estimation accuracy for K-fold random forest with 93.2% prediction cases lying in class A and 6.8% cases in class B. The Random forest technique uses multiple trees which reduces the risk of over fitting. For the system performance validation purpose cross validation is very important as it ensures that each test sample occur in training and testing phase. This eliminates the chance of obtaining very good or bad results. With 10 fold cross validation we obtained R^2 value of 0.95. The accuracy obtained using SVM is also comparable with K-fold Random forest.

5   Conclusion

We investigated the performance of different machine learning models which includes (1) Neural network, (2) SVM, (3) RF and (4) K-fold RF in BGL estimation. We explored two feature extraction techniques as (1) Frame based and (2) Single Pulse Analysis. From the results it is concluded that Single Pulse Analysis technique shows high accuracy as compared with Frame based technique. Using Single Pulse Analysis all the trained models have shown comparable results. Random forest has shown highest accuracy as this method operates by constructing multiple decision trees during training phase. As SVM and RF models shows comparable results, these models can be used for noninvasive BGL real time system implementation.

References

1. So, H.-F., Choi, K.-S.: Recent advances in Noninvasive glucose monitoring. Med. Devices (Auckl) **5**, 45–52 (2012)
2. World Health Organization: Global status report on diabetes 2016. WHO Press, Geneva, Switzerland (2016)

3. Ciudin, A., Hernandrez, C., Simo, R.: Non-invasive Methods of Glucose measurement: Current status and future Perspectives. Curr Diabetes Rev. **8**(1), 48–54 (2012)

4. Fortino, G., Giampa, V.: PPG based methods for noninvasive and continuous blood pressure measurement: an overview and development issues in body sensor networks. In: IEEE Transaction (2010)

5. Poon, C., Zhang, Y.: Cuffless and noninvasive measurements of arterial blood pressure by pulse transit time. In: 27th Annual International Conference on IEEE EMBS (Engineering in Medicine and Biology Society), Shanghai, China, 17–18 January 2005

6. Moreno, E., et al.: Noninvasive estimation of blood glucose and blood pressure from PPG by means of machine learning techniques. J. Artif. Intell. Med. **53**(2), 127–138 (2011)

7. Millasseau, S., Guigui, F., Kelly, R., Prasad, K., Cockcroft, J., Ritter, J.: Noninvasive assessment of the digital volume pulse. Comparison with the peripheral pressure pulse. Hypertension **36**, 952–956 (2000)

8. Jeon, K.J., Hwang, I.D., Hahn, S., Yoon, G.: Comparison between transmittance and reflectance measurements in glucose determination using near infrared spectroscopy. J. Biomed. Opt. **11**, 014022 (2006)

9. Habbu, S., Dale, M., Ghongade, R.: Estimation of blood glucose by noninvasive method using photoplethysmography. Sādhanā **44**, 135 (2019). https://doi.org/10.1007/s12046-019-1118-9

10. Accu-chek Aviva. https://www.accuchek.com/us/glucose-meters/aviva.html. Accessed 2 May 2011

11. Schafer, R.W.: What is a Savitzky-Golay filter? (Lecture notes). IEEE Signal Process. Mag. **28**(4), 111–117 (2011). https://doi.org/10.1109/msp.2011.941097

12. Breiman, L.: Random forests. Mach. Learn. **45**(1), 5–32 (2001)

13. Clarke, W., Cox, D., Gonder-Frederick, L.A., Carter, W., Pohl, S.: Evaluating clinical accuracy of systems for self-monitoring of blood glucose. Diabetes Care **10**(5), 622–628 (1987)

14. Codina, E.G.: Clarke error grid analysis. Matlab central file exchange (2008). http://www.mathworks.com/matlabcentral/fileexchange/20545-clarke-error-grid-analysis. Accessed 2 May 2011

SVD Diagonal Matrix Reconstruction Using OMP

Indrarini Dyah Irawati[1]([⊠]), Ian Joseph Matheus Edward[2],
and Andriyan Bayu Suksmono[2]

[1] Telkom Applied Science School, Telkom University, Bandung, Indonesia
`indrarini@telkomuniversity.ac.id`
[2] School of Electrical and Informatics, Institut Teknologi, Bandung, Indonesia
`{ian,suksmono}@stei.itb.ac.id`

Abstract. We designed a compression and reconstruction system in internet traffic applications by combining Singular Value Decomposition (SVD) and Orthogonal Matching Pursuit (OMP) techniques. This system aims to obtain all internet traffic information from several measured data samples. The problem occurred due to the reconstruction results of the SVD diagonal matrix using OMP are undefined values. This paper presents a solution to fix it. We applied zero forcing and improved it using interpolation. The optimized interpolation is proposed to replace zero value with optimal solution using minimization algorithm. The result shows that the proposed methods is able to overcome zero value problem and improve the accuracy of reconstruction.

Keywords: Internet traffic · Interpolation · Optimization · Orthogonal matching pursuit · Reconstruction · SVD diagonal matrix

1 Introduction

A new method known as Compressive Sensing (CS) has the potential to reconstruct sparse signals in appropriate transformation domain can be reconstructed using fewer sample [1–3]. This scheme requires that the processed signal be sparse and the measurement matrix used must satisfy the Restricted Isometric Property (RIP) [4]. To return a reconstructed signal to its original form needed reconstruction algorithms that have reviewed in [5].

Over the last few years, researches on CS applications for internet traffic matrix began to be developed [6–10, 16]. Some studies have suggested that the right sparse representation of Internet traffic matrix is SVD transformation [6, 11]. According to [12], the measurement matrix is derived from a fixed routing matrix, but it does not satisfy the RIP properties. To meet these requirements, Xu et al. use a Boolean matrix with elements [0,1] as the measurement matrix [13]. Another research generally used random Gaussian matrix [14]. This measurement matrix data is used for the reconstruction process. Presently, there are many reconstruction algorithms which are used properly depending on the application used. The challenge in signal reconstruction is the ability of the

© Springer Nature Singapore Pte Ltd. 2020
S. M. Thampi et al. (Eds.): SIRS 2019, CCIS 1209, pp. 319–331, 2020.
https://doi.org/10.1007/978-981-15-4828-4_26

reconstruction algorithm to work quickly with high accuracy results. One of the groups of reconstruction algorithms is greedy iterative has been widely used in CS [15]. This class is superior because of rapid reconstruction and low mathematical complexity, but if the signal is not sparse the reconstruction is less accurate. The Orthogonal Matching Pursuit (OMP) is one example of reconstruction algorithm in greedy class [16].

In this study, we built CS system to solve the problem of reconstruction on Internet traffic data. We applied the SVD technique to sparse the data and used the routing matrix as the measurement matrix, and took the advantage of the OMP reconstruction algorithm for recovering the compression results. A problem occurs when the results of OMP reconstruction are undefined. This causes reduction the accuracy of the reconstruction results. We solved undefined number by zero forcing and improved the performance of reconstruction accuracy using linear interpolation [10], and optimized interpolation.

2 SVD Sparsing

Internet traffic data as input on the CS system must be sparse. SVD decomposes Internet traffic matrix and utilizes the decomposition result, i.e. SVD diagonal matrix that has sparse properties. SVD mathematical equations is shown as follows:

$$X = U\Sigma V^T. \tag{1}$$

SVD factorization of a matrix $X \in \mathbb{R}^{N \times T}$ yields $U \in \mathbb{R}^{N \times N}$ and $V^T \in \mathbb{R}^{T \times T}$ which are defined as orthogonal matrices, and $\Sigma \in \mathbb{R}^{N \times T}$ denotes a diagonal matrix. The matrix Σ with positive real elements is expressed as follows:

$$\Sigma = \begin{bmatrix} \Sigma_k & 0 \\ 0 & 0 \end{bmatrix}. \tag{2}$$

where $\Sigma_k = \text{diag}(\sigma_1, \sigma_2, \ldots, \sigma_k)$ which is arranged so that $\sigma_1 \geq \sigma_2 \geq \cdots \geq \sigma_k > 0$, and k is called as rank of X.

The rank of X is equal to the number of non-zero singular values, which must be less than or equal to $\min(N, T)$, $k \leq \min(N, T)$ [17]. The SVD diagonal matrix is a sparse matrix that meets the CS requirements. These elements are known as the singular value of X, which represent very significant information from data. The CS uses the SVD diagonal matrix (Σ) for the next process.

3 SVD-OMP for Internet Traffic Reconstruction

3.1 Block Diagram

Figure 1 illustrates SVD-OMP for internet traffic reconstruction. Internet traffic is measured from Abilene network [18]. This traffic is presented temporally as in previous research [8, 9, 16]. SVD sparsing produces a sparse diagonal matrix Σ. We use binary-valued routing matrix as measurement matrix (A).

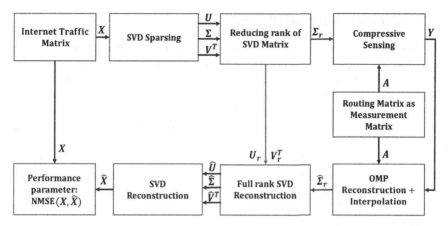

Fig. 1. Block diagram of SVD-OMP Internet traffic reconstruction

SVD sparsing generate sparse diagonal matrix Σ and matrices U and V^T as in Eq. (1). According to [19], Internet traffic matrix has low-level dimension so the traffic matrix X can represented by some rank r, where $r \ll k \leq \min(N, T)$. This term is known as the low-rank singular matrix. The matrix $X \in \mathbb{R}^{N \times T}$ with rank-r has matrix decomposition with new dimension as follows $U_r (N \times r)$, $\Sigma_r (r \times r)$, and $V_r^T (r \times T)$.

The measurement matrix aims to reduce the number of measurements. In this paper, the dimension of measurement matrix is $m \times r$, where m is the number of measurements, and r is the rank, where $m \leq r$. The measurement matrix must satisfy RIP, as in the following equation [4]:

$$(1 - \delta_s) \cdot \|\sigma\|_2 \leq \|A \cdot \sigma\|_2 \leq (1 + \delta_s) \cdot \|\sigma\|_2, \text{ where } 0 < \delta_s \ll 1. \tag{3}$$

Consider for sparse internet traffic matrix Σ_r and measurement matrix A. The measurement of CS can defined as follow:

$$Y = A \times \Sigma_r. \tag{4}$$

The OMP reconstruction algorithms is useful to recover the SVD diagonal matrix ($\hat{\Sigma}$) [16]. Then, SVD reconstruction approximates the Internet traffic matrix ($\hat{X}$) using data U and V, according to the equation:

$$\hat{X} = \sum_{i=1}^{k} \hat{\sigma}_i u_i v_i^T = \hat{\sigma}_1 u_1 v_1^T + \cdots + \hat{\sigma}_k u_k v_k^T, \tag{5}$$

where u_i and v_i^T are the i –th column of U and V.

The performance of CS reconstruction algorithms are evaluated using Normalized Mean Square Error (NMSE) parameter.

3.2 Trade-Off of Compression Results

In this study, we show trade-off of compression results between the number of elements with NMSE. This simulation uses traffic data taken from the Abilene network on March 1, 2004. The data are represented in a matrix of size $(N \times T)$ with N representing the link between nodes and T denoting the measurement time. The number of nodes on the Abilene network is 12 so there are 144 links. The traffic is measured every 5 min so that there are 288 data in a day. We use a size traffic matrix (144×288). The total number of processed elements are 41472.

Figure 2 presents trade-off between the number of elements and NMSE at different rank for random data, while Fig. 3 for the traffic data. Random data are positive values generated randomly. These data are arranged in the form of a matrix with the same size as the traffic matrix $(N \times T)$. While traffic data is obtained from traffic measurements on the Abilene network.

From Fig. 2 and Fig. 3, the solid blue line represents the NMSE between the SVD reconstruction $(\hat{X})$ with original matrix (X) at a certain rank. The solid red line illustrates the total number of matrix elements. The solid green line describes the total number of SVD sparsing. The dash blue line shows maximum rank value for maximum tolerated element number. The dash pink line is minimum rank for tolerated NMSE. The dash red line is maximum tolerated NMSE. Referring to prior research [6, 7, 10], the maximum NMSE is 0.2. The trade-off results is obtained by changing r or rank, then the simulation as described in Fig. 1 is executed. Parameter r is changed from 8 to 144 sample. The output of simulation is the element number (displayed in left-axis in Fig. 2 and Fig. 3) to be transmit to the receiver consisting of the element number of U, V and the compressed matrix Y. Matrix A is assumed to be known in the receiver. The other output of simulation displayed is NMSE displayed on right-axis in Fig. 2 and Fig. 3.

From both figures, it is known that traffic data have tolerated rank region widely than random data. This describes that the number of solutions in traffic data is more than the random data. In addition, traffic data naturally have sparse property compared than random data so that the reconstruction results are more accurate. The element number and NMSE has a trade-off trending when the parameter rank or r is changed. NMSE is decreased, while the element number is increased, when the r is increased.

4 Problem Formation and Solution

In this paper, OMP algorithm reconstructs column-by-column of diagonal matrix $(\hat{\Sigma}_r)$. Each column of $\hat{\Sigma}_r$ is reconstructed by OMP with the same measurement matrix (A) and compression result (Y). The problem of OMP algorithm reconstruction is the appearance of undefined value or NaN (Not a Number) due to the division of a number with a value of zero [10].

NaN value is then replaced with a value of 0 (zero forcing). Zero insertion techniques have not been able to solve the problem because the NMSE value is still high. Figure 4 shows the results of the singular value reconstruction using the OMP algorithm. The x-axis indicates the i-th position of singular value and the y-axis indicates the value of the singular value.

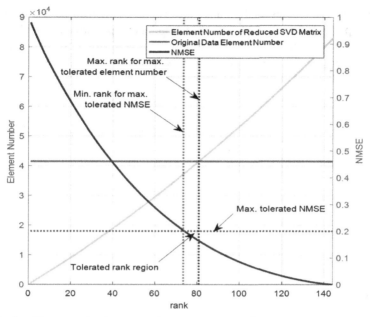

Fig. 2. Trade-off between the element number and NMSE at different rank for random data (Color figure online)

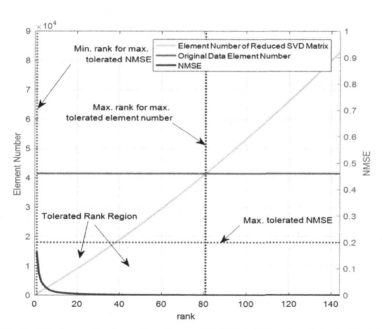

Fig. 3. Trade-off between the element number and NMSE at different rank for traffic data (Color figure online)

From Fig. 4, zero is inserted on undefined values of the reconstruction results, for example at position $i = 1, 2, 4$, etc. Test simulations performed at rank-$r = 60$. Based on the picture, the number of zero insertions is 42 from 60. To overcome this problem, we propose improvements of the reconstruction results in two ways, namely linear interpolation and optimized interpolation.

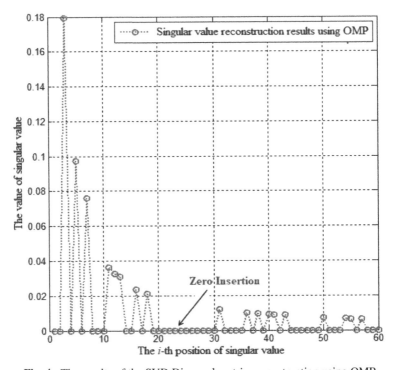

Fig. 4. The results of the SVD Diagonal matrix reconstruction using OMP

4.1 Linear Interpolation

Figure 5 shows the flowchart for solving undefined values using linear interpolation. The solution to the linear interpolation scheme is as follows [16]:

1. If $\hat{\sigma}_{(1)} = 0$, then $\hat{\sigma}_{(i)} = \max(\hat{\sigma}_{(i)})$.
2. In addition, the following equation applies:

$$\hat{\sigma}_{(i)} = \text{mean}(\hat{\sigma}_{(i-1)} + \hat{\sigma}_{(i+1)}), \tag{6}$$

where $\hat{\sigma}_{(i)}$ is singular value solution, $\hat{\sigma}_{(i-1)}$ is singular value at $i - 1$ dan $\hat{\sigma}_{(i+1)}$ is the value at $i + 1$.

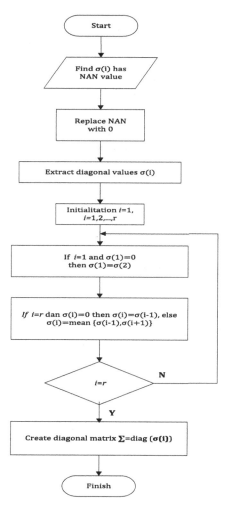

Fig. 5. Flowchart linear interpolation technique

4.2 Optimized Interpolation

The optimized interpolation used in this study is natural logarithm. Singular values have characteristics similar to the values in natural logarithm. Figure 6 shows the flowchart for solving undefined value using optimized interpolation. The function of natural logarithm is shown as follows, for an example:

$$f(x) = e^{-\gamma x}, \tag{7}$$

$$\ln f(x) = -\gamma x \tag{8}$$

For some zero values, the Eq. (7) and (8) can be written as follows:

i. $\sigma_1 = e^{-\gamma x_1}$ then $\ln \sigma_1 = -\gamma \cdot x_1,$ (9)

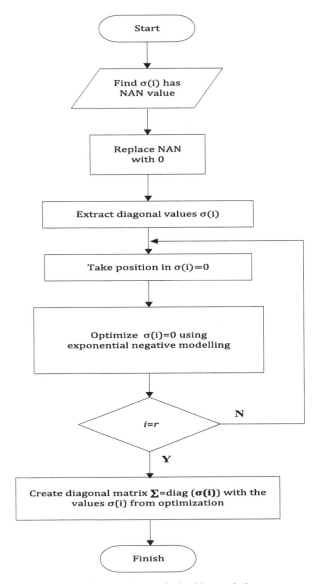

Fig. 6. Flowchart optimized interpolation

ii. $\sigma_2 = e^{-\gamma x_2}$ then $\ln \sigma_2 = -\gamma \cdot x_2,$ (10)

iii. $\sigma_n = e^{-\gamma x_n}$ then $\ln \sigma_n = -\gamma \cdot x_n,$ (11)

From (9), (10), and (11), we can arrange as:

$$
-\gamma \begin{bmatrix} x_1 \\ x_2 \\ \vdots \\ x_n \end{bmatrix} = \begin{bmatrix} \ln \sigma_1 \\ \ln \sigma_2 \\ \vdots \\ \ln \sigma_n \end{bmatrix}, \tag{12}
$$

Least square solution is given as follows:

$$
-\gamma \begin{bmatrix} x_1 \ x_2 \ \cdots \ x_n \end{bmatrix} \cdot \begin{bmatrix} x_1 \\ x_2 \\ \vdots \\ x_n \end{bmatrix} = \begin{bmatrix} x_1 \ x_2 \ \cdots \ x_n \end{bmatrix} \cdot \begin{bmatrix} \ln \sigma_1 \\ \ln \sigma_2 \\ \vdots \\ \ln \sigma_n \end{bmatrix}, \tag{13}
$$

$$
-\gamma X^T \cdot X = X^T \cdot \begin{bmatrix} \ln \sigma_1 \\ \ln \sigma_2 \\ \vdots \\ \ln \sigma_n \end{bmatrix}, \tag{14}
$$

$$
\gamma = -\left(X^T \cdot X \right)^{-1} \cdot X^T \cdot \ln \sigma, \tag{15}
$$

The optimal solution for obtaining γ values is to cover the minimum value according to the equation:

$$
\begin{aligned}
\varepsilon &= arg\, min\, |f(\gamma) \cdot X - f(\sigma)|, \\
\gamma &= \{a, b\} \\
\sigma &= \{c, d\}
\end{aligned} \tag{16}
$$

4.3 Analysis of the Proposed Technique

Figure 7 shows the simulation result of zeros solving with a linear interpolation and optimized interpolation scheme. The experiment was carried out at rank-$r = 60$. The x-axis represents the i-th position of singular value and the y-axis represents the value of singular value. The green line represents SVD diagonal matrix (Σ_r), red line represents solution using linear interpolation, and dash blue line represents solution using optimized interpolation.

The performance parameter is the NMSE which calculated between the original singular values (Σ_r) and the singular values reconstruction result ($\hat{\Sigma}_r$). From the figure, the results of optimized interpolation are closer to the original singular value. The NMSE calculation show that NMSE on linear interpolation is 0.875, while NMSE on optimized interpolation is 0.00060373. The optimized interpolation provides a lower NMSE than linear interpolation. We conclude that both interpolation schemes are able to overcome the problem of zero value in the results of OMP reconstruction. The optimized interpolation has better accuracy than linear interpolation.

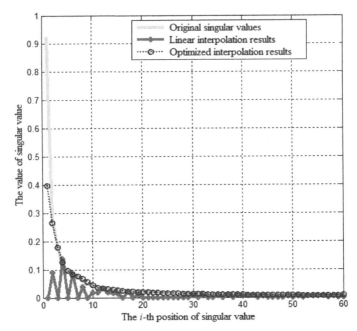

Fig. 7. SVD diagonal matrix improvement using interpolation (Color figure online)

Figure 8 shows the result of Internet traffic signal reconstruction at the 30th link. Simulation is done on traffic data for 1 day, which is measured every 5 min. The x-axis represents the amount of sample time and the y-axis represents internet traffic. From the figure, the optimized interpolation technique produces signal that are close to the original signal compared to linear interpolation technique.

Figure 9 shows the NMSE of reconstruction system. The NMSE is calculated between the original traffic matrix (X) and the reconstructed traffic matrix $(\hat{X})$. NMSE comparisons are performed on various numbers of measurements at rank-$r = 24$. From the figure, the x-axis represents the number of measurement (m) and the y-axis represents NMSE. The simulation results show that the NMSE of optimized interpolation is smaller than linear interpolation. Both interpolation techniques meet the NMSE target, less than 0.2.

Figure 10 shows the effect of measurement number on computational time for linear interpolation and optimized interpolation. The simulation parameter used is rank-$r = 40$. From the Fig. 10 above, it appears that the more the number of measurements used the longer the computation time. At the same number of measurements, the optimized interpolation yields in longer computational time than linear interpolation. This is due to the number of operations on optimized interpolation more than in linear interpolation.

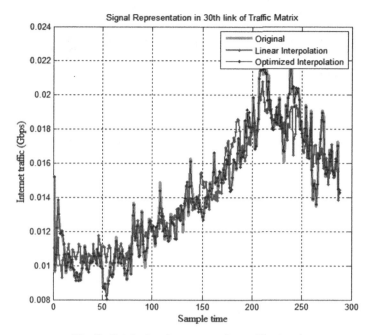

Fig. 8. Original and reconstruction traffic signal

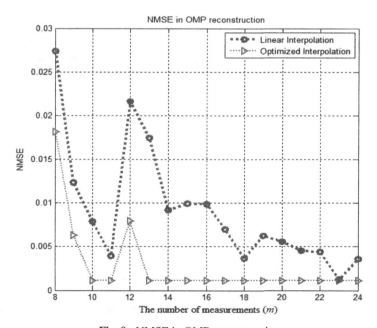

Fig. 9. NMSE in OMP reconstruction

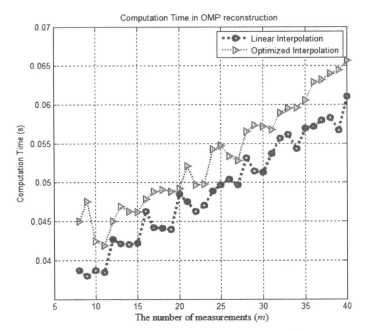

Fig. 10. Computational time in OMP reconstruction

5 Conclusion

We investigates the problem of undefined value from OMP reconstruction for internet traffic data. The proposed technique based on interpolation for optimizing SVD diagonal matrix reconstruction is proposed. Simulation results show that interpolation schemes can reduce errors. Optimized interpolation gives better results than linear interpolation for data reconstruction. In further research, we will create an advanced OMP reconstruction algorithm that can address the problem of undefined values.

References

1. Donoho, D.L.: Compressed sensing. IEEE Trans. Inf. Theory **52**, 1289–1306 (2006)
2. Candes, E.J., Romberg, J., Tao, T.: Robust uncertainty principles: exact signal reconstruction from highly incomplete frequency information. IEEE Trans. Inf. Theory **52**, 489–506 (2006)
3. Candes, E.J., Tao, T.: Near optimal signal recovery from random projections: universal encoding strategies. IEEE Trans. Inf. Theory **52**(12), 5406–5425 (2006)
4. Candes, E.J.: The restricted isometry property and its implications for compressed sensing. Compte Rendus de l'Academie des Sciences **1**, 589–592 (2008)
5. Qaisar, S., Bilal, R.M., Iqbal, M.N.W., Lee, S.: Compressive sensing: from theory to applications, a survey. J. Commun. Netw. **15**(5), 443–456 (2013)
6. Roughan, M., Zhang, Y., Willinger, W., Qiu, L.: Spatio-temporal compressive sensing and Internet traffic matrices (Extended version). IEEE/ACM Trans. Netw. **20**(3), 662–676 (2012)
7. Huibin, Z., Dafang, Z., Kun, X., Xiaoyan, W.: Data reconstruction in Internet traffic matrix. IEEE J. Mag. China Commun. **11**, 1–12 (2012)

8. Irawati, I.D., Suksmono, A.B., Edward, I.J.M.: Low-rank Internet traffic matrix estimation based on compressive sampling. Adv. Sci. Lett. **23**(5), 3934–3938 (2017)
9. Irawati, I.D., Suksmono, A.B., Edward, I.J.M.: Missing Internet traffic reconstruction using compressive sampling. Int. J. Commun. Netw. Inf. Secur. **9**(1), 57–66 (2017)
10. Nie, L., Jiang, D., Guo, L.: A Compressive sensing-based reconstruction approach to end-to-end network traffic. IEEE (2012)
11. Juva, I.: Traffic Matrix Estimation. Helsinki University of Technology, Helsinki (2005)
12. Xu, W., Mallada, E. Tang, A.: Compressive sensing over graph. In: IEEE Proceedings of INFOCOM (2011)
13. Eldar, Y.C., Mishali, M.: Robust recovery of signals from a structured union of subspace. IEEE Trans. Inf. Theory **55**(11), 5302–5316 (2009)
14. Du, L., Wang, R., Wan, W., Yu, X.Q., Yu, S.: Analysis on greedy reconstruction algorithms based on compressed sensing. IEEE-International Conference on Audio, Language and Image Processing, Shanghai, China (2012)
15. Tropp, J.A., Gilbert, A.C.: Signal recovery from random measurement via orthogonal matching pursuit. IEEE Trans. Inf. Theory **53**(12), 4655–4666 (2007)
16. Irawati, I.D., Suksmono, A.B., Edward, I.J.M.: Enhanced OMP for missing traffic reconstruction based on sparse SVD. In: IEEE International Conference on Telecommunications (ICT), Hanoi, Vietnam (2019)
17. Pillow, J.: SVD and Linear Systems, Lecture 3A Note, pp. 1–3. Princeton University, New Jersey
18. Network, The Abilene Research. http://abilene.internet2.edu/.s
19. Ringberg, H.: Sensitivity of PCA for traffic anomaly detection. In: Sigmetrics (2007)

Channel Estimation Using Adaptive Cuckoo Search Based Wiener Filter

Gopala Krishna Mellempudi[1]($\boxtimes$) and Vinay Kumar Pamula[2]

[1] Department of ECE, MIC College of Technology, Kanchikacherla, AP, India
gopal.ece16@gmail.com
[2] Department of ECE, UCEK, JNTUK, Kakinada, AP, India
pamulavk@ieee.org

Abstract. In recent years nature based computing plays an important role in various applications. In this paper, a novel channel estimator is proposed based on nature based computing. The cuckoo search (CS) is a stochastic metaheuristic algorithm is adopted and implemented for the channel estimator due to use of a single control parameter which makes it superior compared with others. The implementation is done at different signal to noise ratio (SNR) condition. M-ary quadrature amplitude modulation (M-QAM) modulation with zero phase offset is used for analysis. Performance of M-QAM scheme, with the proposed channel estimation method over Rayleigh, Rician, Nakagami and Weibull fading is investigated. Performance of channel estimator is determined by mean square error (MSE). On comparison, the proposed channel estimation method gives better MSE when compared to Least Squares (LS) and minimum MSE (MMSE) methods.

Keywords: Adaptive Cuckoo search · Channel estimation · Fading channel · Wiener filter

1 Introduction

The wireless communication has become a leading growth in recent years. This technology will support all needs for a fast internet connection for corporate or home users. Fifth generation (5G) wireless standard is the emerging future cellular network technology to have the massive Multiple-Input Multiple-Output (MIMO) system with millimeter wave (mmWave) communications. To perform with high efficiency and to achieve high gain in both spectral and energy efficiencies, a massive MIMO will employ a large number of antennas. Hence, in the mobile communication the receiving signal become different for different channel response. To obtain particular efficiency, orthogonal frequency division multiplexing (OFDM) technology is incorporated. The OFDM receiver performs channel estimation which is directly related with the bit error rate (BER). In literature, pilot aided channel estimation is performed in frequency domain technique [1]. In OFDM technology, the pilots are 2D-grid type and block type. In

S. M. Thampi et al. (Eds.): SIRS 2019, CCIS 1209, pp. 332–346, 2020.
https://doi.org/10.1007/978-981-15-4828-4_27

fast fading channel environments, reliability in terms of system BER is better for comb type arrangement. Thus, in order to overcome higher computational complexity problems Least Squares (LS) based, minimum mean square error (MMSE) based or maximum likelihood (ML) based techniques have been primarily performed [2].

The OFDM symbol BER which will improve by using the MMSE based techniques. This technique will perform on apriori knowledge of channel statistics and provide optimal performance compared to the techniques based on LS. Hence such kind of drawback suffers from higher computational complexity. This performance can be improved mainly by channel estimator based on LS; many threshold-based denoising strategies have been developed in the literature [3,4]. Thus MMSE technique will enhance the performance of the channel estimator. But with MMSE based techniques, the estimation parameters are not easy to estimate in real-time. The channel impulse response (CIR) algorithm provides an initial estimation to help the pilot symbols using an E-step and a Q-step to obtain the final CIR iterated. In the next section, the past methods are analyzed in detail followed by background methodology, proposed method, and results.

2   Literature Survey

A major challenge affecting wireless communication performance is the recovery of received signals at the receiver end. Accurate channel state information (CSI) is required to estimate the received symbols. These challenges are met using the estimation methods which includes the training based, blind and semi-blind techniques [3]. Estimation in frequency domain is effective in most of the cases [4]. The preliminary 5G standardization activities are the key challenges at all the regulation modes [5,6]. Hence the issues are to be maintained to increase the strength of fourth generation (4G) that will carry the frequencies with the large bandwidth, device densities and extreme base station. Thus, the collective 5G network will gives the high rate of coverage when merged together with Long-Term Evolution (LTE) and wireless fidelity (Wi-Fi). The combined use of more spectrums will have more traffic growth which includes the higher spectral efficiency, more spectrum densification of cells [7,8]. Joint transmission coordinated multipoint (JT CoMP) along with massive MIMO improves the spectral efficiency and coverage [9]. The 5G wireless communication-based design services needs better channel estimation. This method will be launched after 2020 and will have large spectrum and high efficiency of data limits [10]. Such kind of high rate data is been used even in the 4G wireless communication systems.

The 5G network devices are using device-to-device (D2D) communication and yet to overcome some challenges [11]. It enables devices to communicate directly with each other [12]. Hence the user data rate will be lower in older generation than 5G network such as access points (APs) or base stations (BSs). Thus the growth of machine-to-machine (M2M) applications and D2D communication will be more in 5G cellular networks. Gaussian-Mixture Bayesian Learning was presented for channel estimation [13]. The 5G wireless heterogeneous network (Het-Net) is designed to meet the challenges and benefits [14].

This can have a massive MIMO system in which the 5G network will potentially develop a new exploited network. The technology deployed through cellular data network is energy efficient in particular millimeter wave (mmWave) frequencies. The performance is measured using zero forcing receiver (ZF) and maximum ratio combining (MRC) [15,16]. Hence this technique will enhance the communication between the users at very long range that enable efficient single-carrier transmission [17]. The wireless communication is being wide range of usage in the modern days [18,19]. Due to the usage of network, the communication system cannot handle more traffic even for QPSK [20]. This methodology will be constantly used in massive MIMO system for the increased data rates which include the OFDM MIMO [16]. Hence communications technologies give the exponential growth rate in wireless traffic. The new massive MIMO technology is presented to reduce the radiation power and to increase the throughput at a certain signal processing [21,22]. 4G networking technology will improve the efficiency for 3D personal high definition video transmission over a long period of time in millimeter MIMO [23]. Thus, the spectrum efficiency is achieved in time division usage [24]. The massive MIMO system provides asymptotically orthogonal channels [25]. The challenges are on self-interference due to full-duplex mode [26,27]. The iterative detection and decoding (IDD) scheme improves the bit error rate (BER) and the performance of block fading channel [28]. This technique has some novel strategy that takes the root check low density parity check (LDPC) codes in block-fading and fast Rayleigh fading channels.

The frequency-selective fading channel for MIMO is channel estimated using Cramer-Rao lower bounds (CRLBs) along with maximum likelihood estimators (MLEs) [29]. The Frequency Domain- Least Squares (FD-LS) estimator is introduced to improve the performance in time domain multiplexing [30]. The Maximum Like-lihood (ML) detector has the low complexity at the presence of error. The improved FD-LS channel estimator provides 6-dB SNR gain over the FD-LS estimator. Hence the specified FD-LS channel is better when compared to the traditional ML detector at BER [31]. For mmWave FSF channels, channel estimation was carried out using compressive sensing-based channel scheme in angle-domain [32].

3 System Model

Figure 1 shows the OFDM system with K subcarriers, N_t transmit antennas and N_r receive antennas under consideration for testing the channel estimation technique [20]. The conventional QAM/PSK modulator is used at each transmit antenna and demodulator at the receiver. OFDM symbol transmitted at time index n from r^{th} antenna is denoted by the vector $X^r(n)$. After performing inverse fast Fourier transform (IFFT), a cyclic prefix (CP) of length is added. Here, $v \geq L - 1$ is assumed, where L is the maximum length of all channels. In the system model different fading are taken for analysis at the wireless communication system. The CP is removed at the receiving antenna and next FFT will be applied. From the qth receive antenna, after removing CP and FFT calculation, the result is a $K \times 1$ vector represented by $Y\,nolimits^q(n)$.

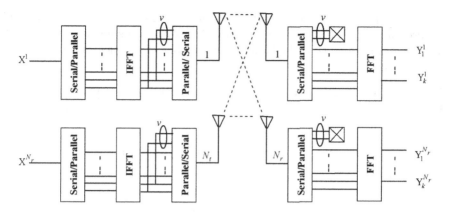

Fig. 1. OFDM system model [20]

4   Background Methodology

Channel estimation plays a crucial role in practically analyzing the characteristics and performance of a channel which fluctuates in time. The different factors which are influenced are natural changes, structures, snags, etc. The system model is considered as:

$$Y = HX + V \tag{1}$$

where H is the channel coefficient matrix, comprising of all the channel parameters and V is additive white Gaussian noise (AWGN). Here, the SNR controls the commotion fluctuation. The transmitted data is indicated by X, where Y is utilized for LS or MMSE estimator.

4.1   LS Estimation

The low complexity is the main characteristics of LS estimation. The simplest algorithm minimizes $\|H_{est}X - V\|^2$, where Y is the received pilot signal in frequency domain, X is transmitted pilot signal in frequency domain and H_{est} is an estimated channel matrix in frequency domain, and $\|\cdot\|$ denotes the Euclidean norm. The corresponding LS channel estimation algorithm in frequency domain is:

$$H_{est} = H_{LS} = (X^H X)^{-1} X^H Y = X^{-1} Y \tag{2}$$

where $(\cdot)^H$ denotes Hermitian transposition. The kth FFT sample of received signal is given by

$$Y(k) = X(k) H(k) + V(k) \tag{3}$$

where $X(k)$, $H(k)$ and $V(k)$ indicate the input samples, channel coefficients and the noise samples, respectively. It is assumed that on each subcarrier there is no inter symbol interference (ISI). Now, for K number of samples, this system can be represented as

$$
\begin{bmatrix} Y(0) \\ Y(1) \\ \vdots \\ Y(K-1) \end{bmatrix} = \begin{bmatrix} X(0) & 0 & \cdots & 0 \\ 0 & X(1) & \cdots & 0 \\ \vdots & \vdots & \ddots & \vdots \\ 0 & 0 & \cdots & X(K-1) \end{bmatrix} \begin{bmatrix} H(0) \\ H(1) \\ \vdots \\ H(K-1) \end{bmatrix} + \begin{bmatrix} V(0) \\ V(1) \\ \vdots \\ V(K-1) \end{bmatrix} \tag{4}
$$

Equation (4) can be written in compact form as

$$
\mathbf{Y} = \mathbf{H}\mathbf{X} + \mathbf{V} \tag{5}
$$

where, $\mathbf{X}$ is input matrix at pilot sub-carrier, $\mathbf{Y}$ is output vector at pilot sub-carriers and $\mathbf{V}$ is the vector of noise samples. Therefore, the LS estimator cost function is given as

$$
C = \|\mathbf{Y} - \mathbf{X}\mathbf{H}\|^2 \tag{6}
$$

The LS solution of (6) is given by

$$
\widehat{\mathbf{H}}_{LS} = (\mathbf{X}^H \mathbf{X})^{-1} \mathbf{X}^H \mathbf{Y} \tag{7}
$$

As $\mathbf{X}$ is an invertible matrix, that is, $(\mathbf{X}^H \mathbf{X})^{-1} = (\mathbf{X}^H)^{-1} \mathbf{X}^{-1}$, (7) can be written as

$$
\widehat{\mathbf{H}}_{LS} = (\mathbf{X}^H)^{-1} \mathbf{X}^{-1} \mathbf{X}^H \mathbf{Y} = \mathbf{X}^{-1} \mathbf{Y} \tag{8}
$$

Hence the LS algorithm does not require any prior information of the channel statistics. The advantage of LS algorithm is its simplicity.

4.2 Wiener Filter Based MMSE Estimation

The Wiener filter based MMSE channel estimation technique has more accuracy than the LS channel estimation. The schematic of Wiener filter based MMSE channel estimation is shown in the Fig. 2, where $\mathbf{M}$ is the weight matrix, $\widehat{\mathbf{H}}_{LS}$ is the LS estimate and $\widehat{\mathbf{H}}_{MMSE}$ corresponds to the MMSE estimate. Now, the standard expression of MMSE estimate at k^{th} sub-carrier is given by,

$$
\widehat{H}_{MMSE}(k) = R_{H(k)Y(k)} R_{Y(k)Y(k)}^{-1} Y(k) \tag{9}
$$

where $R_{H(k)Y(k)}$ is the cross-covariance of $H(k)$ and $Y(k)$ given by

$$
R_{H(k)Y(k)} = L\sigma_h^2 X^*(k) \tag{10}
$$

with L indicating the number of multipaths and σ_h^2 is the variance of channel coefficients, and $R_{Y(k)Y(k)}$ is the auto-covariance of $Y(k)$ given by

$$R_{Y(k)Y(k)} = L\sigma_h^2|X(k)|^2 + N\sigma_n^2 \tag{11}$$

with N indicating the number of sub-carriers and σ_n^2 is the variance of AWGN. Now, by substituting (10) and (11) in (9), we get

$$\widehat{H}_{MMSE}(k) = L\sigma_h^2 X^*(k)\frac{1}{L\sigma_h^2|X(k)|^2 + N\sigma_n^2}Y(k) \tag{12}$$

By rewriting (3) as

$$Y(k) = X(k)\widehat{H}_{LS}(k) \tag{13}$$

and plugging (13) in (12), the kth sample of MMSE estimate can be written as

$$\widehat{H}_{MMSE}(k) = L\sigma_h^2 X^*(k)\frac{X(k)\widehat{H}_{LS}(k)}{L\sigma_h^2|X(k)|^2 + N\sigma_n^2} \tag{14}$$

After some algebraic manipulation, (14) can be written as

$$\widehat{H}_{MMSE} = R_{HH}\left[R_{HH} + (XX^H)^{-1}N\sigma_n^2\right]^{-1}\widehat{H}_{LS} = M\widehat{H}_{LS} \tag{15}$$

where $M = R_{HH}\left[R_{HH} + (XX^H)^{-1}N\sigma_n^2\right]^{-1}$ is the weight matrix.

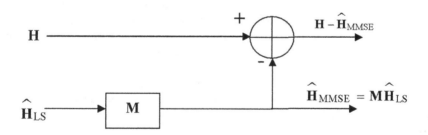

Fig. 2. Block diagram of MMSE channel estimation [2]

5   Cuckoo Search Algorithm

Cuckoo birds may be a nature-inspired rule supported replica. The potential solution with cuckoo eggs laying and latching is considered. This may unremarkably lay their fertile eggs in alternative cuckoos nests with the hope of their off-springs that being their raised proxy oldsters. The cuckoos of foreign eggs if discovered are thrown out of the nests and also the nests can be abandoned. The corresponding basic rules based on cuckoo search (CS) optimization algorithm are the following three rules: (i) The cuckoos are selecting randomly their nest and lay one egg in it, (ii) Among the nest the best eggs are carried over to the next generation and (iii) The probability of is the fixed number of nest by a host cuckoo that can discover a foreign egg. In such case the abandon nest will build a new one somewhere else. The last rule which has the replacing fraction as that is approximated in the n host nests with new random nests solution [33].

Cuckoo breeding behavior: Cuckoos are a group of feathered creatures with novel conceptive procedure progressively forceful contrasted with other winged animal species. Some of the cuckoos winged animal species like Ani and Guira that can lay eggs in collective homes. In such cases they may evacuate other eggs to build forth the likelihood of their own eggs to bring. Different species use brood parasitism strategy for laying their eggs in the homes of different feathered creatures. The parasitic cuckoos are one of the great species in wearing homes. Where eggs have quite recently been laid and they are planning to lay eggs in exact host home which will typically bring forth faster than different eggs. At the point, when this will occurs the outside cuckoo that would expel the non-incubated eggs from the home for driving the eggs out of the home. This conduct is gone for decreasing the likelihood of genuine eggs from brings forth. Moreover, the remote cuckoo chick will access a lot of nourishment by mirroring the decision of the host chicks. There are times once the host cuckoo finds that one of the eggs is outside. All things considered, the cuckoos both dispose of the egg or sur-renders the home by or move to fabricate another home elsewhere [34]. *Lévy flight*: A Lévy flight is a random walk in unexpected directions. The distribution fea-tures are random from which their step lengths are derived. The change in flights of animals and insects are characterized. Hence the comparison of normal random walks is more efficient in exploring large–scale search areas in the Lévy flights. Such kind of flights is mainly due to variances in the flights especially the faster than that of the normal random walk. Thus it can reduce the number of iterations in the optimization algorithm [33]. *Pseudo code*: The corresponding implementation views that the each cuckoo can lay only one egg which is representing one solution at a point. Thus their potential is better than the older egg which can be replaced in it. CS algorithm maintains a balance between local and global random walks. The will be the switching parameters that are balanced between the local and global randomly. Algorithm 1 presents the pseudo code of the CS algorithm given by [33].

Algorithm 1:
begin
Initialize the population of n host nests $x_i(i = 1, \ldots n)$ and define the objective function $f(x), x = (x_1, \ldots x_d)^T$
while $(t < maxit)$ do
Get a cuckoo randomly say i and generate a new solution by Lévy flights, calculate its fitness function and choose a nest randomly say j;
if $(F_i > F_j)$ then
| Replace j by the new solution;
end
Abandon p_a fraction of worse nests and build new ones via Lévy flights, keep the best solution;
end
Post process results and visualization
end

6 Adaptive Cuckoo Search Based Wiener Filter

This section presents the adaptive cuckoo search based Wiener filter (ACSWF) algorithm for channel estimation.

6.1 Concept Behind Proposed ACSWF

This type of adaptive cuckoo search (ACS) based filter depends on 2-D FIR Wiener filtering algorithm [33]. This mode of filter is adaptively modified to minimize the MSE in terms of window weights. This method can ensure the best possible direction as the filter weights used in ACS algorithm are squared errors as compared with other similar meta-heuristic algorithms. Figure 3 shows the ACSWF block diagram. Mathematically MSE can be written as [2]

$$\text{MSE} = \frac{1}{n} \sum_{i=1}^{n} \left(Y_i - \hat{Y}_i \right)^2 \tag{16}$$

6.2 Proposed Adaptive Cuckoo Search Algorithm

A portion of the cuckoo's practices is developed by emulating the commit brood that is a stochastic meta-heuristic calculation in CS. The switching parameters are implemented because they are used as the single control parameter p_a with the use of Lvy flight system as opposed to Brownian arbitrary strolls for arrangement space investigation. This can be contrasted and the key to different elements which make it prevalent contrasted and others in idealized rules. This strategy can stream the correct dependent on CS calculation in three relating rules which are in the single controlling parameter p_a which means the likelihood of finding outsider eggs by the host species. The following rules are practically implemented for replacing a proportion p_a of the current solution set with the new ones. Thus the method of CS algorithm will varies the steps with the size of oversampling at the rate of balanced combination of global and local walk controller in a certain parameter p_a.

Algorithm 2: Proposed Algorithm

---

begin
//initialization;
Population initialization: $W_{k,d}$, **where** $k = 1, 2, ..., N_p$; $l = 1, 2, ..., d$
with $N_p =$ **Size of population;** $d =$**dimensionality;**
Initializing Parameters: Switching parameter (d), **Maximum number of iterations** $(G_{\max})$;
//2D FIR Wiener filtering and MSE calculation 2D
lexicographic ordering of weight vectors W_k **using**
$W_{k+1} = W_k + \alpha L(s, \beta)$, **where** α, s **and** β **are adjustment factors for the Lévy flight; Calculate fitness/MSE values** (F_k) **using Eq. (16); Best fitness/MSE value and the corresponding solution set** (W_{best}) **should be recorded;**
//ACS algorithm
for $(G \leq G_{\max})$ **do**
Estimate new weighting factor and compute their respective fitness values.
if $(F_k\, new \succ F_r)$ **then** $F_r = F_k\, new; W_r = W_{new}$;
else $W_{best} = W_{new}$;
end
if$(randk \geq 1 - G/G_{\max})$ **then**
CS/best/1/bin;$W_{k+1} = W_{r1} + p_a(W_{r2} - W_{r3})$;
else
CS/rand/1/bin;$W_{k+1} = W_{best} + p_a(W_{r1} - W_{r2} + W_{r3} - W_{r4})$;
end
Output estimation, fitness values computation following Eq. (8), (15) and (16); Recording the best (W_{best}) **so far on comparison with the earlier set by newly generated fitness values.**
end

6.3   Proposed ACSWF Implementation

The pseudo code and step-wise (S1 to S7) proposed ACSWF implementation is as follows:

S1: Solution space should be initialized randomly using

$$W_{k,t} = W_{k,t}^{\min} + rand(W_{k,t}^{\max} - W_{k,t}^{\min}) \tag{17}$$

where weights given to 2-D FIR Wiener filter are $W(k,1)$, $k = 1, 2, ..., N_p$, here N_p is the population size and $l = 1, 2, ..., d$; where $d = n^2$ is the required number of coefficients in the filter weight matrix.

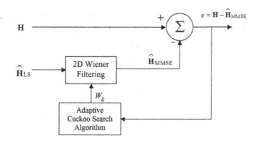

Fig. 3. Proposed ACSWF based Estimator schematic.

S2: Take the linear estimate of the desired signal with the help of adaptive Wiener weight vector as w_k.

S3: Compute the fitness (MSE) values using each possible weight matrix using Eq. (16) for the estimated filter outputs obtained. Repeat this step if the number of iterations $G \leq G_{\max}$.

S4: Retain the most effective double weight matrix (solution) within the previous iteration and generate new random solutions by Lévy flights round the previous solution set. The new population set therefore fashioned follows:

$$W_{k,G+1} = W_{k,G} + \alpha L(s, \beta) \; 0 < \beta \leq 2, \alpha = 0.01$$
$$L(s, \beta) = \frac{\beta \Gamma(\beta) \sin\left(\frac{\pi \beta}{2}\right)}{\pi} \frac{1}{s^l + \beta}; \; (|s| \geq |s_0|) \tag{18}$$

where s represents the step size, s_0 is the smallest step (typical-ly 0.1 to 1) and α, α represents the scaling issue for step size. G represents the iteration/generation count, i.e., $W_{k,G}$ and $W_{k,G+1}$ are the burden vectors developed for the Gth and $(G+1)^{th}$ generation of the ACS formula.

S5: Compute the estimated output using the updated solution set and evaluates their respective fitness values.

S6: Increment the iteration count by 1 i.e., $G = G+1$ until $G = G_{\max}$ (maximum iterations).

S7: The best filter weights W_{best} and $G_{\max}$ should be obtained to get the best possible estimate of the desired signal.

7 Results and Discussion

The analysis of zero phase offset is taken at different SNR conditions with resultant 16-QAM modulation that is used. The performance of QAM scheme over Rayleigh fading, Rician fading, Nakagami fading and Weibull fading channels are plotted here by symbol order which is in binary. The input type is an integer which is taken as the parameters of $m = 3$ and $\Omega = 1$ for Nakagami fading as well as Weibull fading with $\lambda = 100$ and $\beta = 3$.

The LS and MMSE methods are compared with the previous work for performance of channel estimator in terms of MSE. Figure 4 shows comparison of

MSE for Rayleigh fading, Rician fading, Nakagami fading and Weibull fading channels with LS, MMSE and ACSWF techniques over 16-QAM modulation. It shows that the proposed ACSWF estimator gives the better performance in Rayleigh, Rician, Nakagami and Weibull fading channels. ACSWF estimator is not giving remarkable performance in the Weibull fading. MSE decreases as the SNR increases in LS, MMSE estimator and ACSWF estimators. On comparison of Fig. 4 plots, LS estimator is better to MMSE estimator in Rayleigh fading channel. In Rician, Nakagami and Weibull fading channels, MMSE performs better to LS estimator.

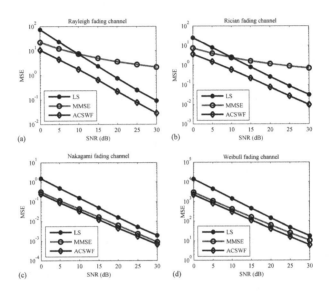

Fig. 4. MSE vs. SNR plot for LS, MMSE-WF and ACSWF Estimators for 16-QAM scheme over (a) Rayleigh fading (b) Rician fading (c) Nakagami fading (d) Weibull Fading channels.

BER for Rayleigh, Rician, Nakagami and Weibull fading channels with LS, MMSE and ACSWF channel estimation techniques for 16-QAM modulation is shown in Fig. 5. Proposed ACSWF estimation gives minimum BER in all fading channels. QAM modulation with M symbols is known as M-QAM. From the results, it is observed that LS, MMSE and ACSWF estimators are giving similar BER vs. SNR performance in Nakagami and Weibull fading channels, MMSE estimator gives the least performance in Rayleigh and Rician fading channels. On comparison of Fig. 5 plots, LS estimator is better to MMSE estimator in Rayleigh, Rician fading channels. In Nakagami and Weibull fading channels, MMSE performs better to LS estimator. In all the four cases, LS gives the similar performance to ACSWF estimator, but at higher SNR value i.e., after 25 dB, undoubtedly ACSWF estimator gives least BER values when compared to LS, MMSE estimators.

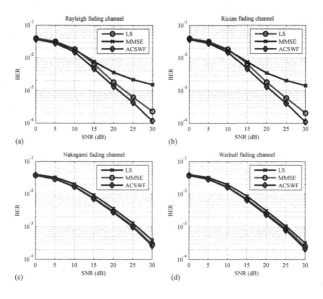

Fig. 5. BER vs. SNR plot for LS, MMSE-WF and ACSWF Estimators for 16-QAM scheme over (a) Rayleigh fading (b) Rician fading (c) Nakagami fading (d) Weibull Fading channels.

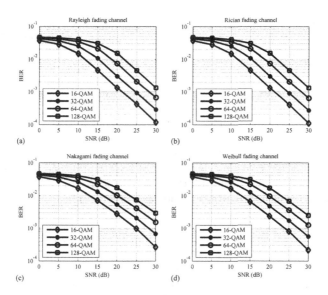

Fig. 6. BER vs. SNR Performance of ACSWF estimator with M-QAM in (a) Rayleigh fading (b) Rician fading (c) Nakagami fading (d) Weibull Fading channels.

Figure 6 gives the BER performance of M-QAM for M = 16, 32, 64 and 128 in different fading channels using proposed ACSWF estimation technique. From the plots, it is observed that 16-QAM gives least BER values for ACSWF channel estimation. Hence, MSE and BER performances are analyzed in different fading channels using LS, MMSE and ACSWF estimators using 16-QAM only. At lower SNR values, there is not much difference among M-QAM modulation schemes. But at higher SNR values i.e., after 20dB, 16-QAM outperforms the remaining schemes.

8 Conclusion

The paper has presented a new channel estimation methodology using nature-based computing. The proposed stochastic meta-heuristic CS algorithm is adopted and implemented for the channel estimator. The main advantage is the control parameters makes it superior when compared to other methods. The implementation is done at different SNR conditions. The modulation scheme used is M-QAM with zero phase offset for analysis. The performance of Rayleigh, Rician, Nakagami and Weibull fading channels are investigated and compared with the ACSWF estimator in terms of BER and MSE. The proposed estimation dominates the performance of LS and MMSE estimators.

References

1. de Figueiredo, F.A.P., Cardoso, F.A.C.M., Moerman, I., Fraidenraich, G.: Channel estimation for massive MIMO TDD systems assuming pilot contamination and flat fading. EURASIP J. Wirel. Commun. Network. **2018**(1), 1–10 (2018). https://doi.org/10.1186/s13638-018-1021-9
2. Almamori, A., Mohan, S.: Improved MMSE channel estimation in massive MIMO system with a method for the prediction of channel correlation matrix. In: 2018 IEEE 8th Annual Computing and Communication Workshop and Conference (CCWC), pp. 670–672. IEEE (2018)
3. Adisa, A., Mneney, S.H., Oyerinde, O.O.: Modified minimized mean value-MMSE algorithm-based semi-blind channel estimator for MC-IDMA systems. In: AFRICON 2015, pp. 1–5. IEEE (2015)
4. Zettas, S., et al.: Performance comparison of LS, LMMSE and adaptive averaging channel estimation (AACE) for DVB-T2. In: 2015 IEEE International Symposium on Broadband Multimedia Systems and Broadcasting, pp. 1–5. IEEE (2015)
5. Andrews, J.G., et al.: What will 5G be? IEEE J. Sel. Areas Commun. **32**(6), 1065–1082 (2014)
6. Albreem, M.A.M.: 5G wireless communication systems: vision and challenges. In: 2015 International Conference on Computer, Communications, and Control Technology (I4CT), pp. 493–497. IEEE (2015)
7. Jungnickel, V., et al.: The role of small cells, coordinated multipoint, and massive mimo in 5G. IEEE Commun. Mag. **52**(5), 44–51 (2014)
8. Kumar, S., Soni, S.K., Jain, P.: Performance analysis of Hoyt-lognormal composite fading channel. In: 2017 International Conference on Wireless Communications, Signal Processing and Networking (WiSPNET), pp. 2503–2507. IEEE (2017)

9. Maurer, J., Matz, G., Seethaler, D.: Low-complexity and full-diversity MIMO detection based on condition number thresholding. In: ICASSP, vol. 3, pp. 61–64 (2007)

10. Jacobsson, S., Durisi, G., Coldrey, M., Gustavsson, U., Studer., C.: One-bit massive MIMO: channel estimation and high-order modulations. In: 2015 IEEE International Conference on Communication Workshop (ICCW), pp. 1304–1309. IEEE (2015)

11. Shen, X.: Device-to-device communication in 5G cellular networks. IEEE Netw. **29**(2), 2–3 (2015)

12. Uddin, M.B., Sarkar, M.Z.I., Faisal, K.N., Ali, M.M.: Performance analysis of Nakagami-m fading massive MIMO channels with linear receivers. In: 2014 17th International Conference on Computer and Information Technology (ICCIT), pp. 510–513. IEEE (2014)

13. Wen, C.-K., Jin, S., Wong, K.-K., Chen, J.-C., Ting, P.: Channel estimation for massive MIMO using gaussian-mixture Bayesian learning. IEEE Trans. Wireless Commun. **14**(3), 1356–1368 (2014)

14. Bogale, T.E., Le, L.B.: Massive MIMO and mmWave for 5G wireless hetnet: potential benefits and challenges. IEEE Veh. Technol. Mag. **11**(1), 64–75 (2016)

15. Muhanned, A.L.-R., Muaayed, A.L.-R.: Performance of massive MIMO uplink system over Nakagami-m fading channel. Radioelectron. Commun. Syst. **60**(1), 13–17 (2017)

16. Singh, R., Soni, S.K., Verma, P.K., Kumar, S.: Performance analysis of MRC combiner output in log normal shadowed fading. In: International Conference on Computing, Communication & Automation, pp. 1116–1120. IEEE (2015)

17. Lu, L., Li, G.Y., Swindlehurst, A.L., Ashikhmin, A., Zhang, R.: An overview of massive MIMO: benefits and challenges. IEEE J. Sel. Top. Signal Process. **8**(5), 742–758 (2014)

18. Björnson, E., Larsson, E.G., Marzetta, T.L.: Massive MIMO: ten myths and one critical question. IEEE Commun. Mag. **54**(2), 114–123 (2016)

19. Rasethuntsa, T.R., Kumar, S., Kaur, M.: A comprehensive performance evaluation of a DF-based multi-hop system over $\alpha-\kappa-\mu$ and $\alpha-\kappa-\mu$-extreme fading channels. arXiv preprint arXiv:1903.09353 (2019)

20. Zhang, X., Li, Y.: Optimizing the MIMO channel estimation for LTE-advanced uplink. In: 2012 International Conference on Connected Vehicles and Expo (ICCVE), pp. 71–76. IEEE (2012)

21. Marzetta, T.L.: Massive MIMO: an introduction. Bell Labs Tech. J. **20**, 11–22 (2015)

22. Zedini, E., Ansari, I.S., Alouini, M.-S.: Performance analysis of mixed Nakagami-m and Gamma-Gamma dual-hop FSO transmission systems. IEEE Photonics J. **7**(1), 1–20 (2014)

23. Park, S., Choi, J.W., Seol, J.-Y., Shim, B.: Expectation-maximization-based channel estimation for multiuser MIMO systems. IEEE Trans. Commun. **65**(6), 2397–2410 (2017)

24. Björnson, E., Hoydis, J., Kountouris, M., Debbah, M.: Massive MIMO systems with non-ideal hardware: energy efficiency, estimation, and capacity limits. IEEE Trans. Inf. Theory **60**(11), 7112–7139 (2014)

25. Coskun, A.F., Kucur, O.: Performance analysis of maximal-ratio transmission/receive antenna selection in nakagami-m fading channels with channel estimation errors and feedback delay. IEEE Trans. Veh. Technol. **61**(3), 1099–1108 (2012)

26. Albu, F., Martinez, D.: The application of support vector machines with gaussian kernels for overcoming co-channel interference. In: Neural Networks for Signal Processing IX: Proceedings of the 1999 IEEE Signal Processing Society Workshop (Cat. No. 98TH8468), pp. 49–57. IEEE (1999)
27. Cirik, A.C., Rong, Y., Hua, Y.: Achievable rates of full-duplex MIMO radios in fast fading channels with imperfect channel estimation. IEEE Trans. Signal Process. **62**(15), 3874–3886 (2014)
28. Uchoa, A.G.D., Healy, C.T., de Lamare, R.C., Li, P.: Iterative detection and LDPC decoding algorithms for MIMO systems in block-fading channels. arXiv preprint arXiv:1404.1981 (2014)
29. Mohammadkarimi, M., Karami, E., Dobre, O.A., Win, M.Z.: Doppler spread estimation in MIMO frequency-selective fading channels. IEEE Trans. Wireless Commun. **17**(3), 1951–1965 (2017)
30. Shu, F., Wang, J., Li, J., Chen, R., Chen, W.: Pilot optimization, channel estimation, and optimal detection for full-duplex OFDM systems with IQ imbalances. IEEE Trans. Veh. Technol. **66**(8), 6993–7009 (2017)
31. Yoshioka, S., Suyama, S., Okuyama, T., Mashino, J., Okumura, Y.: 5G massive MIMO with digital beamforming and two-stage channel estimation for low SHF band. In: 2017 Wireless Days, pp. 107–112. IEEE (2017)
32. Gao, Z., Chen, H., Dai, L., Wang, Z.: Channel estimation for millimeter-wave massive MIMO with hybrid precoding over frequency-selective fading channels. IEEE Commun. Lett. **20**(6), 1259–1262 (2016)
33. Suresh, S., Lal, S., Chen, C., Celik, T.: Multispectral satellite image denoising via adaptive cuckoo search-based Wiener filter. IEEE Trans. Geosci. Remote Sens. **56**(8), 4334–4345 (2018)
34. Yang, X.-S., Deb, S.: Cuckoo search: state-of-the-art and opportunities. In: 2017 IEEE 4th International Conference on Soft Computing & Machine Intelligence (ISCMI), pp. 55–59. IEEE (2017)

Down Link: Error-Rate Performance
of Cognitive D-STTD MC-IDMA System

Asharani Patil[1(✉)], G. S. Biradar[1], and K. S. Vishvaksnan[2]

[1] Department of ECE, PDA College of Engineering, Kalaburagi, Karnataka, India
asharanispatil@gmail.com
[2] Department of ECE, SSN College of Engineering, Chennai, India

Abstract. 5G wireless network has promised to offer high speed data transmission along with quality of service. Cognitive radio network (CRN) with multiple-input-multiple-output system have potential to offer higher bandwidth with high data rate. We present physical layer design of double-space-time-TD (D-STTD)-Transmit Diversity aided MIMO-MC-IDMA system invoking Cognitive spectrum. CRN is a modern device which entreat unused TV-band available spectrum and distribute such spectrum strongly to BS (base station). The fragments subdivided frequency-band of CRN is distributed for secondary users and is utilized for multi-carrier transmission-DSTTD system consists of two STBC block units and we invoke such system to obtain better error-rate performance. We detect signals at each mobile station using block-Nulling detection decoding algorithm. We present error-rate results of CRN defined D-STTD MC-IDMA for standard channel specifications. We observe from the error-rate results that D-STTD assisted MC-IDMA system enhances data rate with better performance while extracting higher bandwidth from CRN network.

Keywords: Block-Nulling detection algorithm · BCJR decoding algorithm · Maximum likely-hood detection (ML) · (MMSE) detection algorithm · Multi-user detection (MUD) · Space-time transmit diversity (S-TTD) · Zero-forcing algorithm (ZF)

1 Introduction

Multiple-Input-Multiple-Output system (MIMO) is a optimistic transmission technique in wireless network by which we can enhance spectral efficiency without expecting additional bandwidth. By incorporating multiple antenna system at receiver and transmitter, we can enhance bit-error rate performance along with data rate and quality of service in wireless transmission. In general, MIMO system is exploited in two possible ways; Space-time transmit diversity (STBC) is aimed to enhance error-rate performance by reducing deep fading effect. Spatial Multiplexing is designed to uplift data rate. In other words, STBC enhances capacity of system while we are invoking higher spectral efficiency using spatial Multiplexing V-BLAST architecture and the advantage of spatial multiplexing is achieving higher gain by scheme. Bell Laboratories introduced V-BLAST architecture which focus on spatial multiplexing and Alamouti introduced the

© Springer Nature Singapore Pte Ltd. 2020
S. M. Thampi et al. (Eds.): SIRS 2019, CCIS 1209, pp. 347–358, 2020.
https://doi.org/10.1007/978-981-15-4828-4_28

concept of STBC. Many detection algorithms have been focused on MIMO detection such as Maximum likely hood algorithm, ZF and MMSE algorithm. Researchers [1] elucidated the benefits of adaptive MIMO scheme by switch over between V-BLAST structure and STBC. Prof. Lee in [2] explained the basic principles of D-STTD detection algorithm using Block-Nulling algorithm. Prof. Iruthayanathan [3] explained D-STTD - OFDM system for mitigation of multi-path fading effects in underwater signal transmission. Spreading code specific CDMA was introduced by Prof. Frenger et al. [4] and such scheme adds advantages over low-rate channel encoding with DSSS techniques by which we can obtain superior performance. Ping in [5] proposed the concept of new accessing scheme called IDMA: Interleave Division Multiple Access scheme to allow more users in a wireless channel. Many research article have addressed the advantages of IDMA that the proposed scheme can mitigate rich scattering effects of multi-path environments thereby offering higher error-rate performance with less E_b/N_0. This scheme is referred as low-rate spreading code CDMA. Hence it exploits all the benefits of CDMA. Prof. Viterbi elucidated in [6] that by allocating full bandwidth for low-rate channel coding, achieves high coding gain along with necessary processing gain with higher capacity in the context of multi-user signal transmission. Research article, authors [7] have addressed benefits of IDMA scheme to get higher error-rate results in multi-path environments. Prof. Hanzo elucidated in [8] Multi-carrier IDMA offers better error-rate results for down-link signal transmission.

The demand for data rate is increasing day by day for specific purpose such as live video telecast system, still image transmission and other relevant multi-media services etc., At present limited frequency-band are accessible for users to transfer data. In the recent past, many research articles focused on Cognitive Radio (CR) un-used TV band spectrum which can be extracted for mobile communication to overcome limitation of bandwidth. It is elucidated in [9] that we can allocate this spectrum for secondary users for huge data transfer application such as human being movements and activities observing system in hospital using IoT etc., Federal Communication Commission (FCC) is being originated to exploit Television-band spectrum for the application of mobile communication.

Turbo code is recommended as preeminent channel encoder for wireless LAN, 5G network. The authors in [10] addressed mathematical model for turbo decoder considering serial concatenated and parallel concatenated convolutional encoder. Prof. Wang and his team [11] explained iterative turbo decoder for CDMA system. Prof. Le Goff and his team [12] explained the benefits of iterative decoding for wireless signal transmission. In literature [13], Benedetto and his team explained bandwidth efficiency of parallel-concatenated style of turbo code in wireless network. In [14], Robertson explained modulation scheme employing turbo code. In [10], Prof. Benedetto explained benefits of channel encoder of serial concatenated type and also decoding technique. In the most of above-mentioned literature, author expressed benefits of MIMO, IDMA scheme along with iterative decoder and Multi-Carrier Communication.

In this treatise, we encapsulate our contribution: -

1. We realize D-STTD style of MIMO structure at Base station to exploit both spatial multiplexing and spatial diversity

2. We implement CRN with TV-Band spectrum of 80 MHz to 800 MHz for utilization of secondary users.
3. At each Mobile station, block-Nulling detection algorithm realization is made to estimate information considering lesser SNR
4. We decode received signals using Maximum Log-Maximum-a-posteriori probability-based detection algorithm.
5. We analyze and present results for CRN assisted D-STTD MC-IDMA system considering SUI Model & LTE channel specifications

Research work is summarized as Sect. 2 describes system model configuration with mathematical modelling. Section 3 explains mathematical structure of D-STTD detection algorithm at the receiver and decoded signals. Section 4 details simulated results observation and exploration. Finally, Conclusion are detailed in Sect. 5.

2 System Model

2.1 Signal Transmission from BS

Considering D-STTD style of MIMO system with N_t - TX from BS-Base Station and N_r - RX antennas at each MS-Mobile Station, we contemplate Cognitive defined Coded MIMO MC-IDMA system with 'K' - DL users in the Fig. 1. We presume that $N_t = N_r = 4$ antennas for calculation in our system. We construct D-STTD using two parallel form of STBC block unit at BS. At each MS, we realize Block-Nulling detection algorithm to decode transmitted sequence in the presence of licensed user interferences known as Primary PR-MUI & Secondary-SE-MUI-Multiple User Interference and noise. We allocate CR frequency-band for secondary users according to the need of users assign SE-users dynamically to utilize CR-Spectrum that ranges 80 MHz– 800 MHz.

Let u_l represents bit-data stream of lth user which undergo turbo code encoding
Let e_l be encoded bit stream where

$$e_l = [e_{l1}, e_{l2}, \ldots, e_{lm}]^T, \quad l = 1, 2, \ldots, L \tag{1}$$

Indicating 'm' as total no. of bits.
Let s_l be the frequency-domain (FD) spreading sequence with respect to l^{th} user

$$\text{Where } s_l = \frac{1}{\sqrt{N}} [s_{l0}, s_{l1}, \ldots, s_{l(N-1)}]^T \tag{2}$$

Indicating N as spreading length.
The FD spreading matrix $Nm \times m$ is represented as S_l where

$$S_l = diag \{s_l, s_l, s_l, \ldots, s_l\} = I_m \otimes s_l, \quad l = 1, 2, \ldots, L \tag{3}$$

$$\text{where } Nm = N. m \tag{4}$$

Consequently, the spreaded sequence is interleaved by user-defined interleaving pattern.

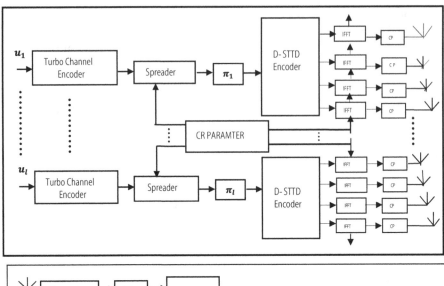

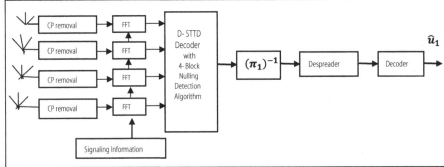

Fig. 1. Transceiver structure for Cognitive defined D-STTD model Coded MIMO MC-IDMA system

Let d_l be interleaved spread sequence and is expressed as

$$d_l = S_l\, e_l \tag{5}$$

The interleaved spread signal of all users is represented as

$$d = \sum_{l=1}^{L} S_l\, e_l \tag{6}$$

We encode bit stream using D-STTD encoder considering each block having four consecutive bit sequences. Then we transmit Mb size of block from each antenna after carrying multi-carrier modulation.

We now represent MC-modulation (Multi-Carrier) matrix as

$$MC = diag\{exp(j2\pi\, f_1 t), exp(j2\pi\, f_2 t), \ldots, exp(j2\pi\, f_{Mb} t)\} \tag{7}$$

We add Cyclic prefixes for 20% to relieve the effects of inter-carrier interference before transmission from each antenna.

We transmit multi-user sum signals in multi-path channel environments with channel specifications as in the Table 1. Considering N - tap delay spread channel matrix connecting r^{th} - RX t^{th}- TX Antenna are represented as

$$h_{rt}(t) = \sum_{n=1}^{N} h_{rt}^n \partial(t - \tau_n) \tag{8}$$

where h_{rt}^n - random process which is Gaussian and zero mean with variance $\psi(\tau_n)$.

2.2 Signal Reception

At each MS, we remove cyclic prefix and carry out multi-carrier demodulation. In this article, we have assumed that the CSI is known to MS. Relation between input and output as assumed to be

$$Y = (H_{\perp})D + \eta \tag{9}$$

Where Y - received vector

$(H_{\perp})$ - Channel Matrix connecting Mobile Station and Base Station and η - indicating both PR-MUI and SE-MUI in addition to the noise. The received matrix at l^{th}. Mobile Station is given by $N_r \times 2$ component matrix received after two symbol durations are expressed.

3 D-STTD Detection Algorithm

$$Y_l = (H_{\perp})_l D + \eta \tag{10}$$

Where

$$Y_l = \begin{bmatrix} y_{11} & y_{12} \\ y_{21} & y_{22} \\ y_{31} & y_{32} \\ y_{41} & y_{42} \end{bmatrix}, \; N_r \times 2 \text{ component received matrix} \tag{11}$$

$$D = \begin{bmatrix} d_1 & -d_2^* \\ d_2 & d_1^* \\ d_3 & -d_4^* \\ d_4 & d_3^* \end{bmatrix} N_t \times 2 \text{ Component transmitted matrix} \tag{12}$$

$$(H_{\perp})_l = \begin{bmatrix} (h_{\perp})_{11} & (h_{\perp})_{12} & (h_{\perp})_{13} & (h_{\perp})_{14} \\ (h_{\perp})_{21} & (h_{\perp})_{22} & (h_{\perp})_{23} & (h_{\perp})_{24} \\ (h_{\perp})_{31} & (h_{\perp})_{32} & (h_{\perp})_{33} & (h_{\perp})_{34} \\ (h_{\perp})_{41} & (h_{\perp})_{42} & (h_{\perp})_{43} & (h_{\perp})_{44} \end{bmatrix}, \; N_r \times N_t \text{ component} \tag{13}$$

$$\boldsymbol{\eta} = \begin{bmatrix} \eta_1 & -\eta_2^* \\ \eta_2 & \eta_1^* \\ \eta_3 & -\eta_4^* \\ \eta_4 & \eta_3^* \end{bmatrix}, \; N_r \times 2 \text{ component noise Matrix} \tag{14}$$

We have to re-arrange the above equation to imply Block-Nulling Signal Estimation Algorithm [2]. Hence the above equation is modified as

$$\left(\overline{Y}_e\right)_l = \left(\overline{H}_e\right)_l \overline{D} + \overline{\eta}_l \tag{15}$$

Where

$$\left(\overline{Y}_e\right)_l = \begin{bmatrix} y_{11} \\ y_{12}^* \\ y_{21} \\ y_{22}^* \\ y_{31} \\ y_{32}^* \\ y_{41} \\ y_{42}^* \end{bmatrix} 2N_r \times 1 \text{ component received vector} \tag{16}$$

$$\left(\overline{H}_e\right)_l = \begin{bmatrix} (h_\perp)_{11} & (h_\perp)_{12} & (h_\perp)_{13} & (h_\perp)_{14} \\ (h_\perp)_{12}^* & -(h_\perp)_{11}^* & (h_\perp)_{14}^* & -(h_\perp)_{13}^* \\ (h_\perp)_{21} & (h_\perp)_{22} & (h_\perp)_{23} & (h_\perp)_{24} \\ (h_\perp)_{22}^* & -(h_\perp)_{21}^* & (h_\perp)_{24}^* & -(h_\perp)_{23}^* \\ (h_\perp)_{31} & (h_\perp)_{32} & (h_\perp)_{33} & (h_\perp)_{33} \\ (h_\perp)_{32}^* & -(h_\perp)_{31}^* & (h_\perp)_{34}^* & -(h_\perp)_{33}^* \\ (h_\perp)_{41} & (h_\perp)_{42} & (h_\perp)_{41} & (h_\perp)_{44} \\ (h_\perp)_{42}^* & -(h_\perp)_{41}^* & (h_\perp)_{44}^* & -(h_\perp)_{43}^* \end{bmatrix} 2N_r \times N_t \text{ component matrix} \tag{17}$$

$$\overline{D} = \begin{bmatrix} d_1 \\ d_2 \\ d_3 \\ d_4 \end{bmatrix} N_t \times 1 \text{ component vector} \tag{18}$$

$$\text{And } \overline{\eta}_l = \begin{bmatrix} \eta_{11} \\ \eta_{12}^* \\ \eta_{21} \\ \eta_{22}^* \\ \eta_{31} \\ \eta_{32}^* \\ \eta_{41} \\ \eta_{42}^* \end{bmatrix} 2N_r \times 1 \text{ component noise vector} \tag{19}$$

We pre-multiply above equation with $\left(\overline{H}_e\right)_l^H$ to express in simple form

$$\left(\overline{Y}\right)_l = \overline{H}_l\,\overline{D} + \overline{\eta}_l \tag{20}$$

Where

$$()^H \text{ - Indicates Hermitian transpose} \tag{21}$$

$$\left(\overline{Y}\right)_l = \left(\overline{H}_e\right)_l^H \times \left(\overline{Y}_e\right)_l = \begin{bmatrix} y_1 \\ y_2 \\ y_3 \\ y_4 \end{bmatrix} N_r \times 1\,\text{component vector} \tag{22}$$

$$\overline{H}_l = \left(\overline{H}_e\right)_l^H \times \left(\overline{H}_e\right)_l = \begin{bmatrix} \upsilon_1 & 0 & \sigma & \omega \\ 0 & \upsilon_1 & -\omega^* & \sigma^* \\ \sigma^* & -\omega & \upsilon_2 & 0 \\ \omega^* & \sigma & 0 & \upsilon_2 \end{bmatrix} N_r \times N_t\,\text{channel gain} \tag{23}$$

Expressing σ, ω, υ_1, and υ_2-channel gain, We have

$$\sigma = \sum_{w=1}^{4} \left(\left(h_{w1}^* \times h_{w3}\right) + \left(h_{w2} \times h_{w4}^*\right)\right) \tag{24}$$

$$\omega = \sum_{w=1}^{4} \left(\left(h_{w1}^* \times h_{w4}\right) - \left(h_{w2} \times h_{w3}^*\right)\right) \tag{25}$$

$$\upsilon_1 = \sum_{w=1}^{4} \left(|h_{w1}|^2 + |h_{w2}|^2 \right) \tag{26}$$

$$\upsilon_2 = \sum_{w=1}^{4} \left(|h_{w3}|^2 + |h_{w4}|^2 \right) \tag{27}$$

And

$$\left(\overline{\eta}\right)_l = \left(\overline{H}_e\right)_l^H \times \left(\overline{\eta}_l\right) = \begin{bmatrix} \eta_1 \\ \eta_2 \\ \eta_3 \\ \eta_4 \end{bmatrix} N_r \times 1\,\text{component noise vector} \tag{28}$$

If STBC block-1 component consisting of d_1 and d_2 are having stronger signals strength than STBC block-2 unit comprising d_3 and d_4, then we detect d_1 and d_2 initially using ML algorithm as given below

$$\hat{d}_i = arg\,min \left| y_i - (\overline{v}_i \times d_i) \right|^2$$
$$d_i \in \{1, -1\} \tag{29}$$

Where

$$\overline{v}_i = v_1 - \left(\frac{|\sigma^2| + |\omega^2|}{v_2} \right) \tag{30}$$

After estimating STBC block-1 unit, we can estimate second STBC block-2 unit using simple expression

$$\begin{bmatrix} \hat{d}_3 \\ \hat{d}_4 \end{bmatrix} = \begin{bmatrix} y_3 \\ y_4 \end{bmatrix} - \begin{bmatrix} \sigma^* & -\omega \\ \omega^* & \sigma \end{bmatrix} \begin{bmatrix} \hat{d}_1 \\ \hat{d}_2 \end{bmatrix} \tag{31}$$

Presuming that STBC block-2 unit having signal component d_3 and d_4 are stronger in signal strength than STBC block-1 unit having d_1 and d_2, we detect signals d_3 and d_4 using ML algorithm based on

$$\hat{d}_i = arg\,min \left| y_i - (\overline{v}_i \times d_i) \right|^2$$
$$d_i \in \{1, -1\} \tag{32}$$

where

$$\overline{v}_i = v_2 - \left(\frac{|\sigma^2| + |\omega^2|}{v_1} \right) \tag{33}$$

Consequently, we detect STBC block-1 unit with signals d_1 and d_2 using expression

$$\begin{bmatrix} \hat{d}_1 \\ \hat{d}_2 \end{bmatrix} = \begin{bmatrix} y_1 \\ y_2 \end{bmatrix} - \begin{bmatrix} \sigma^* & \omega \\ -\omega^* & \sigma \end{bmatrix} \begin{bmatrix} \hat{d}_3 \\ \hat{d}_4 \end{bmatrix} \tag{34}$$

Finally, we de-spread the estimated information and de-interleave followed by turbo decoder [15] to obtain $\hat{u}_l$ - l^{th} user information.

4   Result Discussion and Performance Analysis

Simulation results are represented of coded D-STTD assisted MIMO-MC-IDMA scheme exploiting Cognitive spectrum for SUI-1 [16] and LTE-Vehicular Channel-Model [17]. We summarize channel model in Table 1 and simulation parameter in Table 2.

Table 1. Channel Parameters Model & power delay profiles Specifications for SUI and LTE channels

Path no.	Channel	SUI-1		LTE extended vehicular	
		Delay (μ-microsec)	Power (dB)	Delay (μ-microsec)	Power (dB)
1		0	0	0	0
2		0.4	−15	30	−1.5
3		0.9	−20	150	0
4				310	−1.5
5				370	−0.6
6				710	−9.1
7				1090	−7
8				1730	−12
9				2510	−16.9

Channel desp^n	Doppler shift frequency (Hz)	Antenna correlation	Vehicular speed (km/h)
SUI-1	0.5	0.7	-
LTE-extended vehicular channel model	300	0.7	162

We simulate error-rate results for SUI-1 model with DS-Doppler-Shift of 2.5 Hz along antenna correlation for D-STTD architecture of 0.7 for SUI-1 channel and Doppler shift of 300 Hz with moving vehicle speed of velocity of 162 km/hr considering LTE-extended vehicular model for channel respectively.

Table 2. Parameters for simulation

Parameters	Attributes
Carrier Frequency	Cognitive Spectrum - 80 MHz to 800 MHz
BW-Bandwidth	6 MHz
No's of Channel Realizations	25000
Modulation Technique	BPSK
Channel Models	SUI-1, LTE Extended Vehicular Channel Model Specifications
No's of Transmitter Antenna	4
No's of Receiver Antenna	4
Channel encoder and decoder	Turbo code and log-MAP iterative decoding

We exploit Cognitive spectrum ranging from 80 MHz to 800 MHz Evidently, we have implicated sub-band frequency with bandwidth of 6 MHz of channel for each user. We present simulation of 25000 channel realizations results for each SNR. We have replicated for F-domain spreading sequence length of 8192 with 16-users.

Figure 2 dictates bit error-rate results with SNR for Cognitive D-STTD MC-IDMA system with model of SUI-1 specification. Figure 2 graph, we observe that D-STTD system with MC-IDMA requires 1 Eb/No to generate 10^{-5} of bit error rate. Our considered system offers better bit-error-rate results for lesser SNR while achieving higher processing gain using IDMA system and higher bandwidth using CRN.

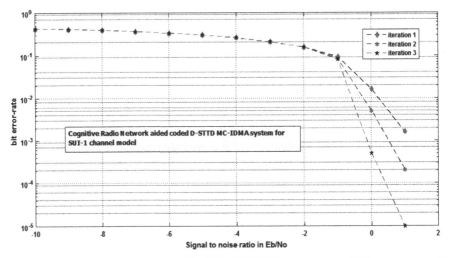

Fig. 2. Bit error performance results of D-STTD MC-IDMA system for SUI-1 channel specification

Figure 3, dictates the error-rate performance with SNR for our system with LTE-vehicular channel specification. The error-rate curve reveals that to reproduce 10^{-5} of bit error rate, the system requires approximately 1 Eb/No. Hence our system provides better results with less number of computation while achieving higher data rate in the rich scattering effects.

From our analysis, we would like to clarify that D-STTD type of MIMO profile achieves both spatial multiplexing and diversity gain. Also the system need three iteration to achieve better results for both channel model. In the first iteration, there is BER of almost 10^{-3} for SUI-1 channel model, 10^{-2} for LTE-vehicular channel model for 1 Eb/No. Clearly, there is no appreciable Bit-Error-Rate performance of our coded system for both channel specifications in first iteration. In the third iteration, iterative turbo decoding algorithm gives appreciable enhancement in terms of Bit-Error-Rate. After three iteration, there is no further improvement in BER results. Hence we have shown up to three iteration results in the graph. Further, we demonstrated from mathematical analysis that Block-Nulling detection techniques [2] offers better estimation technique with less number of computations when compared to MMSE detection algorithm. Furthermore, error-rate curve reveals that coded cognitive radio system achieves better bit

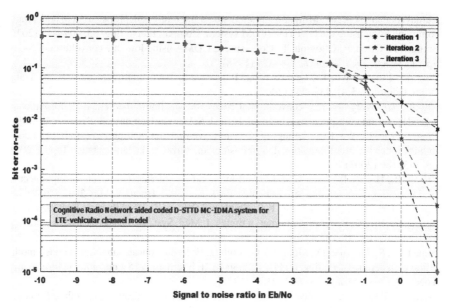

Fig. 3. Bit error-rate performance of D-STTD MC-IDMA system for LTE-vehicular specification

error-rate with lesses SNR even though our Mobile unit is traveling at high speed of 162 km per hour. The error-rate performance reveals that with low value of SNR, we obtain higher performance while exploiting higher bandwidth using Cognitive spectrum.

5 Conclusion

This research work presented performance-analysis of coded cognitive radio network D-STTD aided MIMO MC-IDMA system. The proposed transceiver structure for down-link communication enables to obtain higher performance irrespective of channel condition while offering higher bandwidth by invoking unused TV-band spectrum. Our considered D-STTD MC-IDMA system also supports more user-population by mitigating adverse effects of PR-MUI and SE-MUI with support of D-STTD decoder and iterative style of channel decoder. Further the system reveals that we can confront demand for high data rate using D-STTD structure with minimum number of computation while packing bandwidth efficiency using CRN.

References

1. Muquet, B., Biglieri, E., Sari, H.: MIMO link adaptation in mobile WiMAX systems. In: IEEE Wireless Communications and Networking Conference (WCNC 2007), China, 11–15 March 2007, pp. 1812–1815 (2007)
2. Lee, Y., Shieh, H.-W.: A simple layered space-time block Nulling technique for DSTTD systems. IEEE Commun. Lett. **15**(12), 1323–1325 (2011)

3. Iruthayanathan, N., et al.: Mitigating ambient noise and multi-path propagation in underwater communication using the DSTTD-OFDM system. Comput. Electr. Eng. **53**, 409–417 (2016)
4. Frenger, P., Orten, P., Ottosson, T.: Code-spread CDMA using low-rate convolutional codes. Spread Spectr. Tech. Appl. **2**, 374–378 (1998)
5. Ping, L., Liu, L., Wu, K., Leung, W.K.: Interleave-division multiple-aceess. IEEE Trans. Wirel. Commun. **5**(4), 938–947 (2006)
6. Viterbi, A.J.: Very low rate convolutional codes for Maximum theoretical performance of spread spectrum multiple-access channel. IEEE J. Sel. Areas Commun. **8**(4), 641–649 (1990)
7. Ping, L., Liu, L., Wu, K.Y., Leung, W.K.: Interleave-division multiple-access (IDMA) communications. In: 3rd International Symposium on turbo codes and related Topics 2003, pp. 173–180 (2003)
8. Zhang, R., Hanzo, L.: Iteratively detected multi-carrier interleave division multiple access. In: International conference MICROC0LL 2007, Budapest, Hungary, May 2007 (2007)
9. Lakshmi Dhevi, B., Vishvaksenan, K.S., Senthamil Selvan, K., Rajalakshmi, A.: Patient monitoring system using cognitive internet of things. J. Med. Syst. **42**, 229 (2018). https://doi.org/10.1007/s10916-018-1095-2
10. Benedetto, S., Divsalar, D., Montorsi, G., Pollara, F.: Serial concatenated trellis coded modulation with iterative decoding. In: Proceedings IEEE International Symposium on Information Theory, Ulm, Germany, June/July 1997 (1997)
11. Wang, X., Poor, H.V.: Iterative (turbo) soft interference cancellation and decoding for coded CDMA. IEEE Trans. Wirel. Commun. **47**(6), 225–228 (2007)
12. Le Goff, S., Glavieux, A., Berrou, C.: Turbo-codes and high spectral efficiency modulation. In: Proceedings of IEEE International Conference on Communications, New Orleans, LA, pp. 645–649, May 1994
13. Benedetto, S., Divsalar, D., Montorsi, G., Pollara, F.: Bandwidth efficient parallel concatenated coding schemes. Electron. Lett. **31**(24), 2067–2069 (1995)
14. Robertson, P., Worz, T.: Coded modulation scheme employing turbo codes. Electron. Lett. **31**(18), 1546–1547 (1995)
15. Bahl, L.R., Cocke, J., Jelinek, F., Raviv, J.: Optimal decoding of linear codes for minimizing symbol error rate. IEEE Trans. Inf. Theory **20**(2), 284–287 (1974)
16. Maucher, J., Furrer, J., Heise: IEEE Std. 2007, IEEE Standard for WIMAX 802.16, Publisher, Hannover (2007)
17. GPPP (TR 30.803): Evolved universal Terrestrial radio access (E-UTRA); user equipment (UE) radio transmission and reception (Release 8), Technical specification, Sophia Antipolis, France (2007)

An Approach for Color Retinal Image Enhancement Using Linearly Quantile Separated Histogram Equalization and DCT Based Local Contrast Enhancement

Mayank Tiwari, Riya Ruhela$^{(\boxtimes)}$, and Bhupendra Gupta

Department of Natural Science, PDPM, IIITDM Jabalpur,
Dumna Road, Jabalpur 482005, MP, India
mayanktiwariggits@gmail.com, riyashrama41@gmail.com,
gupta.bhupendra@gmail.com

Abstract. For an accurate diagnosis of medical diseases, medical applications are using computer-aided image processing methods. However, images with poor visibility lead to creating many problems in diagnosis. In this field, it is important to protect meaningful diagnostic information in the image. Image enhancement is an applicable issue found in diverse image processing and computer vision problems. The important property of a good image enhancement approach is that it should enhance the image without creating any distortion in the meaningful information of the image. Here we propose a method for visibility improvement of color retinal images. The structure of the proposed method is divided into two basic steps. Initially, overall visibility of the given color retinal image is improved using a quantile segmentation based histogram equalization method. In the second step, discrete cosine transform (DCT) based tool is applied for local contrast improvements of the image. From the subjective and quantitative measures, it is shown that the proposed method enhances global contrast and preserves the local information of the color retinal images.

Keywords: Color retinal image · Histogram equalization · Contrast improvement · Transformed domain based local contrast enhancement · Mean brightness preservation

1 Introduction

Dealing with color retinal images (CRIs) in digital image processing (DIP) is a highly challenging task. This is due to the reason that the CRIs include the complex structure and major investigation information, and poor perceptibility (contrast) of CRIs hides valuable diagnostic information. As a result, the medical inspector is not able to make proper decisions from the given CRI. To solve such problems CRI improvement methods have been developed. These methods improve the visibility of the given CRI in such a way that the processed image shows meaningful information and thus improves the efficiency of diagnosis.

© Springer Nature Singapore Pte Ltd. 2020
S. M. Thampi et al. (Eds.): SIRS 2019, CCIS 1209, pp. 359–367, 2020.
https://doi.org/10.1007/978-981-15-4828-4_29

1.1 Literature Review

Many CRI enhancement methods have been developed by various groups of researchers [1–10]. However, classical CRI enhancement methods suffer from the following problems.

1. Most of CRI enhancement methods over-enhance the image. As a result, the processed image may have saturated regions. Also, the information in those regions is completely distorted [1].
2. Most of the CRI enhancement methods shift mean brightness in the processed image. This brightness change distorts meaningful diagnosis information of the CRI. This means brightness change is a problem termed as the 'mean-shift' problem in the literature. Due to the 'mean-shift' problem further post-processing becomes difficult [2–4].
3. Some local contrast improvement methods enhance the foreground (part of the image contains meaningful information required for diagnosis) and background part (rest of the part of the image after foreground) of the image at an equal level. This leads to an increase in the false-positive rate during the diagnosis [5–8].

1.2 Research Gap

As it is well known that the image enhancement methods are problem-specific. This means that the best method developed for CRIs of 'type-A' will not work effectively for CRIs of 'type-B'. Also, the CRI enhancement scenario requires methods that are capable of improving the visibility of given CRI without creating noise amplification, saturation, and mean-brightness shift. We have found that none of the methods discussed in [1–10] are capable of doing this.

1.3 Motivation

In this work, we have developed a method for solving all such problems. The proposed method is based on the quantile separation based histogram equalization method [11] and transformed domain local contrast enhancement [12]. We termed this method as linearly quantile separated histogram equalization followed by transformed domain-based local contrast enhancement (LQSHE-LCE) method.

1.4 Organization of the Work

The remaining portion of this paper is written as follows: Sect. 2 describes in detail the proposed method LQSHE-LCE. For performance evaluation of the proposed method Sect. 3 contains experimental results. Section 4 concludes the proposed method. At last authors acknowledged the reviewers.

2   The Proposed Method

A complete description of the proposed LQSHE-LCE method is presented in this section. For a better understanding of the proposed method, we are initially showing its flowchart in Fig. 1. The proposed method initially converts the given CRI to Lab color space from red, green, blue (RGB) color space. The Lab color space is widely used for color-preserving contrast improvement of all types of images. Then the global contrast of this image is enhanced using a 'linearly quantile separated histogram equalization' method. In the next step for local details enhancement of the global contrast-enhanced image is performed using transformed domain processing. As a result of global and local contrast enhancement, we get an image that has sufficient contrast with good interpretation of local details.

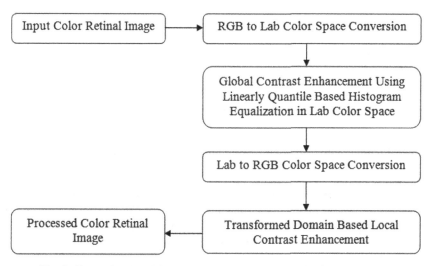

Fig. 1 Shows functional block diagram of proposed method LQSHE-LCE.

2.1   Linearly Quantile Separated Histogram Equalization

The linearly quantile separated HE method is a well-known method for contrast improve-ment. This method was initially proposed by Tiwari et al. [11] and later modified by Gupta et al. [15]. In the proposed work we have only used the quantile tool for his-togram segmentation. This is because the quantile based segmentation does not require a recursive method for histogram segmentation. We have observed that the quantile val-ues generation algorithm used in [11] and [15] is not optimized. As for quantile value q, in the worst case, this algorithm requires calculations. We have found that by optimizing this algorithm, its computational complexity can be improved up to. In the proposed work we have used the improved method for quantile based histogram segmentation (see Algorithm 1).

Algorithm 1: improved_quantile_values_generations(.)

Required: P[];
Required: q;
  X=256;
  quantile = zeros(q,1);
  sum = 0;
  count = 0;
  for i = 1 to q **do**
    limit = 1.0/q;
    while count <= X-1 **do**
      sum = sum + P[count];
      if sum >= limit **then**
        quantile[i] = count;
        break;
      end if
      count = count + 1;
    end while
  end for
  quantile[q] = X - 1;
array quantile[] is output of this algorithm.

2.2 Transformed Domain Based Local Contrast Enhancement

In this work, our main objective is to develop a method that is capable of preserving features of the image. To achieve this we have used a method of scaling the coefficients of the contrast-enhanced image in the transformed domain. For transformation we have used the discrete cosine transform (DCT) [13, 14]. In the transformed domain we have scaled only those coefficients that have lower values than a pre-defined threshold [12].

Let $x(a, b)$ be the transformed domain coefficient value at location (a, b). Now the improved coefficient value $x'(a, b)$ is obtained as:

$$x'(a, b) = x(a, b) \text{ if } |x(a, b)| > 0.01 \times x(1, 1)$$

$$x'(a, b) = \alpha \times x(a, b) \text{ \textbf{Otherwise}.} \tag{1}$$

The level of enhancement is controlled by the scaling factor $\alpha > 1$. Automatic selection of parameter alpha can be defined as below:

$$\alpha = 1 + \sqrt{\frac{k}{255}}, \tag{2}$$

and the value of k is calculated as:

$$k = std(Y) - std(X). \tag{3}$$

Where X is input image and Y is the image obtained after applying LQSHE method.

At last, the processed image is built back from the transformed domain using inverse DCT (IDCT). A surprisingly huge value of parameter would produce an improvement in terms of edges and details damages. The automated parameter setting contributes an immeasurable trade-off in the level of improvement and detail preservation.

Now it is clear that the LQSHE method improves the contrast of the given image and the DCT based local contrast enhancement method improves the local details in the image. As a result, we get an image with sufficient global and local contrast.

3 Experimental Results

In this section, numerical and visual experiments have been performed to check the effectiveness of the proposed image. As test images, we have selected 400 CRIs [16]. We have compared the performance of the proposed method with contrast limited adaptive histogram equalization (CLAHE) in different color spaces namely Lab ($CLAHE_{Lab}$), RGB ($CLAHE_{RGB}$), and histogram equalization (HE) method in different color space namely RGB (HE_{RGB}), Lab (HE_{Lab}) and method proposed by Zhou et al [4]. Each of these methods have their special feature for CRI enhancement. The HE method is the standard method used for contrast improvement of all type of images. The CLAHE method is the widely accepted method for local contrast improvement of all type of CRIs. The method proposed by Zhou et al [4] is the best method of CRI enhancement in the given literature [4]. Also, these methods do not distort meaningful information of the given CRIs.

3.1 Image Quality Assessment

In Fig. 2, we are showing results produced by various methods on '20051020_62461_0100_PP' image. Here we observe that the original low-contrast CRI does not provide clear information about inner muscles structures of the image. Clear information about muscles structure of the image is necessary for performing different type of medical treatment. One such situation arise when we are interested to find the location of infected region in the given low-contrast CRI. The accuracy of the further treatment depends on the precise detection of the infected region. Otherwise the medical examiner may need to find this based on his/her experience, which will be an inefficient and time consuming procedure. To find the accurate location of the infected region, the given CRI can be enhanced by using various image enhancement methods. For comparison the efficiency of different methods we are showing resultant images obtained by these methods. The HE method improves contrast of the CRI, however result of this method suffers from the 'mean-shift' problem. Next result generated by the CLAHE method enhances muscle structure and other part of the image at equal level. In many cases this leads to rise of false positive rate. Result of the method proposed by Zhou et al [4] is capable in enhancing the overall details of the given CRI; however this result is not able to maintain the color information in the processed image. On the other hand result produced by the proposed LQSHE-LCE method has sufficient global contrast with good interpretation of local details.

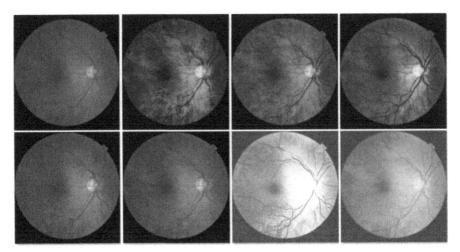

Fig. 2. Shows results produced by various methods on '20051020_62461_0100_PP' image. Here from left to right and top to bottom the images are as: input image, result of Zhou et al method [4], result of $CLAHE_{Lab}$ method, result of $CLAHE_{RGB}$, result of proposed method (q = 3), result of proposed method (q = 5), result of HE_{RGB} method, and result of HE_{Lab} method respectively.

Next in Fig. 3, we are showing results produced by various methods on an image patch cropped from '20060523_50003_0100_PP'. The low contrast nature of CRIs creates the problem in the further post-processing. Due to extreme low-contrast present in the CRI, diagnosis of the infected region becomes a difficult task. Hence in situations like case-(a) a healthy person is marked as ill. The person will have to take the heavy dosage of medicines for a disease from which he/she is not suffering at all. While in case-(b) an ill person is marked as healthy. The person may loss his/her eyes, if the infection increases suddenly. To avoid such situations proper diagnosis of eye disease from a given low-contrast CRI is necessary. For improving the accuracy of diagnosis color retinal image enhancement methods are applied. For doing this we are showing image enhancement results of different methods on the given image. A careful observation of Fig. 3 shows that the input CRI does not provide information about the infected region. Using the HE method the results are improved. However, it is clearly visible that the HE method is over enhancing the contrast of the image. The CLAHE method is enhancing infected region and other parts of the image at the equal level. This situation leads to raising the cases (a) and (b) as discussed earlier. The result produced by the LQSHE-LCE method has sufficient contrast and this method does not enhance infected region and another region at the equal level. As result chances of raise in the cases like (a) and (b) is very low, which shows that the proposed method is a good choice for contrast improvement of all type of color retinal images.

For numerical evaluation, we have used three widely used image quality measures. These measures are peak signal to noise ratio (PSNR), structural similarity index measure (SSIM), and absolute mean brightness error (AMBE). These measures have been widely used for image quality assessment [11, 15, 17–23]. The PSNR is used to measure contrast improvement in all type of images. A PSNR value in the range 20 to 40 shows better

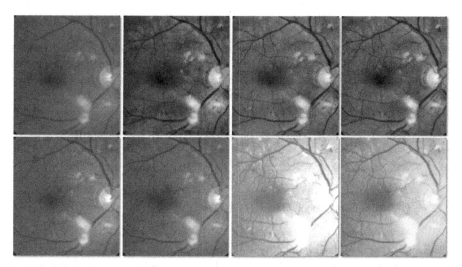

Fig. 3. Shows results produced by various methods on an image patch cropped from '20060523_50003_0100_PP' image. Here from left to right and top to bottom the images are as: input image, result of Zhou et al method [4], result of $CLAHE_{Lab}$ method, result of $CLAHE_{RGB}$, result of proposed method (q = 3), result of proposed method (q = 5), result of HE_{RGB} method, and result of HE_{Lab} method respectively.

contrast improvement. Next, the SSIM is a good statistical measure to check overall structural similarity of the image [22]. The range of SSIM is any fractional value in the range [0, 1]. An SSIM value near to 1 shows better results. At last for brightness preservation, AMBE measure is used. A small value of AMBE shows better brightness preservation.

In Table 1, we are showing average values of PSNR, SSIM and AMBE measures. It is clear from the numerical results that the proposed method is enhancing sufficient contrast in the resultant CRI at the same time able to preserve structural similarity and brightness more accurately than other methods.

Table 1. Results of average-PSNR, average-SSIM and average-AMBE.

Methods	A-PSNR	A-SSIM	A-AMBE
HE_{RGB}	8.17	0.28	97.68
HE_{Lab}	8.47	0.30	94.90
$CLAHE_{RGB}$	25.47	0.57	9.26
$CLAHE_{Lab}$	25.77	0.54	7.86
Zhou et al [4]	23.39	0.59	5.43
Proposed (q = 3)	32.56	0.76	4.80
Proposed (q = 5)	33.46	0.76	4.32

3.2 Computational Complexity

Table 2 shows analysis of computational complexity of the proposed method at various parts of its working. Here we have assumed all the single line arithmetical statements (statements performing $+$, $-$, $*$ and $/$ on simple variables) to take constant time. Also we have assumed the image size as $m \times n$ and the image has X grey levels.

Table 2. Results of average-PSNR, average-SSIM and average-AMBE.

Methods	Computational complexity
LQSHE	$2 \times O(m.n) + O(X)$
Transformed Domain Based LCE	$O(m.n)$

The total computational complexity of the proposed method is $3 \times O(m.n) + O(X)$. By this computational complexity the proposed method can be easily implemented for real time CRI enhancement applications.

4   Conclusion

In this work we have proposed a new method for contrast improvement of all types of CRIs. The proposed method enhances global contrast of the given CRI in the Lab color space by using quantile based histogram equalization method and then it enhances local details of the image in transformed domain. We have experimentally shown that processed images produced by the proposed method have least amount of distortion. Also these images do not suffer from the common issues such as the 'mean-shift' problem, color distortion, and noise enhancement in some visually important areas. In future the proposed method can be tested for all types of images too.

Acknowledgement. The authors thank the reviewers, who helped us in improving the quality of this work. Without their help we were not able to present many interesting concepts related to this work.

References

1. Hsu, W.Y., Chou, C.Y.: medical image enhancement using modified color histogram equalization. J. Med. Biol. Eng. (Springer) **35**(5), 580–584 (2015)
2. Singh, N., Kaur, L., Singh, K.: Histogram equalization techniques for enhancement of low radiance retinal images for early detection of diabetic retinopathy. Eng. Sci. Technol. Int. J. **22**, 736–745 (2019)
3. Soomro, T.A., Gao, J., Khan, M.A.U., Khan, T.M., Paul, M.: Role of image contrast enhancement technique for ophthalmologist as diagnostic tool for diabetic retinopathy. In: 2016 Conference, DICTA, pp. 1–8. IEEE (2016)
4. Zhou, M., Jin, K., Wang, S., Ye, J., Qian, D.: Color retinal image enhancement based on luminosity and contrast adjustment. IEEE Trans. Biomed. Eng. **65**(3), 521–527 (2018)

5. Setiawan, A.W., Mengk, T.R., Santoso, O.S., Suksmono, A.B.: Color retinal image enhancement using CLAHE. In: 2013 Conference on International Conference on ICT for Smart Society, pp. 1–3. IEEE (2013)
6. Sonali, Sahu, S., Singh, A.K., Ghrera, S.P., Elhoseny, M.: An approach for de-noising and contrast enhancement of retinal fundus image using CLAHE. Opt. Laser Technol. **110**, 87–98 (2019). Elsevier
7. Jintasuttisak, T., Intajag, S.: Color retinal image enhancement by Rayleigh contrast-limited adaptive histogram equalization. In: 2014 14th International Conference on Control, Automation and Systems (ICCAS 2014), pp. 692–697. Seoul (2014)
8. Setiawan, A.W., Mengko, T.R., Santoso, O.S., Suksmono, A.B.: Color retinal image enhancement using CLAHE. In: International Conference on ICT for Smart Society, pp. 1–3. Jakarta (2013)
9. Jebaseelia, T.J., Duraib, A.D., Petera, D.: Segmentation of retinal blood vessels from ophthalmologic Diabetic Retinopathy images. Comput. Electr. Eng. **73**, 245–258 (2019)
10. Wang, X., Jiang, X., Ren, J.: Blood vessel segmentation from fundus image by a cascade classification framework. Pattern Recogn. **88**, 331–341 (2019)
11. Tiwari, M., Gupta, B., Shrivastava, M.: High-speed quantile-based histogram equalization for brightness preservation and contrast enhancement. IET Image Proc. **9**(1), 80–89 (2014)
12. Fu, X., Wang, J., Zeng, D., Huang, Y., Ding, X.: Remote sensing image enhancement using regularized-histogram equalization and DCT. IEEE Geosci. Remote Sens. Lett. **12**(11), 2301–2305 (2015)
13. Khayam S.A.: The Discrete Cosine Transform (DCT): Theory and Application. Department of electrical & computing engineering (2003)
14. Gupta, B., Tiwari, M.: Improving performance of source-camera identification by suppressing peaks and eliminating low-frequency defects of reference SPN. IEEE Signal Process. Lett. **25**(9), 1340–1343 (2018)
15. Gupta, B., Agarwal, T.K.: Linearly quantile separated weighted dynamic histogram equalization for contrast enhancement. Comput. Electr. Eng. **62**, 360–374 (2017)
16. Decenciere, E., et al.: Feedback on a publicly distributed image database: the Messidor database. Image Analysis & Stereology **33**(3), 231–234 (2014)
17. Gupta, B., Tiwari, M.: Minimum mean brightness error contrast enhancement of color images using adaptive gamma correction with color preserving framework. Int. J. Light Electron Opt. **127**, 1671–1676 (2015)
18. Tiwari, M., Gupta, B., Lamba S.S.: Performance improvement of image enhancement methods using statistical moving average histogram modification filter. In: 2018 CONFERENCE, ICDSP, pp. 65–69. New York (USA) (2018)
19. Tiwari, M., Gupta, B.: A consistent approach for image de-noising using spatial gradient based bilateral filter and smooth filtering. In: Proceedings SPIE 10011, First International Workshop on Pattern Recognition, 100110Q (2016)
20. Tiwari, M., Gupta, B.: Brightness preserving contrast enhancement of medical images using adaptive gamma correction and homomorphic filtering. In: 2016 CONFERENCE, SCEECS, pp. 1–4. Bhopal (2016)
21. Tiwari, M., Gupta, B.: Maximum absolute relative differences statistic for removing random-valued impulse noise from given image. Circ. Syst. Signal Process. **37**(5), 2098–2116 (2018)
22. Wang, Z., Bovik, A.C., Sheikh, H.R., Simoncelli, E.P.: Image quality assessment: From error measurement to structural similarity. IEEE Trans. Image Process. **13**(4), 600–612 (2003)
23. Tiwari, M., Lamba, S.S., Gupta, B.: A software-supported approach for improving visibility of backlight images using image threshold-based adaptive gamma correction. In: Bhatia, S.K., Tiwari, S., Mishra, K.K., Trivedi, M.C. (eds.) Advances in Computer Communication and Computational Sciences. AISC, vol. 760, pp. 299–308. Springer, Singapore (2019). https://doi.org/10.1007/978-981-13-0344-9_26

Recurrence Network-Based Approach to Distinguish Between Chaotic and Quasiperiodic Solution

Ardhana Mohan, V. Vijesh, Drisya Alex Thumba,
and K. Satheesh Kumar[✉]

Department of Futures Studies, University of Kerala,
Kariavattom PO 695 581, Kerala, India
kskumar@keralauniversity.ac.in

Abstract. The quasiperiodic route to chaos is observed in many complex dynamical systems. Quasiperiodic solutions have many similar features of a chaotic system, making it difficult to distinguish between them. In this paper, we propose a complex network-based method for differentiating quasiperiodic from chaotic solutions. The efficiency of the method is demonstrated using chaotic and quasiperiodic realisations generated from two dynamical systems, Chua's circuit and Lorenz system involving complex variables. Statistical analysis of the networks generated from time series shows that there is a significant difference between the network characteristics of quasiperiodic and chaotic dynamics.

Keywords: Quasiperiodicity · Chaoticity · Recurrence plot · Complex networks

1 Introduction

Studies on nonlinear dynamical systems have got significant applications in different fields ranging from simple physical systems to complex social systems. The quantitative understanding of the dynamical behaviour of any nonlinear system is important as it can explain the inherent complexities and thereby modelling the system efficiently and effectively. There are many methods available in literature and practice to quantify the different characteristics of nonlinear dynamical systems. Fractal geometry is one of the typical attributes of chaotic systems and its presence explains the self-similarity structure. Various dimension estimates such as box-counting dimension, Hausdorff dimension, correlation dimension *etc.* are used for measuring the structure and self-similarity of chaotic systems [1–3]. Sensitivity to initial condition is another important characteristic of the systems with chaotic dynamics as the nearby trajectories diverge in an exponential rate later in time. Lyapunov exponent is the best-known technique to quantify this rate of divergence and the positive value of Lyapunov exponent indicates the system is very much sensitive [4]. Chaotic systems are deterministic in nature,

© Springer Nature Singapore Pte Ltd. 2020
S. M. Thampi et al. (Eds.): SIRS 2019, CCIS 1209, pp. 368–375, 2020.
https://doi.org/10.1007/978-981-15-4828-4_30

and hence the evolutionary trajectories may revisit the previous state later in time. Recurrence plot introduced by Eckmann, Oliffson-Kamphorst, and Ruelle is an efficient tool to visualise the repeated occurrence of a state in a trajectory [5]. However, one has to visually observe the plot for interpreting the revisits to similar states and the quantification of recurrence plot structure, such as how frequently a state recurs and how long it follows the former state is therefore necessary to get an insight to the complexities in a system [6]. *Theil et al.* successfully estimated dynamical invariants such as correlation dimension and entropy with a recurrence quantification analysis [7]. Recurrence quantification analysis using the concept of complex networks, where the recurrence matrix is considered as the adjacency matrix, has gained much attention these days. *Donner et al.* has done a detailed analysis of chaotic systems through recurrence network and suggested that dynamical complexities can be quantitatively well explained with the network characteristics such as centrality measures and clustering coefficients [8]. Quantifying the dynamical characteristics has a wide range of applications, especially in systems that exhibits multiple dynamical behaviours corresponding to different initial conditions. In this work, we attempt to quantitatively analyse the distinct dynamics of a system to discriminate the quasiperiodic behaviour from chaotic behaviour.

Analysing the quasiperiodicity in a dynamical system is important as it can be an intermediate step to chaos through a period-doubling process [9]. Quasiperiodic solutions are originated from disproportionate frequencies in phase space, and usually, the attractor takes the form of a torus. This behaviour of a dynamical system is caused by the distortion of stable periodic attractor and well-known for its weak nonlinearity with the Lyapunov exponent as zero. As this type of behaviour is a route to the chaos, it is very important to identify some suitable empirical methods for differentiating quasiperiodic from chaos. As of now, only limited works are available in literature in this perspective. *Zou et al.* discussed the use of recurrence plot for distinguishing quasiperiodic from chaotic attractor [10]. They used the concept of Slater's theorem for identifying the torus in the quasiperiodic system. In this work, we present a recurrent network-based quantification method to differentiate quasiperiodic attractor from a strange chaotic attractor. For analysis, we used the time series data obtained from two different dynamical systems, Lorenz system and Chua's circuit solutions for both quasiperiodic and chaos.

2   Recurrence Quantification Using Complex Networks

Analysis of non-linear systems using the time series arose of complex networks is a growing field as it efficiently characterizes the dynamical properties of a system. The first step into this analysis is the proper construction of networks from the historical time series data. There are numerous methods available in the literature for the proper creation of networks from the available time series. Methods for generating complex networks from time series can be broadly classified into three namely, proximity networks, visibility graphs and transition networks [11]. Proximity network uses the similarity information of different states in evolution

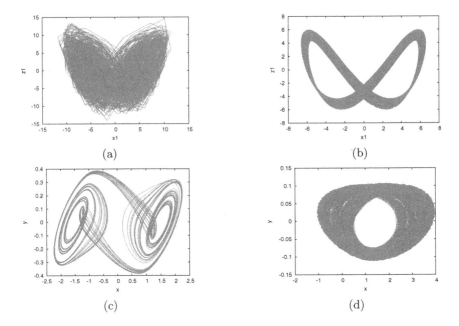

Fig. 1. The cross-sectional view of attractor of (a) chaotic (b) quasiperiodic behaviour of Lorenz system with complex variables and (a) chaotic (b) quasiperiodic behaviour Chua's circuit.

trajectory for network construction, and it considers the characteristics such as Euclidean distance, correlation and pseudo-periodicity [12–14]. Visibility of two points in state space is considered to form edges in visibility networks and two points are visible each other if there is no obstacle between them. Such networks and its applications are reported in various literature [15–17]. In the transition networks, individual dynamics are identified, and the transition probabilities are calculated [18,19]. For network construction, we used the concept of proximity by identifying the recurring states of a dynamical system and the network is known as recurrence network [8,20].

The basic concept of recurrence network is based on the recurrence plot, a visualisation tool to picture the revisit of the system to its previously occurred states. For a network $G = (V, E)$, each state in the trajectory, $\mathbf{x}_i$, is considered as node-set V and the neighbouring nodes $U_\epsilon(\mathbf{x_i})$ are connected to form edge set E. The neighbourhood of a point $\mathbf{x}_i$ in state space is identified with a threshold distance of ϵ. If any point $\mathbf{x}_j$ is at a distance less than ϵ from $\mathbf{x}_i$ then they are considered as connected in the network structure. With this, we can form a recurrence matrix as,

$$R_{ij} = H(\epsilon - \|\mathbf{x}_i - \mathbf{x}_j\|) \tag{1}$$

where H is the Heaviside function and the metric calculated is the Euclidean distance. Here the values of i and j depend on the maximum number of points in vector space $\mathbf{x}$. Once recurrence matrix $R_{i,j}$ is generated adjacency matrix

$A_{i,j}$ can easily be attained by subtracting $R_{i,j}$ from unit matrix $I_{i,j}$. Once the adjacency matrix is defined, we can build the corresponding network to study the time series characteristics.

3   Results and Discussion

We have carried out a detailed analysis, based on empirical data generated from the basic realisation of the Lorenz system involving complex variables and Chua's circuit. Edward Lorenz introduced the concept of nonlinear deterministic systems with a three-variable mathematical model for atmospheric convection which is commonly known as Lorenz equation. Apart from the standard chaotic attractor, more distinguished behaviours such as quasiperiodic and hyper-chaotic can be achieved by changing Lorenz variables to complex forms [21]. This higher dimensional Lorenz system can be mathematically expressed as

$$\dot{x} = a(y - x)$$
$$\dot{y} = cx - y - x^* z \qquad (2)$$
$$\dot{z} = xy - bz$$

where x, y, z are the complex variables and $x = x_1 + ix_2, y = y_1 + iy_2, z = z_1 + iz_2$. Along with chaotic solutions, for a specific range of c values the system also generates quasi-periodic behaviours.

Chua's circuit is known as the universal circuit for studying chaotic behaviour as it is capable of generating multiple attractors by varying the parametric values. The basic Chua's circuit can be modelled with a set of nonlinear differential equations

$$\dot{x} = \alpha(y - x - g(x))$$
$$\dot{y} = x - y + z \qquad (3)$$
$$\dot{z} = -\beta y$$

where $g(x) = m_1 x + (m_0 - m_1)/2(|x + 1| - |x - 1|)$ is the nonlinear realisation. To capture the different states in the evolutionary dynamics of both Lorenz system with a complex variable and Chua's circuit, we used a *Matlab* code with a fourth-order Runge Kutta integration solution. Since the behaviour of Chua's circuit is highly dependent on the chosen initial condition of voltages x, y across the capacitors, current z through the inductor and circuit parameters α, β, we used different combinations of these values to generate time series of ten strange chaotic attractors and ten quasiperiodic attractor containing 2000 data points. Similarly, ten sets of chaotic and quasiperiodic solutions of the Lorenz system is also obtained by varying the values of c ranging $3.1 - 3.5$ and $3.6 - 4.2$ respectively. An example of the dynamical behaviour of Chua's system and Lorenz system with complex variables, both chaos and quasiperiodic, are plotted in Fig. 1 and it is evident from the figure that compared to the chaotic attractor, the quasiperiodic phenomenon has much simple structure.

For any systems, if one or more signals can be measured as a function of time, then the dynamics of the system can be unfolded by embedding the signals into a higher dimensional space. For further analysis, we used only the time series values of x_1 from Lorenz system with complex variables and x from Chua's circuit obtained with the Matlab simulation of Eqs. 2 and 3 respectively.

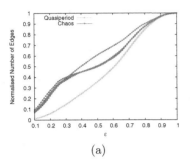
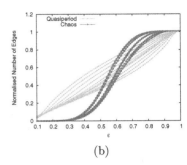

(a)   (b)

Fig. 2. Variation in normalised number of edges in relation to ϵ (a) Chua's circuit and (b) Lorenz system with complex variables.

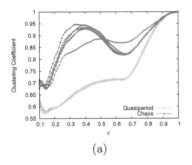
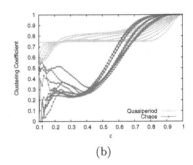

(a)   (b)

Fig. 3. Variation in clustering coefficients with varying ϵ (a) Chua's circuit and (b) Lorenz system with complex variables.

According to Takens Embedding Theorem, with appropriate embedding delay (τ) and dimension (m), a historically measured scalar time series data $x(i)$ can be projected into higher dimension vector space as a series [22] $\mathbf{x_i} = (x_i, x_{i-\tau}, x_{i-2\tau}, ..., x_{i-m\tau})$. Suitable embedding dimension (m) and delay (τ) can be obtained by the method of false nearest neighbours and mutual information [23,24]. As a first step, in order to change the values into a common scale the signals under consideration are normalized. Appropriate measures of embedding dimension (m) and delay (τ) are then calculated with aforementioned techniques for attractor reconstruction. Embedding values obtained are $m = 12, \tau = 9$ for Lorenz system with complex variable and $m = 6, \tau = 7$ fo Chua's circuit. For generating the associated complex networks, as discussed earlier in Eq. 1, each vector point $\mathbf{x_i}$ in the phase space is considered as the nodes, and edge set is formed by connecting neighbouring points at a distance less than ϵ (recurring states). It has been noted in the literature that fixing a suitable ϵ value is crucial for revealing the dynamical characteristics [25]. However, we constructed a series of networks for a range of ϵ values for each time series. These networks were analysed for its structural properties such as the number of edges and clustering coefficient. Network characteristics obtained from ten sets of each quasiperiodic and chaotic solutions of Lorenz system with complex variables and Chua's circuit are plotted in Figs. 2 and 3. It can be noted that for a wide range of ϵ values, these structural properties of the periodic solution is significantly different from that of chaotic solutions.

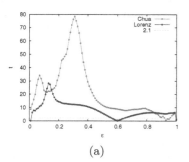
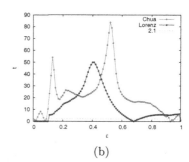

(a)   (b)

Fig. 4. Plot of ϵ versus significance of difference t between the means of (a) clustering coefficient and (b) normalised number of edges calculated from chaotic and quasiperiodic solution.

In order to validate the results obtained, the network characteristics of quasiperiodic and chaotic solutions were further analysed with t-$test$ to see whether they differ significantly. The hypothesis assume that the signals originated from quasiperiodic and chaotic realisations of a nonlinear system exhibits similar network characteristics and the test statistic t is given by

$$t = \frac{\mu_c - \mu_q}{\sqrt{\frac{s_c^2}{N_c} + \frac{s_q^2}{N_q}}} \tag{4}$$

where μ_c and s_c are the mean and variance of network characteristics obtained for signals from chaotic solutions, μ_c and s_c are the mean and variance of network characteristics obtained for quasiperiodic solutions and N_c, N_q the number of samples in each set which 10 in this case. In order to prove the alternative hypothesis with 95% the critical for t is calculated as 2.1. That means if $t > 2.1$ the probability is 95% or more that the network characteristics quasiperiodic and chaotic solutions of a dynamical system are different. Figure 4 plots the values of t versus ϵ obtained from quasiperiodic and chaotic solutions of both the Lorenz system with complex variables and Chua's circuit. We can observe that the significance of difference t for both normalised number of edges and clustering coefficients reaches upto 85, and hence the null hypothesis can be safely rejected.

Recurrence plots have already been reported in the literature as a means of distinguishing periodicity from chaos [10]. However, recurrence plots are generally visual detection tool. The recurrence network method proposed in this work provides a range of ϵ values for network construction to quantify the difference.

4   Conclusion

We propose a recurrent network approach for differentiating quasiperiodicity from chaotic dynamics. In order to generate the recurrent network from the measured time series, the attractor is reconstructed using the concept of high-dimensional embedding. Each point in the phase space is considered as the nodes in the network, and the connections are formed if, in phase space, two points are at a distance less than ϵ. We constructed a series of networks from chaotic and periodic time series data by varying the distance of ϵ. The characteristics of the obtained complex network are then analysed. For the analysis, we used a set of forty different signals generated from quasiperiodic and chaotic solutions of the Lorenz system with complex variables and Chua's circuit realisations. Comparing the recurrent network characteristics, it is evident that the network features of chaotic and quasiperiodic solutions are remarkably different. The statistical analysis shows that recurrence network approach is efficient in discriminating chaotic and periodic behaviour quantitatively.

Acknowledgement. The authors wish to thank the campus computing facility of University of Kerala set up under DST-PURSE programme for providing computational facilities. The third author (DAT) would like to thank Department of Science and Technology (India), for financial assistance as post doctoral research fellowship through PURSE programme.

References

1. Hausdorff, F.: Dimension und äußeres maß. Math. Ann. **79**, 157–179 (1919). eudml.org/doc/158784
2. Liebovitch, L.S., Toth, T.: A fast algorithm to determine fractal dimensions by box counting. Phys. Lett. A **141**(8–9), 386–390 (1989)
3. Grassberger, P., Procaccia, I.: Measuring the strangeness of strange attractors. Physica D **9**(1–2), 189–208 (1983)
4. Ott, E.: Chaos in Dynamical Systems. Cambridge University Press, Cambridge (2002)
5. Eckmann, J., Kamphorst, S.O., Ruelle, D., et al.: Recurrence plots of dynamical systems. World Sci. Ser. Nonlinear Sci. Ser. A **16**, 441–446 (1995)
6. Marwan, N.: A historical review of recurrence plots. Eur. Phys. J. Spec. Top. **164**(1), 3–12 (2008). https://doi.org/10.1140/epjst/e2008-00829-1
7. Thiel, M., Romano, M.C., Read, P., Kurths, J.: Estimation of dynamical invariants without embedding by recurrence plots. Chaos Interdisc. J. Nonlinear Sci. **14**(2), 234–243 (2004)

8. Donner, R.V., Zou, Y., Donges, J.F., Marwan, N., Kurths, J.: Recurrence networks–a novel paradigm for nonlinear time series analysis. New J. Phys. **12**(3), 033025 (2010)

9. Ding, M., Grebogi, C., Ott, E.: Evolution of attractors in quasiperiodically forced systems: from quasiperiodic to strange nonchaotic to chaotic. Phys. Rev. A **39**(5), 2593 (1989)

10. Zou, Y., Pazó, D., Romano, M., Thiel, M., Kurths, J.: Distinguishing quasiperiodic dynamics from chaos in short-time series. Phys. Rev. E **76**(1), 016210 (2007)

11. Zou, Y., Donner, R.V., Marwan, N., Donges, J.F., Kurths, J.: Complex network approaches to nonlinear time series analysis. Phys. Rep. **787**, 1–97 (2018)

12. Zhang, J., Sun, J., Luo, X., Zhang, K., Nakamura, T., Small, M.: Characterizing pseudoperiodic time series through the complex network approach. Physica D **237**(22), 2856–2865 (2008)

13. Yang, Y., Yang, H.: Complex network-based time series analysis. Phys. A **387**(5–6), 1381–1386 (2008)

14. Gao, Z.K., Jin, N.D.: A directed weighted complex network for characterizing chaotic dynamics from time series. Nonlinear Anal. Real World Appl. **13**(2), 947–952 (2012)

15. Lacasa, L., Luque, B., Ballesteros, F., Luque, J., Nuno, J.C.: From time series to complex networks: the visibility graph. Proc. Natl. Acad. Sci. **105**(13), 4972–4975 (2008)

16. Luque, B., Lacasa, L., Ballesteros, F., Luque, J.: Horizontal visibility graphs: exact results for random time series. Phys. Rev. E **80**(4), 046103 (2009)

17. Donner, R.V., Donges, J.F.: Visibility graph analysis of geophysical time series: potentials and possible pitfalls. Acta Geophys. **60**(3), 589–623 (2012)

18. Nicolis, G., Cantu, A.G., Nicolis, C.: Dynamical aspects of interaction networks. Int. J. Bifurcat. Chaos **15**(11), 3467–3480 (2005)

19. Kulp, C.W., Chobot, J.M., Freitas, H.R., Sprechini, G.D.: Using ordinal partition transition networks to analyze ECG data. Chaos Interdisc. J. Nonlinear Sci. **26**(7), 073114 (2016)

20. Marwan, N., Donges, J.F., Zou, Y., Donner, R.V., Kurths, J.: Complex network approach for recurrence analysis of time series. Phys. Lett. A **373**(46), 4246–4254 (2009)

21. Moghtadaei, M., Golpayegani, M.H.: Complex dynamic behaviors of the complex lorenz system. Sci. Iranica **19**(3), 733–738 (2012)

22. Takens, F.: Detecting strange attractors in turbulence. In: Rand, D., Young, L.-S. (eds.) Dynamical Systems and Turbulence, Warwick 1980. LNM, vol. 898, pp. 366–381. Springer, Heidelberg (1981). https://doi.org/10.1007/BFb0091924

23. Fraser, A.M., Swinney, H.L.: Independent coordinates for strange attractors from mutual information. Phys. Rev. A **33**(2), 1134 (1986)

24. Kennel, M.B., Brown, R., Abarbanel, H.D.: Determining embedding dimension for phase-space reconstruction using a geometrical construction. Phys. Rev. A **45**(6), 3403 (1992)

25. Jacob, R., Harikrishnan, K., Misra, R., Ambika, G.: Uniform framework for the recurrence-network analysis of chaotic time series. Phys. Rev. E **93**(1), 012202 (2016)

Non Linear Analysis of the Effect of Stimulation on Epileptic Signals Generated at Right Hippocampus

Siri Dhathri Kataru and Sunitha R.[✉]

Department of Electronics and Communication Engineering, Amrita School of Engineering,
Amrita Vishwa Vidyapeetham, Bengaluru, India
r_sunitha@blr.amrita.edu

Abstract. Epilepsy is a neurological disorder which is a result of excessive electric discharge between the neurons of brain. Epileptic seizures can be treated with Anti Epileptic Drugs (AEDs), surgery and stimulation. In this study, Epileptor a dynamical system model is used to generate epileptic seizure signals in The Virtual Brain-a simulation framework and electrical Deep Brain Stimulation (DBS) has been used to reduce the effect of epilepsy. The non- linear dynamics of the simulated signals with and without Deep Brain Stimulation has been analyzed using various entropy methods and Hurst Exponent. It is observed that the entropy values of the signals stimulated with DBS are greater than those of the non-stimulated epileptic seizure signals emphasizing an increase in randomness. Also, it is observed that the effect of DBS reduces the predictability of the signal as the Hurst exponent reduces.

Keywords: Epilepsy · Seizure · Epileptor · Deep brain stimulation · Entropy · Hurst exponent

1 Introduction

Epilepsy is a disorder in which the central nervous system is affected due to abrupt electric discharges between the neurons of brain and which is characterized by spontaneous recurrences. An Epileptic seizure is defined as a "transient occurrence of symptoms due to abnormal and excessive neuronal activity in the brain" [1]. Most of the times, Epilepsy occurs due to genetic defects or due to injuries in brain [2] and may sometimes cause death of the individual. Symptoms of seizures depend upon from where the abnormal signals originate in the brain and it has been observed that thirty percent of the Epileptic patients suffer from medically refractory epilepsy [2, 5, 6]. Epileptic seizures can be identified through brain scans such as EEG, CT, MRI, fMRI etc.

Treatments for Epilepsy include Anti Epileptic Drugs (AEDs), abscission, Neurostimulation etc. Neurostimulation be of different forms including Vagus Nerve Stimulation (VNS), Spinal cord stimulation, Transcutaneous brain stimulation and Deep Brain Stimulation (DBS) [5]. It was found that stimulation frequencies less than 12 Hz or

© Springer Nature Singapore Pte Ltd. 2020
S. M. Thampi et al. (Eds.): SIRS 2019, CCIS 1209, pp. 376–384, 2020.
https://doi.org/10.1007/978-981-15-4828-4_31

greater than 70 Hz are more efficient in diminishing the seizures [6] and also higher amplitude stimulus were found to be better in abating the seizures [7]. DBS has been found to be effective in cerebellum, hippocampus, subthalamus, anterior thalamus, hypothalamus and brain stem [3, 5, 6]. A few complications as part of DBS include site pain, infection at the implant site, dizziness, migration etc. [3].

Different types of models that are already available in the literature help us to mimic the already present data or analyse the demeanor of a system under new conditions. A neural model has been portrayed as a group of equations which depict the activity of a neuron or a network of neurons [8]. Such models could be used to find out which therapies most suitably can treat epilepsy [8]. Neuronal models have been categorized into different sub divisions as spatial scale models (micro and macro scale) and deterministic or stochastic models [2]. Macro-scale models are used to provide interactions between networks of neurons whereas micro scale models are concerned with modeling the activity of individual neurons such as ion channel, neuronal architecture and communication between neurons. Micro-scale models give highest accuracy of communication between neurons [2].

The FitzHugh–Nagumo model which has two state variables as membrane potential and recovery variable has been implemented as a micro scale neuron model [2, 9]. Hodgkin- Huxley is also found to be a low dimensional dynamic system model which describes the generation and propagation of the action potential in neuron. Deterministic models gave precise predictions unlike stochastic models which give probabilities. Stochastic modeling could be used when the system is levied to sudden quirks and could predict the seizures [9]. Stochastic models were also used to predict the seizure onset and to observe the dynamics during ictal and inter-ictal states of seizure [10]. Wilson-Cowan model is found to be a low dimensional lumped deterministic model with two state variables describing excitable population firing and inhabitable population firing [2, 10]. Markov model is found to be stochastic in nature which has three states: Normal, Pre-seizure and seizure states. This could predict the probability of transitions between the states [10]. Another attempt was also made to model the EEG rhythm based on fractional Brownian motion and fractional Gaussian noise once the multirate filterbank could separate the various frequency bands in the EEG [22].

In this study, the Epileptor model from "The Virtual Brain" simulation platform has been used to model the epileptic signals with and without stimulation. Non linear methods like entropy and Hurst Exponent have been used to analyze the dynamics of epileptic signal with and without stimulation. Entropy gives the measure of randomness of a system. It also measures the amount of chaos a signal possesses. A signal is said to be more informative if the entropy of the signal is greater [11].

2 Methodology

2.1 The Virtual Brain

The Virtual Brain (TVB) is a large scale neuroinformatics online simulation platform which allows multiple users to handle the data [12, 13] and can be installed for various operating systems such as Linux and Windows. It has been installed in the Windows 10 operating system for this study. TVB uses a blend of both tractoraphic data and existing cortical connectivity information to create connectivity matrix to construct the brain networks. TVB supports two types of stimulations [7, 12], surface based stimulation and region based stimulation. In surface based stimulation, each region of the cortex is taken as a node and each node is modeled by neural mass models. This includes short range and long range connectivity [7]. In region based stimulation, each region is considered as a node which results in inclusion of only short range connectivity [7].

TVB also has the Brain Visualizer where one can see how the epileptic signal is being transferred from one region to the other. Figure 1 shows the Brain Visualizer, in which the dark red coloured regions represents the epileptic zones, light red represents the regions with less epileptic intensity and the pale red represents the regions with no epileptic activity.

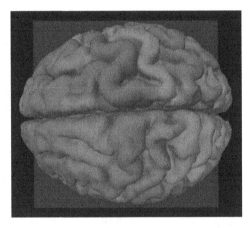

Fig. 1. Brain Visualizer generated in TVB (Color figure online)

The simulation results of TVB are stored in HDF5 file format and have been analysed in the Python using various packages.

2.2 Epileptor

The Epileptor model is a dynamic model in TVB which has been used to generate an epileptic signal in this study. Using five state variables in three different time scales, Epileptor generates epileptic seizure signals. The state variables x_1 and y_1 describes the fast discharges on the fastest time scale and the state variables x_2 and y_2 describe the spike and wave events on the intermediate time scale. The permittivity variable z describes the slow processes like ion concentration, tissue oxygenation and energy consumption. Switching between the ictal and inter-ictal states is governed by the permittivity variable [14–16]. The following equations describe the coupling between the state variables of the Epileptor model.

$$\dot{x}_2 = -y_2 + x_2 - x_2^3 + I_{rest2} + 0.002g(x_1) - 0.3(z - 3.5) \tag{1}$$

where $g(x_1) = \int_{t_0}^{t} e^{-\gamma(t-\tau)} x_1(\tau) d\tau$

2.3 Analysis of Epileptic Signals

Non-linear methods of analysis such as Approximate Entropy, Sample Entropy and Hurst Exponent have been used to analyze and validate the effect of Deep Brain Stimulation on right Hippocampus.

Approximate Entropy. Approximate Entropy (ApEn) is described by Pincus as the complexity of a time series signal [4, 14, 17] and it gives the measure of the similarity level of the signal [18]. The following equations represent the Approximate entropy [4, 14].

$$ApEn(M, m, r) = \Phi^m(r) - \Phi^{m+1}(r) \tag{2}$$

where $\Phi_m(r) = \dfrac{\sum_i^{M-m+1} \ln C_i^m}{M-m+1}$

Here r is the tolerance specified as 20% of the standard deviation, M is the number of data points of the signal, m is the number of samples each data point is divided into and $\ln C_i^m$ is the conditional probability for m segments corresponding to i^{th} data point.

Moorman and Richman detailed ApEn as a self matching algorithm in which each data point is compared with itself [18]. It was observed that this aspect of self match led to bias in results because of which the approximate entropy showed more resemblance than which is actually present. Also, ApEn is highly dependent on length of records [18].

Sample Entropy. Sample Entropy (SampEn) has been coined to abolish the short comings observed in the Approximate Entropy [18]. It is found that sample entropy works better for short data sets and is less effective to noise [18, 19]. The following equation governs the Sample Entropy.

$$SamEn = \ln \frac{B^m(r)}{A^m(r)} \tag{3}$$

Where $B^m(r)$ represents the probability that two data points of the signal would match for m samples and $A^m(r)$ represents the probability that two data points of the signal are identical for $m + 1$ samples with a tolerance of r which is found to be twenty percent of the standard deviation of the signal.

Hurst Exponent. Hurst Exponent indicates the long term memory of the system which is generating the signal. The value of Hurst Exponent is expected to be in the range amidst 0 and 1 for bio signals. The value of Hurst exponent in each moment will determine the demeanor of the signal in the successive moment. Value of Hurst exponent between 0 and 0.5 indicates an anti- persistent system i.e. the system behaves disparately in two consecutive moments. A value between 0.5 and 1 indicates a persistent system i.e. any two consecutive moments depict a similar behaviour. A Hurst exponent of 0.5 emphasizes that the system is highly random [20, 21].

3   Results and Discussion

The "Epileptor" model in "The Virtual Brain" has been used to model the epileptic signals of the right Hippocampus (rHc) with and without stimulation. The stimulation feature in TVB has been used to generate pulse and Gaussian stimuli and is applied to the right Hippocampus. The approximate entropy, sample entropy and the Hurst exponent of different signals simulated in TVB with and without stimulation (DBS) have been calculated using the Python platform. The value of tolerance (r) and m (number of samples each of the M segments the signal is divided) for sample entropy and approximate entropy is chosen to be 20% of the standard deviation of the signal and 2 respectively. Sigmoidal and an improvised version of the Sigmoidal coupling by name Jansen-Rit coupling have been used to generate the epileptic signals in which the difference coupling function between pre and post synaptic activity is used. The following are few of the simulations and observations inTVB.

Figure 2 shows the epileptic signal generated at rHc with difference coupling. An integration step size of 0.01220703125 ms and a sampling period of 0.98 ms have been used for simulating the signal. Figure 3 shows the pulse stimulated rHc signal with an onset of 200 ms, pulse width of 2000 ms, time period of 2000 ms, amplitude of 2 and a scaling factor of 20. Figure 4 shows the Gaussian stimulated rHc signal with a sigma of 3000, midpoint of 3500 ms, amplitude of 2 and an offset of 1.

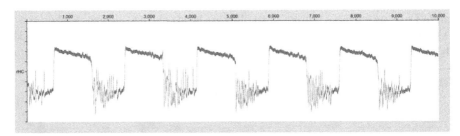

Fig. 2. Epileptic signal generated at right Hippocampus with Difference coupling

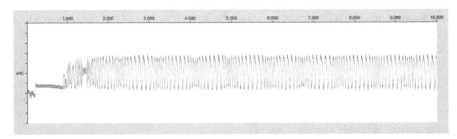

Fig. 3. Pulse stimulated rHc signal with onset = 200 ms, pulse width = 2000 ms, time period = 2000 ms, amplitude of 2, scale = 20

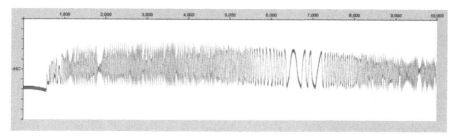

Fig. 4. Gaussian stimulated rHc signal with sigma = 3000 ms, midpoint = 3500 ms, amplitude = 2, offset = 1, scale = 50

From the above three figures, we can see that the stimulated signal is observed to be more irregular and unpredictable than the normal signal and the Gaussian pulse has been observed to have more effect in subsiding the seizures than pulse stimulus. The sample entropy, approximate entropy for m = 2 and Hurst exponent of the signals shown in Figs. 2, 3 and 4 are tabulated in Table 1, which emphasizes the fact that application of DBS made the signal more random.

Table 1. ApEn, SampEn and Hurst Exponent of signals with Difference coupling shown in Figs. 2, 3 and 4

Epileptic signal	ApEn	0.046
	SampEn	0.011
	Hurst Exponent	0.874
Pulse Stimulated signal	ApEn	0.207
	SampEn	0.089
	Hurst Exponent	0.361
Gaussian stimulated signal	ApEn	0.224
	SampEn	0.094
	Hurst Exponent	0.341

Figure 5 shows the epileptic signal generated at right Hippocampus with Sigmoidal JansenRit coupling. An integration step size of 0.01220703125 ms and a sampling period of 0.98 ms have been used for simulating the signal. Figure 6 shows the pulse stimulated rHc signal with an onset of 0 ms, pulse width of 1600 ms, time period of 1865 ms and amplitude of 2 with a scaling factor of 20. Figure 7 shows the Gaussian stimulated rHc signal with a sigma of 1000 ms, midpoint of 3700 ms, amplitude of 2 and an offset of 0.

Again, It can be observed that the stimulated signal is more random than the normal signal can also be noticed that Gaussian stimulated signal is more random than Pulse stimulated signal.

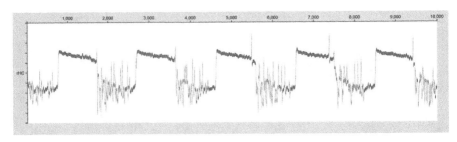

Fig. 5. Epileptic signal generated at right Hippocampus with Sigmoidal Jansen Rit coupling

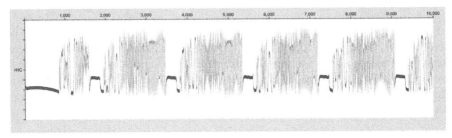

Fig. 6. Pulse stimulated signal at rHc with onset = 0 ms, pulse width = 1600 ms, time period = 1865 ms, amplitude = 2, scale = 20

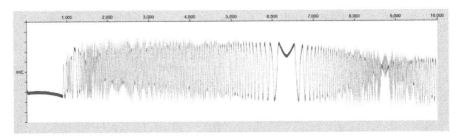

Fig. 7. Gaussian stimulated signal with sigma = 1000 ms, midpoint = 3700 ms, amplitude = 2, offset = 0, scale = 20

These results have been quantified in Table 2 using the Approximate entropy and the Sample entropy for m = 2. The Hurst Exponent of these signals has also been calculated and shown in Table 2.

From the tabulated results, it can be understood that entropy can be used as one of the measures to explore the effect of DBS. Stimulation with appropriate parameters is found to increase the entropy and thereby reduces predictability of the signal. Decrease in Hurst Exponent indicates that the signal is more irregular.

Table 2. ApEn, SampEn and Hurst Exponent of signals with SigmoidalJansenRit coupling shown in Figs. 5, 6 and 7

Epileptic signal	ApEn	0.065
	SampEn	0.014
	Hurst Exponent	0.832
Pulse stimulated signal	ApEn	0.184
	SampEn	0.061
	Hurst Exponent	0.725
Gaussian stimulated signal	ApEn	0.230
	SampEn	0.084
	Hurst Exponent	0.321

4 Conclusion

The Epileptor model in TVB has been used to generate epileptic signals with and without DBS. The entropy values and the Hurst exponents of the signals found reveal that the right Hippocampus could generate more random and unpredictable signals when stimulated with DBS than the epileptic signals generated without stimulation. The study also could reveal the dependency of DBS on the width of the stimulating pulse, as increase in the width of the triggering pulse could increase the randomness. The effect of stimulus can further be analyzed in detail to explore better ways of applying DBS to reduce the effect of Epilepsy and can be validated using clinical data.

References

1. Baier, G., Goodfellow, M., Taylor, P.N., Wang, Y., Garry, D.J.: The importance of modeling epileptic seizure dynamics as spatio-temporal patterns. Front. Physiol. **3**, 1–7 (2012)
2. Stefanescu, R.A., Shivakeshavan, R.G., Talathi, S.S.: Computational models of epilepsy. Seizure **21**(10), 748–759 (2012). Foster, I., Kesselman, C.: The Grid: Blueprint for a New Computing Infrastructure. Mor-gan Kaufmann, San Francisco (1999)
3. Klinger, N., Mittal, S.: Deep brain stimulation for seizure control in drug-resistant epilepsy. Neurosurg. Focus **45**(2), E4 (2018)
4. Acharya, U.R., Molinari, F., Sree, S.V., Chattopadhyay, S., Ng, K.H., Suri, J.S.: Automated diagnosis of epileptic EEG using entropies. Biomed. Signal Process. Control **7**(4), 401–408 (2012)
5. Fisher, R.S., Velasco, A.L.: Electrical brain stimulation for epilepsy. Nat. Rev. Neurol. **10**(5), 261–270 (2014)
6. Laxpati, N.G., Kasoff, W.S., Gross, R.E.: Deep brain stimulation for the treatment of epilepsy: circuits, targets, and trials. Neurotherapeutics **11**(3), 508–526 (2014)
7. Costers, L.: Second semester examination period dynamical diseases : modelling the effect of stimulation-based interventions on intrinsic brain dynamics. An application to epilepsy Master Thesis II submitted to obtain the degree of Master of Science in Psychology (2016)

8. Holt, A.B., Netoff, T.I.: Computational modeling of epilepsy for an experimental neurologist. Exp. Neurol. **244**, 75–86 (2013)

9. Jarray, R., Jmail, N., Hadriche, A., Frikha, T.: Innovations in Bio-Inspired Computing and Applications, vol. 735. Springer, Cham (2018). https://doi.org/10.1007/978-3-319-01781-5

10. Soltesz, I.: Computer modeling of epilepsy. Epilepsy **9**, 1–18 (2012)

11. Dhanya, E., Sunitha, R., Pradhan, N.: Power spectral scaling and wavelet entropy as measures in understanding neural complexity. In: 12th IEEE International Conference Electronics, Energy, Environment Communication Computation Control (E3–C3), INDICON 2015, vol. 2015-Janua, pp. 1–6 (2016)

12. Jirsa, V., et al.: The virtual brain: a simulator of primate brain network dynamics. Front. Neuroinform. **7**, 10 (2013)

13. Woodman, M.M., et al.: Integrating neuroinformatics tools in TheVirtualBrain. Front. Neuroinform. **8**(April), 1–9 (2014)

14. Jirsa, V.K., Stacey, W.C., Quilichini, P.P., Ivanov, A.I., Bernard, C.: On the nature of seizure dynamics. Brain **137**(8), 2210–2230 (2014)

15. Proix, T., Bartolomei, F., Chauvel, P., Bernard, C., Jirsa, V.K.: Permittivity coupling across brain regions determines seizure recruitment in partial epilepsy. J. Neurosci. **34**(45), 15009–15021 (2014)

16. Jirsa, V.K., et al.: The virtual epileptic patient: individualized whole-brain models of epilepsy spread. Neuroimage **145**, 377–388 (2017)

17. Kumar, Y., Dewal, M.L., Anand, R.S.: Features extraction of EEG signals using approximate and sample entropy. In: 2012 IEEE Students' Conference Electrical Electronics and Computer Science Innovation for Humanity SCEECS 2012, pp. 1–5 (2012)

18. Richman, J.S., Moorman, J.R.: Physiological time-series analysis using approximate entropy and sample entropy. Am. J. Physiol. Heart Circ. Physiol. **278**(6), H2039–H2049 (2000)

19. Song, Y., Liò, P.: A new approach for epileptic seizure detection: sample entropy based feature extraction and extreme learning machine. J. Biomed. Sci. Eng. **03**(06), 556–567 (2010)

20. Namazi, H., et al.: A signal processing based analysis and prediction of seizure onset in patients with epilepsy. Oncotarget 7(1), 342–350 (2015)

21. Stan, C., Cristescu, C.M., Cristescu, C.P.: Computation of hurst exponent of time series using delayed (log-) returns. Application to estimating the financial volatility. UPB Sci. Bull. Ser. A Appl. Math. Phys. **76**(3), 235–244 (2014)

22. Gupta, Anubha, et al.: A novel signal modeling approach for classification of seizure and seizure-free EEG signals. IEEE Trans. Neural Syst. Rehabil. Eng. **26**(5), 925–935 (2018)

What Are 3GPP 5G Phase 1 and 2 and What Comes After

Valerio Frascolla[✉]

Director Research and Innovation, Intel Deutschland, Lilienthalstraße 15,
85579 Neubiberg, Germany
valerio.frascolla@intel.com

Abstract. This paper describes the ongoing standardization of the new 5G system and the planned next steps in the 3rd Generation Partnership Project (3GPP) groups. Starting with an overview of how 3GPP works, what is meant with 5G Phase 1 and 2 and a clear timeline for ongoing and planned activities are provided. Finally an interesting innovation vector for beyond 5G systems, i.e. dynamic (in time) spectrum management, is briefly mentioned.

Keywords: 3GPP · 5G · Beyond 5G · Dynamic spectrum management · Standardization · Funded research projects

1 Introduction

The continuous development, generation after generation [1], of the telecommunication system allows mobile users to benefit from new services and applications almost with a yearly cadence. Starting with the second half of 2019, the first commercial deployments of the latest generation of the cellular communication system, known as the 5th generation (5G), are being rolled out in several countries.

The new 5G system aims at two main targets: First, to be a game changer in the way mobile users can experience nomadic communications, by introducing brand new technologies and functionalities that on the one hand can enhance several system key performance indicators (KPI), and on the other hand manage to relax some system constraints. Second, to provide for the first time in a unique framework a real synergy and a seamless integration of the legacy systems with the so called '5G verticals', i.e. domains and businesses different from the telecommunication domain, like Automotive and more in general Intelligent Transport Systems (ITS), E-health, or Industrial Internet of Thing (IIoT).

In order to fulfil its set targets, 5G has been defined along three main so called 'usage scenarios' (also known as 5G 'use cases'), which, as shown in Fig. 1, are supposed to cluster, along three main innovation vectors, similar requirements and constraints, among the numerous new ones coming out of the long list of 5G expected improvements. Such usage scenarios, which are detailed and broadly discussed in [2–4], have been jointly defined by both the research community and the industry as:

© Springer Nature Singapore Pte Ltd. 2020
S. M. Thampi et al. (Eds.): SIRS 2019, CCIS 1209, pp. 385–398, 2020.
https://doi.org/10.1007/978-981-15-4828-4_32

- enhanced Mobile BroadBand (eMBB),
- Ultra Reliable and Low-Latency Communications (URLLC),
- massive Machine Type Communications (mMTC).

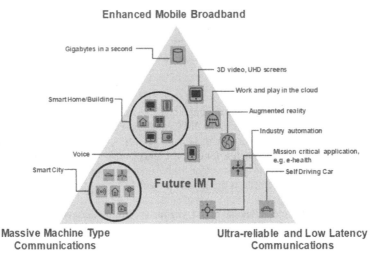

Fig. 1. 5G usage scenarios (Source: Recommendation of ITU-R M.2083-0 [4]).

In a nutshell, *eMBB* is supposed to take care of data hungry applications and how they can exchange from and to terminals a huge amount of data per second. For instance, mmWave access is considered a new enabling technology under this usage scenario. *URLLC* covers a variety of use cases, like Mission critical services, E-Health and Autonomous driving, where reliability and low latency are both equally important. Finally, *mMTC* covers a broad spectrum of IoT-related services and applications, usually concerning a huge number of small or cheap devices communicating a tiny or small amount of data. mMTC is usually relevant for the different Smart-x paradigms (with x = city, grid, etc.).

Following the very recent launch of 5G commercial services, which finally allow mobile users to experience the new services and applications based on the 5G system, it is now a timely move to provide, thanks to this paper, an updated insight on the current status and planned work of the 3rd Generation Partnership Project (3GPP) [5], i.e. the most relevant standard for the 5G system definition.

The rest of the paper is organized as follows: Sect. 2 surveys some key works that provide background on why the 5G system is needed and hints at some promising enabling technologies for 5G and beyond 5G (B5G) systems. Section 3 explains how the 3GPP standardization groups work and describes what the content of the so called 5G Phase 1 and Phase 2 is. Section 4 provides an overview of what comes after 5G Phase 2, and hints at a new topic for B5G systems, which was recently proposed to the European Telecommunications Standards Institute (ETSI) [6] work. Section 5 gives a glimpse of what could be the content of the 3GPP work in the timeframe 2022–2023. Finally, Sect. 6 concludes the paper.

2 5G Background and Some Key Seminal Works

Several papers have provided in the last few years interesting overviews on the potential impact of newly defined technologies for the forthcoming 5G era. Several of those works are the outcome of international collaborative research projects, funded by the European Union or other international institutional agencies, under the 'Framework Programme 8' (also called 'FP8' or 'Horizon 2020'), which started in 2013 and is going to be completed in 2020. This paper therefore takes some main outcome of those projects, but not only, to provide a short overview of main 5G aspects and related work, planned in the 3GPP standardization body. Some of the touched-upon outcomes of those projects will be instrumental to build the foundation for the new technology enablers that will characterize the successor of FP8, i.e., the forthcoming new research programme FP9, at the moment tentatively also called 'Horizon Europe', which will run in the time window 2020–2027.

Concerning works on the 5G usage scenarios, regarding the impact of the 5G *eMBB*-related technologies in the society, authors in [7], from the *5G-CHAMPION* project, elaborate on the benefits brought to the Winter Olympic games held in Korea in 2018, and authors in [8], from the *5G-MiEdge* project, on the potential impact on the 2020 Summer Olympic games in Japan. Regarding the *URLLC* usage scenario, authors in [9] elaborate on how merging together different technologies like Wi-Fi and SDN can help in reducing the communication latency, so to more effectively steer a robot from remote, whereas authors in [10] provide an overview of several different interesting aspects of URLLC services and applications. Regarding the *mMTC* usage scenario authors in [11, 12] detail constraints and aspects of effective 5G mMTC deployments.

In addition to the 5G usage scenarios, several other papers focus on the so called enabling technologies, i.e. newly developed means that will allow the 5G system to fulfill its system KPIS and reach its ambitious goals. Among others, one of the most promising new 5G enabler is *millimeter waves (mmWave) access and backhaul* enhancements: authors in [13, 14], from the *MiWaves* project, elaborate on what could be impactful contributions in standards regarding a smooth introduction of mmWave access and backhaul, up to 70 GHz bands. Authors in [15], from the *mmMAGIC* project, move even further and broadly elaborate on different aspects of newly proposed technologies, mainly at the lower layer of access networks, up to the 100 GHz bands. Specifically on the backhaul topic, an interesting overview of key Radio Resource Management (RRM) aspects are discussed in [16].

Another vector of innovation, very important for an effective 5G deployment, is the *usage of new spectrum bands*, or *the more enhanced usage of existing spectrum bands*. Interesting works in frequencies below 6 GHz are described in [17], from the *ADEL* project, on extending the concept of Licensed Spectrum Access (LSA) with time dynamicity, and even further in [18], from the *SPEED-5G* project, with the introduction of the extended dynamic spectrum allocation (eDSA) concept. For bands in the mmWave domain a disruptive approach is sketched in [19] from the *5GENESIS* project, and for a more forward looking perspective for B5G on THz bands, a very recent survey is presented by authors in [20].

The *integration among wireless and optical networks* is a very appealing target of future networks, as discussed in [21, 22] from the EU-Brazil co-funded project *FUTE-BOL*, and very recently in [23] regarding its positive impact on the backhaul topology optimization.

The *integration between terrestrial and satellite networks*, as discussed in [24, 25] is a topic gaining more and more traction in 3GPP. That stems from the fact that real ubiquitous and disaster-resilient communications will be more and more looked for in the years to come, and a seamless integration of the satellite link coverage, also in remote areas, together with the features and performance of terrestrial networks can offer a valuable solution to ease the always more numerous natural disasters caused by the warming up of the average air temperature on earth.

Finally, an important field of activities is the integration of cellular technologies with the 5G verticals, like *IIoT* or *Automotive and the more general ITS*. Among the huge existing literature on those topics, it may be worth mentioning, regarding IIoT, the work in [26], where authors provide a recent overview of applications and advances on 5G and B5G mobile technologies in that domain. Regarding the Automotive and the ITS domains, particularly interesting is the work in [27], where authors deliver a comprehensive survey on the standardization status and the open issues of the broader vehicular communication topic.

3   3GPP 5G Phase 1 and Phase 2

In this section we provide a short description of the way 3GPP bodies works, and then summarize what is the meant with and what is the content of '5G Phase 1' and '5G Phase 2'.

3.1   3GPP Working Procedures

3GPP organizes its work splitting the topics to be standardized in numerous parallel activities and in a hierarchical manner. The focus area of communication systems is organized in three high level groups, called *Technical Specification Groups (TSG)*, which are:

- Radio Access Network (RAN).
- Service and System Aspects (SA).
- Core Network and Terminals (CT).

Each TSG is made of a different number of so called TSG Working Groups (WG), i.e., six each in RAN and SA, four in CT. The number of WGs can increase or decrease with the time, thus adapting in a dynamical way to the changes of the ecosystem, so to be always capable of coping with the raise of new needs or technologies. A WG can be closed if its focus area is considered completed, a new one can be opened if a new topic deserves a major effort and a focused activity spanning more years.

Each one of those WGs organize independently (in different times and locations) meetings with a specific cadence, during which a consensus-based decision mechanism is used to define the kind of and the content of the newly discussed features, and changes, corrections or amendments of the already existing ones. The participants to the 3GPP work usually physically meet each quarter co-locating the attendees of the three TSG in the same place, and on an almost monthly cadence for the work of the several WGs. Some WGs may decide, on special occasions, e.g. when cross-WGs decisions are to be taken, or when the expertise of different teams need to converge to solve a difficult technical issue, to co-locate their meetings as well.

When a new feature or functionality is to be added to the standards, the WGs operate in a 3-steps cycle of activities called *Stage 1 - Stage 2 - Stage 3*, each one partially overlapping in time with the others and feeding the next one in the row with its outcome:

- *Stage 1*: first use cases and scenarios for the new services and applications are introduced, together with related (new) requirements on the system architecture.
- *Stage 2*: Then changes are proposed to the different part of the lower (in the ISO/OSI layer sense) communication system architectural blocks, protocols and messages. If needed new architectural blocks can be proposed to take into account new needed functionalities.
- *Stage 3:* Finally, also the upper layers of the protocol stacks, together with the related new communication protocols and messages, are enhanced in order to fulfil the new services in the enhanced system architecture.

There is also a fourth short stage called '*ASN.1*', however we do not have enough room in this paper to enter in too many details.

During the work of each stage a set of documents is produced, each one of them usually focusing on a new single feature to be added to the system. Depending on their content and scope, they are called be **Feasibility Study (FS)**, **Technical Report (TR)**, or **Technical Specification (TS)**, which is the most important one as it contains normative text, i.e. a description of a functionality or a feature that is to be literally fulfilled, if compliancy with 3GPP standards is looked for when a new device is commercially launched into the market.

The full-fledged implementation of a new feature therefore is obtained through the overall outcome of a cycle (Stage 1 to Stage 3), as shown in Fig. 2. When a feature is particularly complex, or when it implies a huge re-work of the existing standards, it is usually split in smaller activities called **Work Items (WI)**, which are the smallest pieces of work that can be independently tracked and discussed at WG level.

A set of parallel activities on different features, and spanning a defined time window, is the so-called **Release-x (Rel-x)**, which is composed of a set of new or enhanced architectural blocks, protocols, functionalities, and messages, to be added to the previously existing system, where '*x*' is a progressive number that distinguishes among old and new generations.

A Release is '*frozen*' when no new feature can be added to it, only essential corrections to the standards are allowed; that happens usually with roughly a yearly cadence. A Release can be '*closed*', when it no longer is maintained, and no changes or amendments are allowed any more. Finally, it is worth mentioning that TR and TS, once created, can evolve and be enhanced through several Releases, each one of them adding new functionalities to the specific feature in focus in that TR or TS.

Technical discussions happen primarily during WGs meetings, where new documents, amendments, enhancements and new features are discussed and agreed-upon within the different partners participating to the work, thus creating so called 'normative documents', i.e., the mentioned above *TS*. Overall alignment, resolution of controversial cases, and assessment of the compliancy with the working procedures of 3GPP of the work produced by WGs is usually the scope of the TSG meetings.

The following two subsections explain what the content of the first specifications on 5G are, what is already finalized and what are the still ongoing activities on the 5G system definition.

3.2   Release 15 (5G Phase 1)

3GPP bodies started to work on the definition of the 5G system with Rel-15 in 2016. Figure 2 shows a timeline of Rel-15 and Rel-16 work, where the x axis is a time axis based on the increasing number of the TSG meeting, taking place on average each three months. The first set of 5G features is composed of several TSs, which were completed under Rel-15, and that is what is called '*5G Phase 1*'.

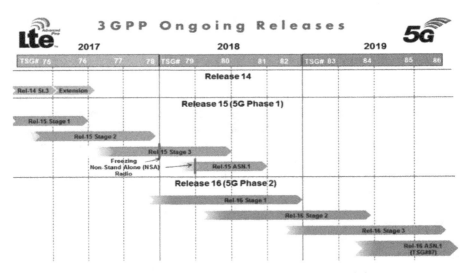

Fig. 2. Phase 1 and Phase 2 timeline (Source: 3GPP website).

Rel-15 mainly focused on eMBB services and worked on building the basic blocks of the new 5G system, describing the new 5G system access part, also called New Radio (5G NR) and the new 5G core network (5GC). Other new concepts have been introduced, like network slicing and virtualization mechanisms, a functional interworking with legacy (2G, 3G, 4G/LTE) cellular system and other wireless technologies (Wi-Fi).

Several proposals were made and discussed on the new 5G architecture, however only two so-called 'operations modes' were defined in Rel-15, i.e. the *Standalone mode*

(SA) and the *Non-Standalone (NSA) mode*. The former assumes an independent operation of the 5G NR with the 5GC, whereas the latter assumes that an existing LTE network, on top of which the new 5G network overlaps, provides the control plane for the 5G network to operate. The NSA work was prioritized as the 3GPP participant companies, as well as the overall ecosystem, identified that one as the mode which would create most added value; its new use cases and scenarios focus on features and services that are supposed to have a smoother and more successful market launch.

In a nutshell, the main new features of Rel-15 are the introduction of the new 5G NR access, and the first enhancements to the whole communication network in order to take more into account the requirements of the so-called 5G verticals, i.e. sectors of the industry that so far have not been seamlessly integrated with the mobile network, or of which only a partial support for their requirements was taken into consideration in previous 3GPP releases. Among the numerous 5G vertical one can mention Media and entertainment, ITS, IIoT, or the different paradigms that look towards the future evolution of our society called Smart-City, Smart-Grid, Smart-x, ...

It is important to mention that currently available commercial networks that run 5G services offer only a subset of all the defined 5G Phase 1 features. More will be deployed in the forthcoming quarters in an incremental manner. That is always the case when a new technology is launched, as it has to be ensured that first a smooth integration with the legacy system is possible, and then that the newly proposed services and applications are economically viable; indeed all of the above takes time.

3.3 Release 16 (5G Phase 2)

Rel-16, fulfilling the time-scheduled indicated in early 2016 as shown in Fig. 1, started at the end of 2017, its Stage-2 and Stage-3 definition is currently in full swing and principally *takes care of the other two usage scenarios, i.e. mMTC and URLLC*.

The content of the new features and functionalities introduced by Rel-16, considered an evolution of the 5G system defined in 5G Phase 1, is what is called *'5G Phase 2'*,

In general terms Rel-16 can be split in two sets of main activities or features, i.e. *Efficiency Items*, listed in Fig. 3, and *Expansion Items*, listed in Fig. 4. The former are a mix of system improvements and enhancements on different domains (positioning, MIMO, dual connectivity, etc.), which can be seen as an evolution of the previous releases and aim at overall improving the core 5G system functionalities, as defined by the 5G Phase 1. The latter aim at providing functions to serve (and better interact with) 5G verticals, like Automotive (in the 3GPP flavor of interaction between vehicles and the environment called Vehicle-to-everything (V2X)) and IIoT, as well as at handling other features like URLLC and unlicensed spectrum operations, which were not yet taken into consideration in the first set of features composing 5G Phase 1.

In both figures lighter coloured lines indicate set of features that got lower priority when compared to others of the same Release. Finally, it is worth mentioning that there are numerous other features under work that cannot be easily mapped on the listed items shown in both figures. However, The intention of this paper is not to provide a comprehensive report of the activities per release, rather to summarize and briefly elaborate on the main activities that drive the work of 3GPP WGs.

Looking at the content of Fig. 3, one can notice from the names that much attention is given to enhance the status quo, and to reduce the impact of the new introduced functionalities on the legacy system, specifically regarding the interference issues that might arise in using adjacent spectrum bands between 5G NR and legacy access.

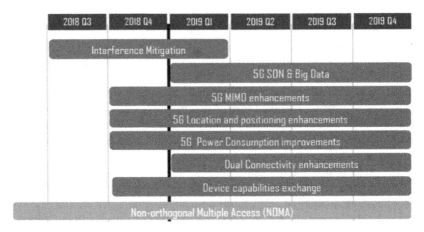

Fig. 3. 3GPP Release 16 Efficiency Items (Source: 3GPP website).

Looking instead at the content of Fig. 4, one can notice that a more disruptive approach is used on introducing new features, mainly targeting the mentioned above new interaction with 5G verticals, and with introducing the support for 5G NR to operate in unlicensed bands, a rather promising new feature for industrial deployments willing to operate by their own a cellular network (e.g. a vast industrial production environment).

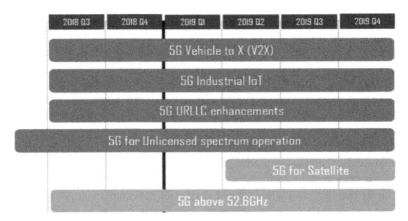

Fig. 4. 3GPP Release 16 Expansion Items (Source: 3GPP Website).

It is worth noting that a smoother interworking with satellite networks and work needed to operate on mmWave bands above 52.6 GHz have been given lower priority compared to the other items. That might stem from the fact that the telecommunication community is waiting to analyze the outcome of the work done during the World Radio-communication Conference (WRC-19), expected to be finished by the end of November 2019. The WRC is driven by the International Telecommunication Union (ITU) council and governs the usage of both terrestrial and satellite networks, including the allocation of new spectrum bands, e.g. in the lower mmWave range, for worldwide deployment. Therefore, decision taken there will have an impact on both satellite-terrestrial interworking and on the usage of mmWave access.

The same lower priority in the 3GPP work applies also to the introduction of more disruptive technologies like the Non-orthogonal Multiple Access (NOMA), a promising approach to improve the spectral efficiency while taming the interference experienced at receiver side [28].

In conclusion, what composes the Phase 1 and the Phase 2 of the 5G system is very well defined, part is being commercially launched starting with June 2019, part will be in an incremental manner commercialized in the forthcoming quarters.

4 Beyond 5G Phase 1 and Phase 2: Release 17

As Rel-16 is getting close to its completion, work has already started in defining the content of Rel-17, as sketched in Fig. 5.

Fig. 5. 3GPP Release 17 timeline (Source: 3GPP website).

During 2019 a long list of candidate features for Rel-17 has been drafted, 15 from the RAN groups and 29 from the SA groups, summing up to 44 entries, too many for the planned duration of Rel-17. It was therefore decided that a prioritization exercise will take place in the December 2019 plenary TSG meeting, and those entries that will not make it in Rel-17 will be shifted into Rel-18. It is important to stress that the list of Rel-17 features is now 'closed', i.e. no new items can be added to it. The latest news on Rel-17 is that during the November SA1 WG meeting, the Stage-1 of Rel-17 has been completed, and first discussions have just started on Rel-18 topics.

In the following, the list of features under discussion are reported, to give an idea of where in 2020 the focus of the 3GPP WGs will concentrate on.

The following SA features seem to have a common agreed-upon relevance and will therefore be treated within Rel-17 [29]:

- 5G system enhancement for Advanced Interactive Services,
- Architectural enhancements for 5G multicast-broadcast services,
- Architecture enhancements for 3GPP support of advanced V2X services - Phase 2,
- Enablers for Network Automation for 5G-Phase 2,
- Enhanced support of Industrial IoT,
- Enhanced support on Non-Public Networks,
- Enhancement of Network Slicing Phase 2,
- Enhancement of support for Edge Computing in 5GC,
- Integration of satellite in 5G systems,
- Support for Multi-USIM devices,
- System enhancement for Proximity based Services in 5G.

There is no shared agreement on the importance of the following SA entries, which will therefore be good candidate to be included in Rel-17 work during the December TSG meeting:

- Enhancement of support for 5G LAN-type services,
- Enhancement to the 5GC location services - Phase 2,
- Extended Access Traffic Steering, Switch and Splitting support in the 5G system architecture,
- Multimedia Priority Services (MPS) Phase 2,
- Study on enhancement of support for 5G wireless and wireline convergence,
- Supporting Unmanned Aerial Systems Connectivity, Identification, and Tracking,
- UPF enhancement for control and SBA.

And finally, the SA features that will be most probably shifted to Rel-18 are:

- Application awareness Interworking between LTE and NR,
- Cellular IoT enhancement for the 5G system,
- Enhancement of 5G UE policy,
- Service-based support for SMS in 5GC,
- Smarter User Plane,
- Support for minimization of service interruption,
- Supporting Flexible Local Area Data Network,
- System architecture for next generation real time communication services,
- Usage of User Identifiers in the 5G system.

For completeness, also the features planned to be discussed by the RAN groups are listed here below.

- Coverage enhancements,
- Generic enhancements to NR-U,
- Integrated Access ad Backhaul enhancements,
- MIMO enhancements,
- Multi SIM operation,
- NB-IoT and eMTC enhancements,

- NR above 52.6 GHz (incl 60 GHz unlicensed),
- NR for Non Terrestrial Networks,
- NR light,
- NR Multicast broadcast,
- Positioning enhancements,
- Power saving enhancements,
- RAN data collection enhancements,
- Sidelink enhancements,
- Small data transfer optimization.

In summary, even if the work has not started on the definition of the Pahse 2 and Phase 3 of Rel-17 features, we already know the content of the 5G system in the 2021/2022 timeframe, i.e. the dates by when Rel-17 systems will be commercially launched. This is important information, as knowing how standards work, people have the possibility to understand when and how the 5G system will be deployed, and which features are expected in which year. Such information is very relevant to both academia and industry to plan future activities, and push at the right time their innovative ideas, patents, or products into standards.

5 Beyond 5G or 5G Long Term Evolution? Release 18

According to a recent document discussed during the September plenary meeting [29], the real work on defining the Rel-18 content is supposed to start at the beginning of 2020 (Stage 1 work). However, we already know that some of the items listed above will not make it in Rel-17 and are therefore the best candidates to be among the features composing what will be called Rel-18, or at least a part of it.

Will the content of Rel-18 still be called '5G Phase X' or will it get a more appealing name, like for instance the rumored '5G Long-Term-Evolution', a name that has been used by the European Union for some of the text of its recent calls for new funded projects that target a long-term vision of 5G capabilities? We can only speculate on such topics, and we will know the answer probably in the next couple of quarters.

Even if the list of features that will be shifted out of Rel-17 into Rel-18 is already rather long, there are several other interesting new topics that could be taken into consideration for Rel-18 work. For instance, the feature of adding a ***dynamic (in time) spectrum management*** to the RRM functionalities. That is one of the innovation vectors that the 5GENESIS project [19] aims at pushing into the ecosystem. The proposal is to extend the work done in the finalized EU-funded research projects ADEL [30] and SPEED-5G [301], aiming at extending the 5G system architecture with logical blocks, functionalities, and messages that would make possible to allocate spectrum (disregarding if using Wi-Fi, 5G NR, or legacy cellular access) in a timely dynamical manner, say on minutes base. Several benefits can be mentioned if such a feature would be implemented, e.g., a much more granular and effective utilization of all the access technologies that surround a mobile phone, with the consequence of a much more energy-efficient operation of telecommunication networks. Moreover, having the possibility to allocate spectrum chunks with minute-granularity would increase of an important factor the capability of making the best usage of all the available spectrum for access, disregarding of the underlying access technology, thus allowing for an overall increased spectral efficiency.

To that end the 5GENESIS project has already started to provide contributions to a standards body, i.e., ETSI Technical Committee Reconfigurable Radio System Work Group 1 (TC RRS WG1) [32, 33] so to engage with the ecosystem and prepare the discussions and the contributions for potentially impacting also the 3GPP bodies, ideally within the Rel-18 timeframe.

6   Conclusion

This paper provides an overview of the current status in 3GPP bodies of features and usage scenarios of 5G systems. International collaborative research projects and 5G Verticals are taken as examples of the new features and services that 5G can bring as added values for the final mobile phone users. After explaining what 3GPP 5G Phase 1 and 2 are, we have elaborated on what comes after, i.e., Rel-17, also providing for all the 5G releases a timeline of the planned work. Finally a glimpse of an interesting new item to be discussed in Rel-18 is provided, as part of the ongoing and future work of the 5GENESIS research project.

Acknowledgment. Part of the research in this work has received funding from the European Commission H2020 programme, grant agreement No. 815178 (5GENESIS project).

References

1. Raaf, B., et al.: Key technology advancements driving mobile communications from generation to generation. Intel Technol. J. **18**(1), 12 (2014)
2. NetWorld2020 ETP: 5G: Challenges, research priorities, and recommendations, White Paper, September 2014. https://www.networld2020.eu/
3. Marsch, P. et al.: 5G System Design, Wiley, June 2018
4. Series, M.: IMT vision-framework and overall objectives of the future development of IMT for 2020 and beyond. Recommendation ITU-R M.2083-0, September 2015
5. GPP Website. https://www.3gpp.org
6. ETSI Website. https://www.etsi.org
7. Won, S.H., et al.: Development of 5G CHAMPION testbeds for 5G services at the 2018 Winter Olympic Games. In: 2017 IEEE 18th International Workshop on Signal Processing Advances in Wireless Communications (SPAWC), Sapporo, pp. 1–5 (2017)
8. Frascolla, V., et al.: 5G-MiEdge: design, standardization and deployment of 5G phase II technologies: MEC and mmWaves joint development for Tokyo 2020 Olympic games. In: 2017 IEEE Conference on Standards for Communications and Networking (CSCN), Helsinki, pp. 54–59 (2017)
9. Martínez, V.M.G., et al.: Ultra reliable communication for robot mobility enabled by SDN splitting of WiFi functions. In: 2018 IEEE Symposium on Computers and Communications (ISCC), Natal, pp. 527–530 (2018)
10. Soldani, D., et al.: 5G for ultra-reliable low-latency communications. IEEE Netw. **32**(2), 6–7 (2018)

11. Bockelmann, C., et al.: Towards massive connectivity support for scalable mMTC communications in 5G networks. IEEE Access **6**, 28969–28992 (2018)
12. Ket, M., et al.: Compressive massive random access for massive machine-type communications (mMTC). In: 2018 IEEE Global Conference on Signal and Information Processing (GlobalSIP), Anaheim, CA, USA, pp. 156–160 (2018)
13. Frascolla, V., et al.: MmWave use cases and prototyping: a way towards 5G standardization. In: 2015 European Conference on Networks and Communications (EuCNC), Paris, pp. 128–132 (2015)
14. Frascolla, V., et al.: Challenges and opportunities for millimeter-wave mobile access standardisation. In: 2014 IEEE Globecom Workshops (GC Wkshps), Austin, TX, pp. 553–558 (2014)
15. Tercero, M., et al.: 5G systems: the mmMAGIC project perspective on use cases and challenges between 6–100 GHz. In: 2016 IEEE Wireless Communications and Networking Conference, Doha, pp. 1–6 (2016)
16. Shariat, M., et al.: Enabling wireless backhauling for next generation mmWave networks. In: 2015 European Conference on Networks and Communications (EuCNC), Paris, pp. 164–168 (2015)
17. Morgado, A., et al.: Dynamic LSA for 5G networks the ADEL perspective. In: 2015 European Conference on Networks and Communications (EuCNC), Paris, pp. 190–194 (2015)
18. Herzog, U., et al.: Quality of service provision and capacity expansion through extended-DSA for 5G. In: 2016 European Conference on Networks and Communications (EuCNC), Athens, pp. 200–204 (2016)
19. Koumaras, H., et al.: 5GENESIS: the genesis of a flexible 5G facility. In: 2018 IEEE 23rd International Workshop on Computer Aided Modeling and Design of Communication Links and Networks (CAMAD), Barcelona, pp. 1–6 (2018)
20. Huq, K.M.S., et al.: Terahertz-enabled wireless system for beyond-5G ultra-fast networks: a brief survey. IEEE Netw. **33**(4), 89–95 (2019)
21. Marquez, P., et al.: Experiments overview of the EU-Brazil FUTEBOL project. In: 2017 European Conference on Networks and Communications (EuCNC), Oulu, pp. 1–6 (2017)
22. Marques, P., et al.: Optical and wireless network convergence in 5G systems – an experimental approach. In: 2018 IEEE 23rd International Workshop on Computer Aided Modeling and Design of Communication Links and Networks (CAMAD), Barcelona, pp. 1–5 (2018)
23. Frascolla, V., et al.: Optimizing C-RAN backhaul topologies: a resilience-oriented approach using graph invariants. Appl. Sci. **9**(1), 136 (2019)
24. Gineste, M., et al.: Narrowband IoT service provision to 5G user equipment via a satellite component. In: 2017 IEEE Globecom Workshops (GC Wkshps), Singapore, pp. 1–4 (2017)
25. Koumaras, H., et al.: 5G experimentation facility supporting satellite-terrestrial integration: The 5GENESIS approach. In: 2019 European Conference on Networks and Communications (EuCNC), Valencia, pp. 1–5 (2019)
26. Mumtaz, S., et al.: Guest editorial special issue on 5G and beyond—mobile technologies and applications for IoT. IEEE Internet of Things J. **6**(1), 203–206 (2019)
27. Zhao, L., et al.: Vehicular communications: standardization and open issues. IEEE Commun. Stan. Mag. **2**(4), 74–80 (2018)
28. Ding, Z., et al.: Application of non-orthogonal multiple access in LTE and 5G networks. IEEE Commun. Mag. **55**(2), 185–191 (2017)
29. GPP doc SP-190949 "SA#85 agreements on SA R17 prioritization & Content"

30. Frascolla, V., et al.: Dynamic licensed shared access - a new architecture and spectrum allocation techniques. In: 2016 IEEE 84th Vehicular Technology Conference (VTC-Fall), Montreal, QC, pp. 1–5 (2016)
31. Frascolla, V., et al.: Breaking the access technologies silos by enhancing MAC and RRM in 5G+ networks. In: 2018 European Conference on Networks and Communications (EuCNC), Ljubljana, Slovenia, pp. 1–9 (2018)
32. ETSI document RRSWG1(18)044009: Adding dynamicity to spectrum management
33. ETSI document RRSWG1(18)044010: EU-funded project 5GENESIS high-level description

Author Index